W.-G. Forssmann, D. W. Scheuermann, J. Alt (Eds.)

Functional Morphology of the Endocrine Heart

Steinkopff Verlag Darmstadt
Springer-Verlag New York

CIP-Titelaufnahme der Deutschen Bibliothek

Functional morphology of the endocrine heart
W.-G. Forssmann ... (eds.). − Darmstadt: Steinkopff; New York: Springer, 1988
ISBN-13: 978-3-642-72434-3   e-ISBN-13: 978-3-642-72432-9
DOI: 10.1007/978-3-642-72432-9

NE: Forssmann, Wolf G. (Hrsg.)

Preface

In 1956, Bruno Kisch (90) discovered a special form of myocyte in the guinea pig heart atrium which contained peculiar, dense inclusions, but it was not before the early 1980s that cardiac hormones were isolated and characterized independently and almost simultaneously by several working groups. In 1964 Jamieson and Palade (84) were the first to postulate the secretory nature of the atrial myocytes and their specific granules. In 1976, Marie et al. (103) revealed the relation of the granular index of these atrial cells with the water- electrolyte balance of the body fluid. The biological effects of purified atrial extracts, i.e., the diuretic and vasorelaxant effects, were detected in the early 1980s, (26,35,38,56). Now, a series of polypeptides, of the same family, and all derived from a homologous precursor in the different species are know to exert the diuretic and vasorelaxant effects. This development of the discovery of a peptide hormone is unparalleled in the sense that it was mostly morphologists who contributed to the final characterization of these cardiac hormones. Thus, it was a great honor for us to be given the opportunity on the occasion of the 8th European Anatomical Congress to organize a special satellite symposium on the rapidly growing field of research in cardiac endocrinology. The symposium "Functional Endocrinology of the Endocrine Heart" was planned as an interdisciplinary meeting to enable anatomists to follow recent advances in cardiac endocrinology. A number of internationally respected scientists were willing to review their fields of research. The meeting also included several original communications and a poster session. The high quality of the presentations encouraged us to edit the present volume in cooperation with the same publishers as that of Kisch's first ultrastructural-morphological monograph on the heart. We assume that this edition represents a consequent contribution which accounts for the rapidly growing interest in the field of cardiac endocrinology. Although a book in German language on the same subject was recently published by Steinkopff, (Kreye VAW, Bussmann, WD (Hrsg) (1987) ANP - Atriales Natriuretisches Peptid und das kardiovaskuläre System. Steinkopff, Darmstadt), we suppose that the present volume covers complementary subjects because it contains reviews rather than original contributions; the latter are enculded in the Kreye-Bussmann edition. Thus, the two books are based on different intentions − one giving a general account of cardiac endocrinology, the other detailing current research projects.

We are grateful to Bissendorf Peptide GmbH (Wedemark, FRG) and Schwarz GmbH (Monheim, FRG) for supporting the symposium. Without the help of Bissendorf Peptide GmbH the edition of this book would not have been possible. We would further like to acknowledge the help of K. Grützner and C. Behrends for preparing the manuscripts, and of J. Sis and G. Sürig, for editing assistance. The particular interest and enthusiasm of Prof. Dr. Klaus Döhler and Dr. Jean-Pierre Timmermanns greatly contributed to the organization of the symposium.

We hope that this book will serve not only to update advances in cardiac endocrinology, but that it will also stimulate further activities in this new and rapidly expanding field of bio-medical research. (for references please see pp. 33)

September, 1988

W.G. Forssmann, Heidelberg
D.W. Scheuermann, Antwerp
J. Alt, Hannover

Contents

VIII

Author Index

Brabant, G., Dr.
Depts. Klinische Endokrinologie, Nephrologie und Gastroenterologie, Medizinische Hochschule, 3000 Hannover 61, FRG

Brand, T.
Max-Planck-Institut for Physiological and Clinical Research, Dept. of Experimental Cardiology, Benekestr. 2, 6350 Bad Nauheim, FRG

Currie, Mark G., Ph.D.
Department of Cell and Molecular Pharmacology and Experimental Therapeutics, Medical University of South Carolina, 171 Ashley Avenue, Charleston, SC 29425, USA

Eisenhauer, T., Dr. med.
Dept. of Nephrology, University Clinic, Robert-Koch-Str. 40, 3400 Göttingen, FRG

Forssmann, W.-G., Prof. Dr. med.
Institut für Anatomie und Zellbiologie der Universität Heidelberg, Im Neuenheimer Feld 307, 6900 Heidelberg, FRG

Gagelmann, M., Dr. rer. nat. habil.
Anatomisches Institut III, Universität Heidelberg, Im Neuenheimer Feld 307, 6900 Heidelberg, FRG

Gerbes, A.L., M.D.
Dept. of Medicine II, Klinikum Großhadern, University of Munich, Marchioninistraße 15, 8000 Munich 70, FRG

Gutkowska, J., Ph.D.
Laboratory of Biochemistry of Hypertension, Clinical Research Institute of Montreal, 110 Pine Avenue West, Montreal, Quebec, Canada, H2W 1R7

Hock, D., Dipl. Biol.
Anatomisches Institut III, Universität Heidelberg, Im Neuenheimer Feld 307, 6900 Heidelberg, FRG

Jungmann, E., Prof. Dr.
Centre of Internal Medicine, Johann Wolfgang Goethe-University, Theodor-Stern-Kai 7, 6000 Frankfurt am Main 70, FRG

Reinecke, M., Dr. rer. nat.
Dept. Anatomy III, University of Heidelberg, Im Neuenheimer Feld 307, 6900 Heidelberg, FRG

Riegger, A.J.G., Prof. Dr.
Medizinische Universitätsklinik Würzburg, Josef-Schneider-Str. 2, 8700 Würzburg, FRG

Rippegather, G., Dipl.-Biol.
Anatomisches Institut III, Universität Heidelberg, Im Neuenheimer Feld 307, 6900 Heidelberg, FRG

Scheuermann, D.W., Prof. Dr. med.
Institute of Histology and Microscopic Anatomy, University of Antwerp, Groenenborgerlaan 171, B-2020 Antwerp, Belgium

Schnermann, J., Prof. Dr. med.
University of Michigan, Medical School, Department of Physiology, Medical Science Building II, 7712 Ann Arbor, Michigan 48109, USA

Stumpf, W.E., Prof. MD.
Department of Anatomy, University of
North Carolina, 111, Swing Building,
Chapell Hill NC 27514, USA

Tarlartschik, J., Dr. med.
Medical University Clinic, Robert-
Koch-Str. 40, 3400 Göttingen, FRG

Thompson, R.P., Ph.D.
Professor of Anatomy, Department
of Anatomy, Medical University of
South Carolina, 171 Ashley Avenue,
Charleston, SC 29425, USA

Weidmann, P., M.D.
Professor of Medicine, Medizinische
Universitätspoliklinik, Freiburgstr. 3,
CH-3010 Berne, Switzerland

Weil, J., M.D.
Haunersche Kinderklinik der Universität München, Lindwurmstraße 4,
8000 München 2, FRG

Vollmar, A.M., Dr. med.
Institut für Pharmakologie, Toxikologie
und Pharmazie der Tierärztlichen
Fakultät der Universität München,
Königinstr. 16, 8000 München 22, FRG

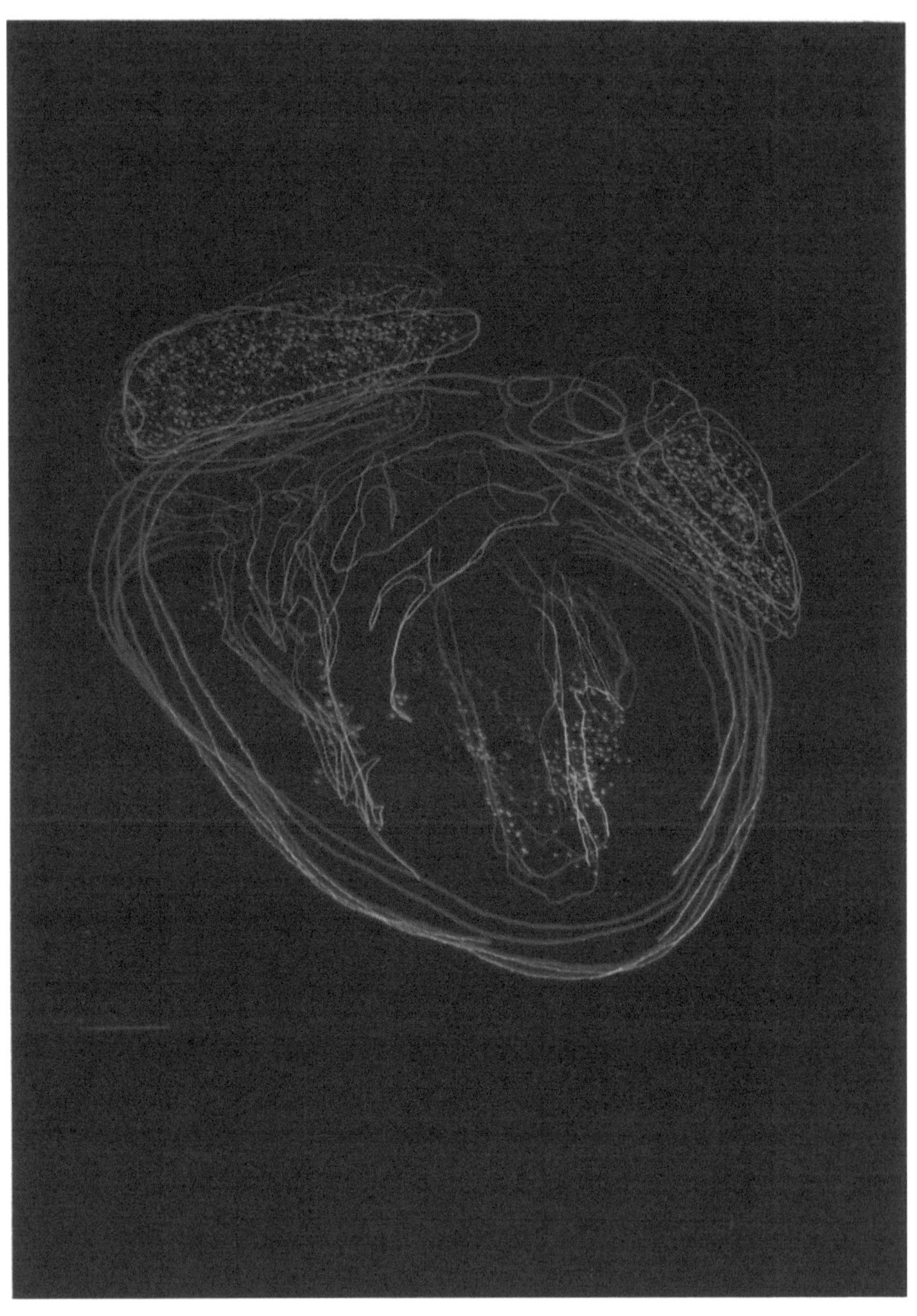

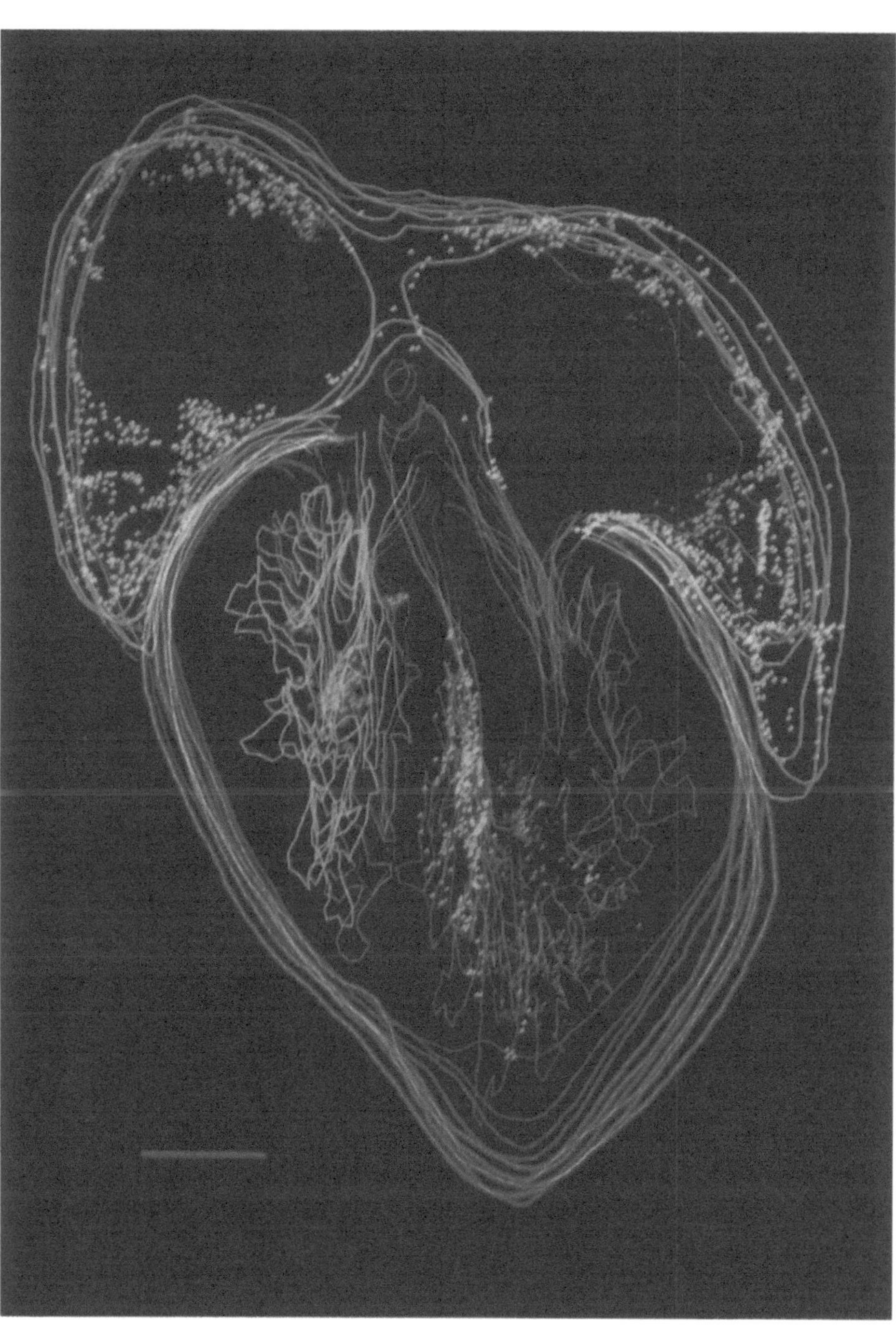

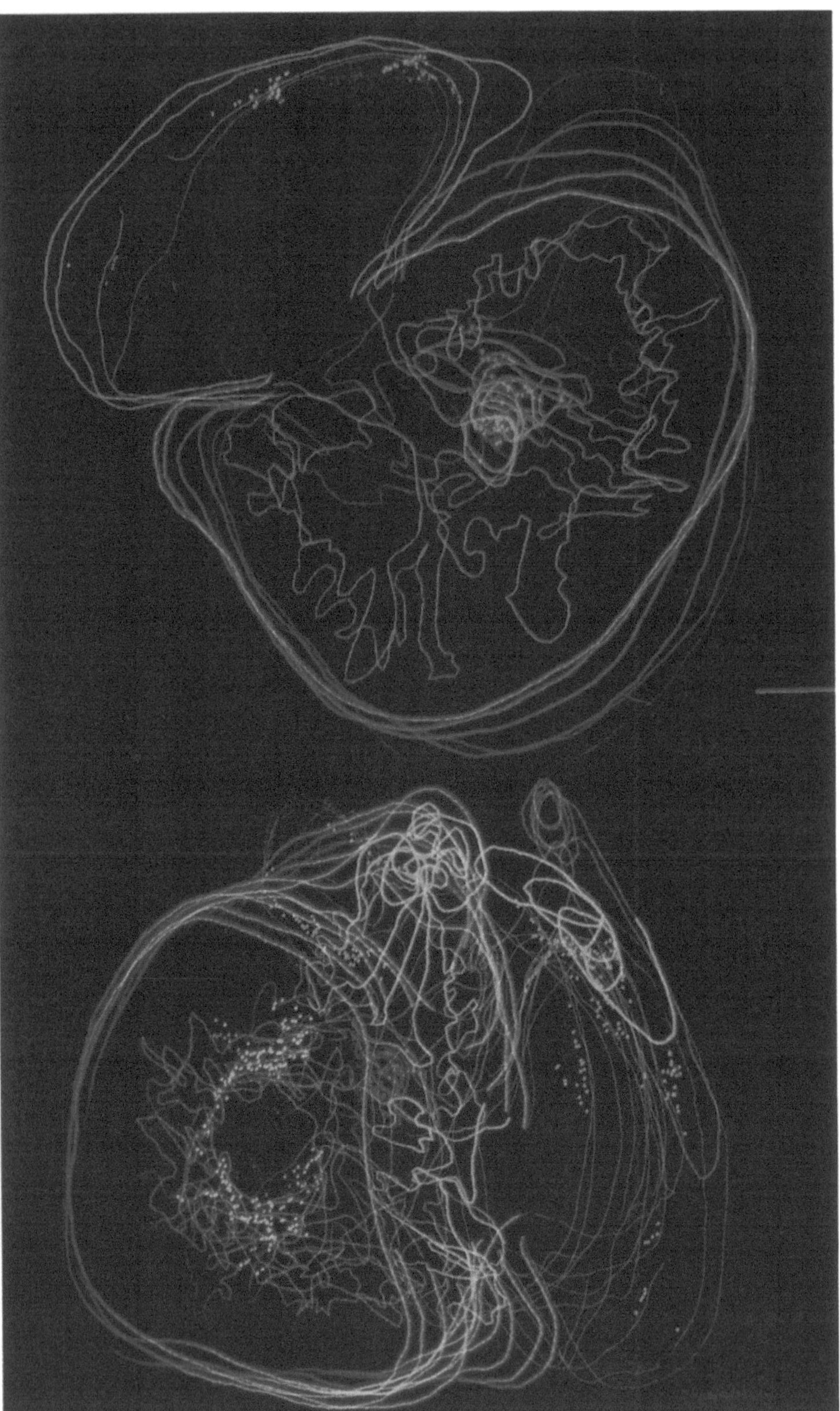

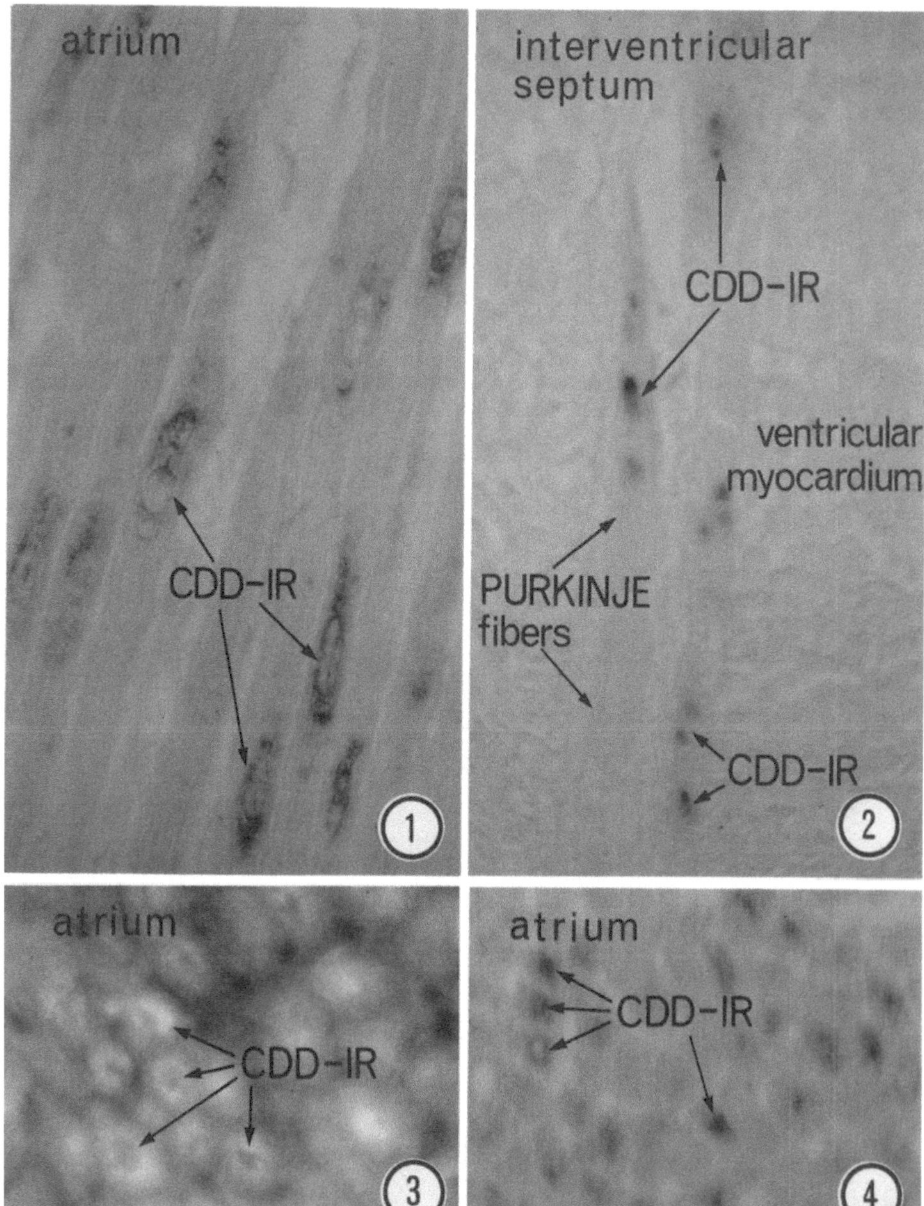

atrium
CDD-IR
interventricular
septum
CDD-IR
ventricular
myocardium
PURKINJE
fibers
CDD-IR
1
2
atrium
CDD-IR
3
atrium
CDD-IR
4

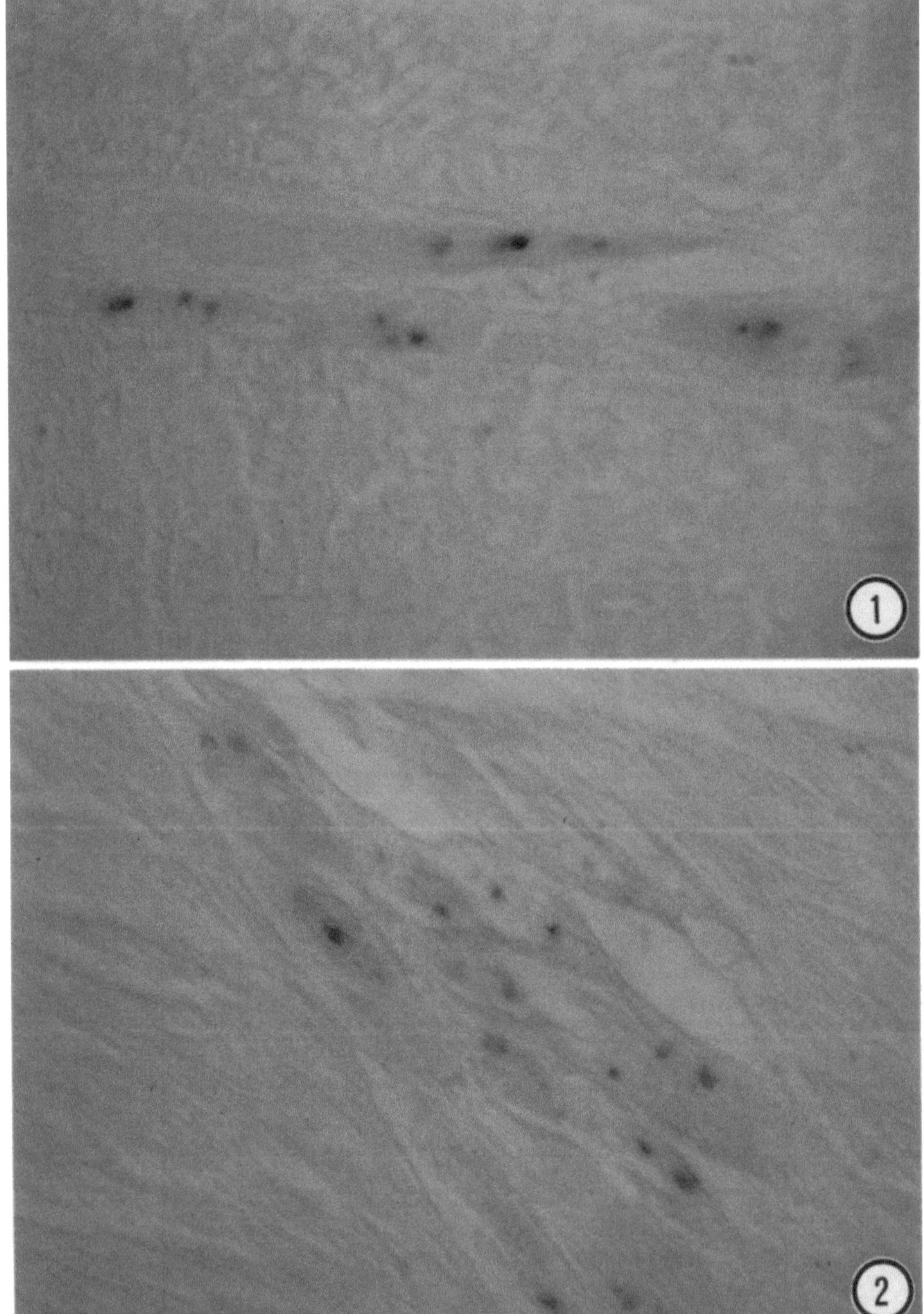

Embryology of the endocrine heart

R.P. Thompson, J. Göbel, J.R. Lindroth[1], M.G. Currie[2]

[2]Department of Pharmacology Medical University of South Carolina, Charleston, South Carolina, USA; and [1]Department of Anatomy and Cell Biology, University of Heidelberg, FRG

Summary

Recent interest in the storage, release, and action of atrial natriuretic peptide (ANP) in adult mammals has directed our attention to the early appearance, distribution, and possible function of this cardiac hormone during embryonic development. This report summarizes efforts to map and quantify cardiac ANP in developing and neonatal rats, rabbits, and pigs.

Levels of ANP immunoreactivity in fetal and neonatal hearts were measured by radio-immunoassay of atrial and ventricular homogenates. Histochemical localization of ANP deposits was accomplished by immunoperoxidase staining of paraffin embedded tissues. The antibodies used were raised in rabbits or guinea pigs against purified rat pro-ANP or synthetic fragments of porcine pro-ANP.

In series of rat and rabbit hearts spanning the period from cardiac septation (gestational day 15) to birth, RIA-detectable ANP levels in atrial homogenates rose steadily to maximum values near birth. In ventricular homogenates, lower levels of ANP immunoreactivity declined gradually throughout the embryonic and neonatal period from highest levels in the early embryonic period toward extremely low levels in the adult. During the morphogenetic period of cardiac septation, ANP immunoreactivity intensified within the inner layers of atrial myocardium and along ventricular trabeculae in apparent association with the coalescing cardiac conducting system, especially apparent in the embryonic pig.

Introduction

Adult mammalian atrial myocardium features a cell population characterized by specific atrial granules (2,12,13). Recent efforts in numerous laboratories represented at this symposium have established that these granules contain a family of peptide hormones referred to as atrial natriuretic peptide (ANP) (atriopeptin, cardiodilatin, and others), which cause natriuresis, diuresis, inhibition of aldosterone secretion, and vasodilation in adults (7,10,11).

Despite considerable interest in the role of ANP in the adult, little is known concerning the appearance of ANP in embryonic life or its possible role in development. Early ultrastructural studies noted secretory or neurosecretory granules in atrial and ventricular myocardium of rat and mouse embryos (6,21), in the ventricular wall of the eight-week human embryo (14), in ventricular muscle of chick embryos (18), and lower vertebrates (15), and in conducting tissue of adult rat (2). These dense-core, membrane-bound granules were similar in appearance to the more abundant granules reported in adult mammalian atria, now thought to be the principal storage site for ANP (5,20). Recent morphological studies of ANP during embryonic and neonatal development of the rat (26,28) have demonstrated ANP accumulation in the cardiac ventricles, as well as atria, early in development. The distribution of ANP storage sites within the ventricles suggest that ANP may be a structural feature, perhaps a functional participant, in coalescence of peripheral portions of the cardiac conducting system, recently shown to contain ANP in the adult rat (1). High tissue and plasma levels during the perinatal period in the rat have also now been reported (4,31). Together, these studies suggest participation of ANP during development and the need for futher systematic mapping and quantitation of ANP accumulation in developing mammals, especially larger mammals with well-defined conducting tissues.

This study provides such new data concerning the appearance and distribution of ANP in the rat, rabbit, and pig heart during the embryonic and perinatal period as detected by immunohistochemistry and radioimmunoassay.

Materials and methods

Embryos of known gestational age were removed by laparotomy under sodium pento-barbital from timed-pregnant Long-Evans rats (Charles River Laboratories, Raleigh, North Carolina), and from white New Zealand rabbits (Rabbits, Ltd., Summerville, South Carolina). Yorkshire-cross commercial pig embryos were obtained at slaughter (Marvin's Meats, Hollywood, South Carolina).

For *immunohistochemical tests*, cardiac and thoracic tissues were dissected in saline and immersed in Carnoy's fixative overnight at 4 °C. After embedding in paraffin, serial sections were cut at 5 or 10 microns. Immunostaining methods were adapted from methods previously described by (19,25). The antibodies used were raised in rabbits or guinea pigs against purified rat pro-ANP or synthetic fragments of porcine pro-ANP, with primary antibody dilutions ranging from 1:50 to 1:1000. Secondary and tertiary antibodies were goat anti-rabbit immunoglobulin, rabbit anti-horse radish peroxidase (HRP) and goat anti-guinea pig immunoglobulin-peroxidase conjugate at 1:10 to 1:100 dilutions. HRP (1%) was added when rabbit anti-HRP was used.

Adjacent control sections were either incubated with normal rabbit or normal guinea pig serum in place of immune serum or stained for glycogen with periodic acid/Schiff reagent (PAS). The specificity of the antisera was assessed as described elsewhere (1,24).

Radioimmunoassay for ANP was performed as reported elsewhere (8).

Computer images were reconstructed on an Evans and Sutherland PS340 Colour Graphics System interfaced with an IBM PC/AT for data input. Outlines of selected tissue structures were traced, aligned, and digitized manually, essentially as described previously (32). The locations of immunoreactive cells were determined by alignment of such contours with a video image from a Dage M68 video camera followed by auto-mated cluster analysis of the digitized image as described elsewhere (16). Following adjustment of color, rotation, clipping and depth cueing, photographs were taken directly

2

from the Evans and Sutherland screen using 4 x 5 Fujichrome 100 transparency film and direct positive printing.

Results

Radioimmunoassay: Levels of ANP in the dissected atria and ventricles of developing rats and rabbits are shown in Fig. 1, expressed as nanograms of ANP per milligram of fresh, blotted tissue. In both species, highest reactivity was observed in atrial homo-

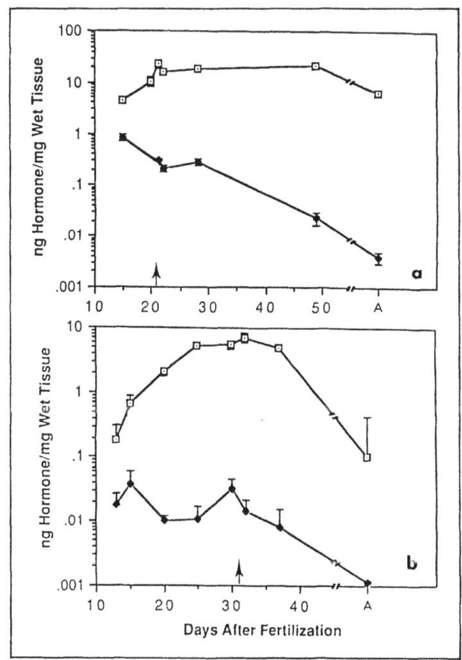

Fig. 1. ANP levels in homogenized whole atria (open boxes) or ventricles (closed boxes) in developing rats (a) and rabbits (b). Ordinate: ng hormone/mg blotted tissue pooled from left and right chambers. Abscissa: days after fertilization, with birth indicated by arrows; A = adult. Each point represents mean (S.E.M.) of 3-8 samples.

genates from the neonatal period, with lower levels present during the critical development period of cardiac septation (gestational days 14-16) and in the adult.

Ventricular homogenates from both species contained considerably less radioimmunoassayable ANP throughout life, with less dramatic fluctuations during embryonic and fetal development and decreases in the first few weeks following birth decreasing to extremely low levels in the adult.

Immunohistochemistry of the developing rat heart: The two principal cardiac sites of ANP immunoreactivity in the developing rat are clearly illustrated in Fig. 2. Intense staining was routinely observed within the atrium and trabeculated regions of the left ventricle in embryos of gestational age 14, 16, and 19 days. Lesser sites of focal staining (not shown) were seen in the right ventricular trabeculae and the muscular walls of both ventricles at these ages and along the primitive trabeculae of the tubular heart as early as 11 days of gestation.

In the neonatal rat (Fig. 3), ANP-immunoreactivity was strongest in the well-developed atrial appendages, especially along the inner layers of atrial myocardium. In the neonatal left ventricle, ANP-positive cells were found along the minor trabeculae and

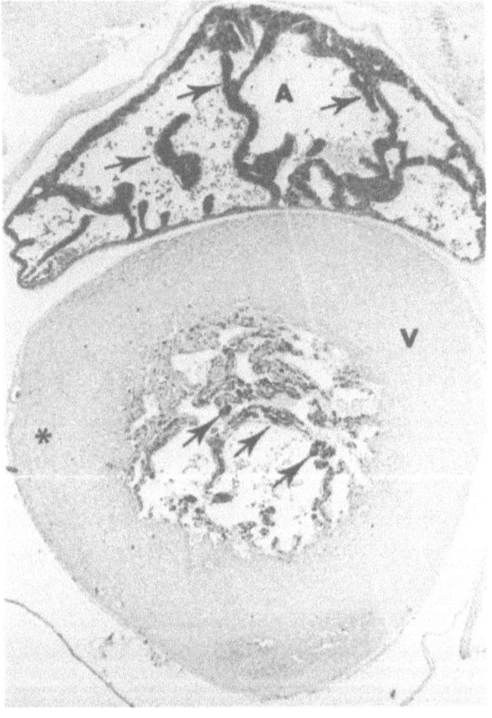

Fig. 2. Parasagittal section of embryonic rat heart (19 days of gestation) illustrating intense ANP-immunoreactivity (arrows) within left atrium (A) and trabeculated regions of the left ventricle (V) compared with the ventricular free wall (*). (x50)

subendocardial layer of the interventricular septum, in apparent association with the left bundle branch of the cardiac conducting tissue and with the base of the principal papillary muscles. Scattered foci of immunoractivity were also seen along the trabeculae of the right ventricle of neonatal rats. No identifiable foci of ANP-reactivity were seen within central regions of the conducting system in dozens of fetal and neonatal rat hearts examined.

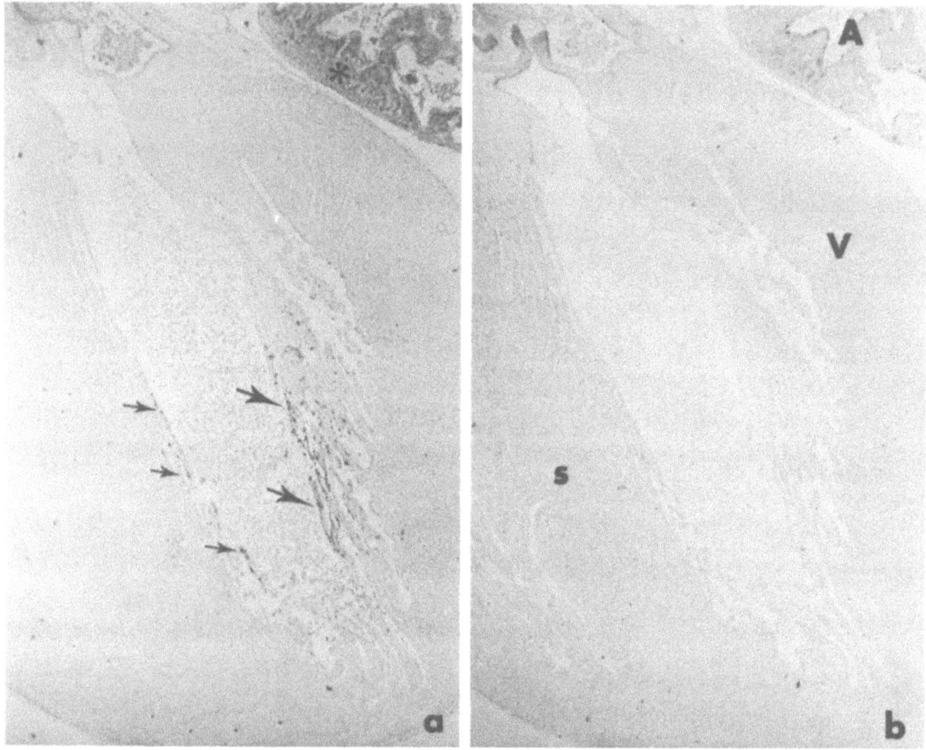

Fig. 3. Cardiac frontal view of neonatal rat heart reacted with ANP-antiserum (a) or normal rabbit serum (b). Note immunostaining in atrium (A,*), subendocardial tracts and minor trabeculae along the septal wall of the left ventricle (small arrows) and papillary muscle (large arrows) and relative lack of staining in left free wall (V) and interventricular septum (S) (x40) (cf. Plate 1)

A three-dimensional reconstruction of ANP-positive sites from serial sections through the neonatal rat heart shown in Fig. 3 is presented in Plate 1 in front of the book. Note the widespread distribution of ANP in both atria, along the interventricular septum and anterior papillary muscle of the left ventricle, and lesser scattered foci along the interior of the right ventricle.

Immunohistochemistry of developing rabbit hearts: Under similar staining conditions as those used for rat tissues, foetal rabbit tissues displayed high background staining, due probably to non-specific binding of secondary goat-anti-rabbit-peroxidase conjugate with ubiquitous rabbit antigens. Using primary antibody raised in guinea pigs against rat ANP, in conjunction with a commercial goat-anti-guinea pig secondary antibody, we observed specific ANP-positive reactions within the atria and at isolated subendocardial sites within the ventricles of rabbit embryos at 15 and 30 days of gestation (data not shown). Staining of rabbit tissues was neither as intense nor as widespread as the reactions observed in the rat, due perhaps to reduced cross-reactivity of antibody with rabbit ANP.

5

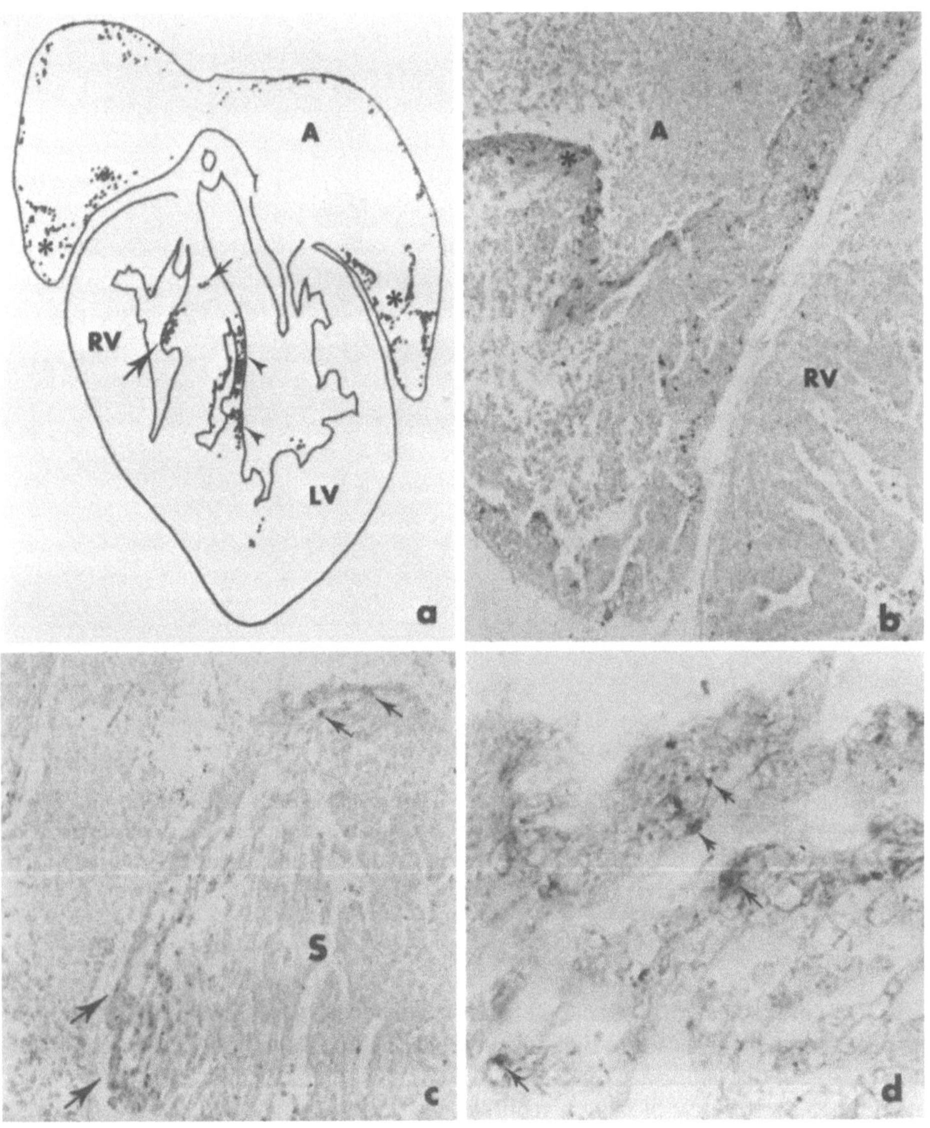

Fig. 4. Cardiac frontal view of 18mm pig embryo heart (ca. 25 days of gestation) showing ANP-immunoreactivity along atrial trabeculae (*, A in a,b), developing right bundle branch (large arrows in a,c) and fainter staining in a few cells (small arrows a,c,d) along the crest of the interventricular septum (S). Arrowheads (a) indicate location of ANP-positive cells along septal surface of the left ventricle; (a= x25; b,c= x150; d= x1000) (cf. Plate 2)

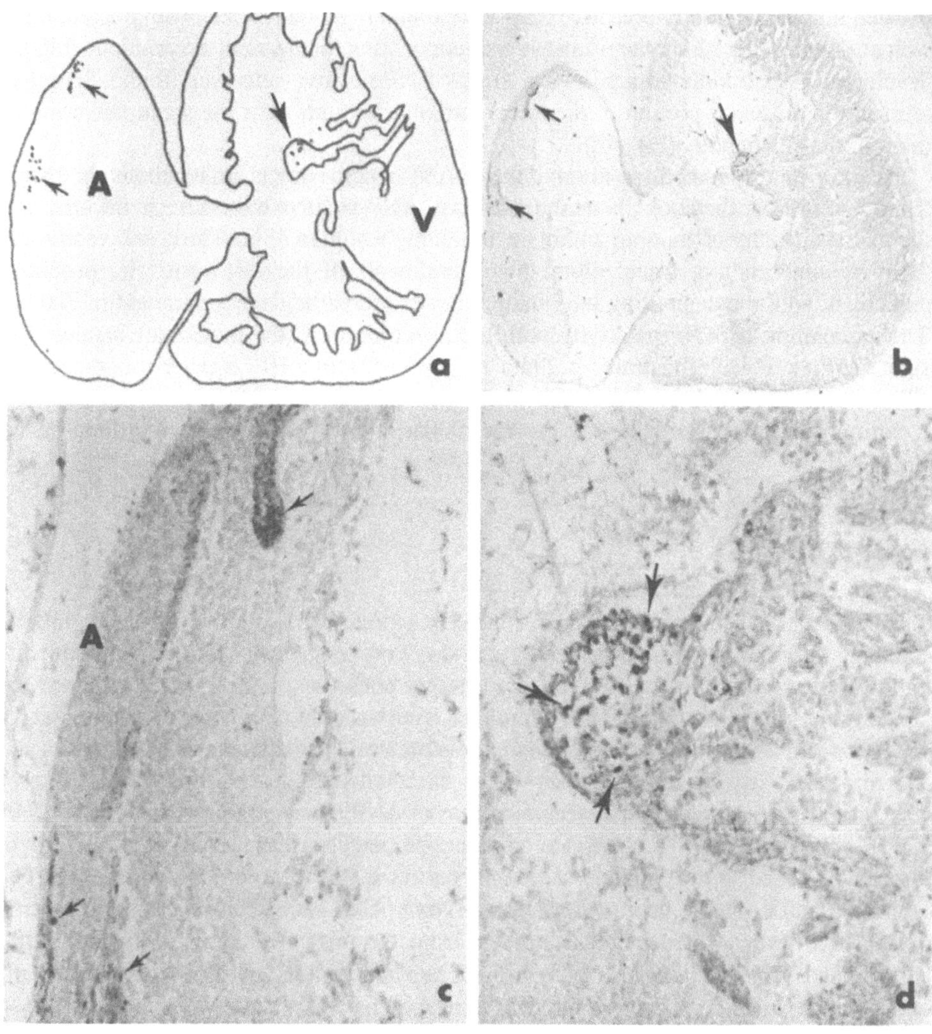

Fig. 5. Parasagittal section of 17mm pig embryo heart showing ANP-immunoreactivity along atrial myocardium (A, small arrows, a,b,c) and within the primordial moderator band of the right ventricle (large arrows, a,b,d). Small arrows (a,c,d) indicate faintly ANP-positive cells along crest of interventricular septum (s). (a,b= x30; c,d= x210) (cf. Plate 3)

Immunohistochemistry of embryonic pig heart: In the hearts of 17-18 mm pig embryos (Figs.4 and 5), distinct ANP-immunoreactivity was found in several atrial and ventricular sites. Within the incompletely divided atria, positive staining was particularly prominant along the folded primordia of the pectinate muscles. In the left ventricle, ANP-immunoreactivity was seen along the septal wall and adjacent minor trabeculae, in apparent association with the developing left bundle branch and more peripheral fibers ramifying to the free wall and base of the papillary muscles. In the right ventricle, distinct and

7

isolated bundles of immunoreactive cells were found high in the septal wall in continuity with additional sites along the primitive moderator band, in apparent association with the developing right bundle branch. A few faintly ANP-positive cells were found along the crest of the muscular portion of the interventricular septum, near the presumed primordium of the His bundle (Fig. 4d).

The three dimensional disposition of these ANP-reactive tracts are illustrated in Plates 2 and 3 in front of the book . Note the strands of ANP-positive trabeculae in the atria and the relative paucity of staining within the mural myocardium of both atria and ventricles. ANP-immunoreactive tracts along the septal wall of the left ventricle, probably coincident with the coalescing left bundle branch, are particularly prominant in Plate 1. The association of ANP-reactivity with the primordium of the moderator band of the right ventricle is well illustrated in Plate 2a. Viewed from a left superior oblique angle (Plate 2b), immunoreactive trabeculae within the left ventricle of this same heart are seen to form a ring of tissue encircling the ventricular interior below the primordium of the mitral valve.

Discussion

In summary, ANP immunoreactivity is present in the hearts of embryonic rats, rabbits and pigs during critical developmental periods of organogenesis, cardiac septation, and birth. These results confirm and extend to larger mammals, our original finding in the developing rat (26). Others (28) reported similar results for the rat and clearly demonstrated immunohistochemical and ultrastructural localization of ANP-deposits and electron dense cytoplasmic granules within the tubular rat heart as early as 11 days of gestation, with immunoreactive accumulations of ANP demonstrable in such granules by day 14. The morphogenetic period of cardiac septation spans gestational days 13.5-16 in the rat (27) and approximately the same period in the developing rabbit (23). The secretion and probable action of such embryonic ANP stores can be strongly inferred from such immunohistochemical studies, from the present demonstration that atrial tissue levels of ANP peak near birth in both rats and rabbits, and from its presence and responsiveness in the plasma of the near term rat fetus, less than a week later (4,31). As noted by those authors, however, specific and possibly important roles of ANP in the regulation of fetal or postnatal vascular resistance, fluid and electrolyte balance, or other systemic actions, remain open to speculation and further experiment.

Ventricular ANP storage sites have also proved of interest in the embryology of the endocrine heart. The present study confirms and extends to rabbits and pigs the recent finding in rats that ANP accumulates within the cardiac ventricles during the embryonic and fetal period (4,26,28,29,31). The relative abundance of atrial ANP stores in developing rats and rabbits renders it unlikely that such ventricular ANP stores contribute significantly to circulating ANP levels or actions in the fetal period of higher vertebrates. Early in development, however, or in lower vertebrates, ventricular ANP may contribute a larger fraction of circulating hormone levels, since it is found in the walls of the tubular rat heart (28) and similar neurosecretory-like granules have been reported in the ventricles of embryonic rodents (6,21), chicks (18), and adult fish (15).

The distribution of ventricular ANP in the rat and pig during or soon after formation of distinct cardiac chambers leaves little doubt that ANP is indeed a salient feature of early conducting tissues in several mammals, probably in the human as well. The embryo and adult pig hearts share important similarities in cardiac development with those of humans and other large mammals (22,23), as well as in conducting tissue organization (30). Apart from a possible role in regulating early coordination of cardiac function, the three dimensional distribution of ANP within the newly formed ventricles, as presented here, constitutes strong evidence that peripheral portions of the cardiac conducting system coalesce during development from early trabecular tissue (3,9,17). The effect of such ANP accumulations upon the strength and timing of contractility, and the opposing question of whether such ANP sites organize in response to uneven physical stress or reflect the predetermined fate of specific ventricular myocyte populations, remain important questions for further experimental study.

Acknowledgements: We thank Dr. W.G. Forssmann for support of this collaboration and for antibodies to synthetic porcine pro-ANP fragments, Howie Schomer for technical assistance; Mathew Wong for the design and supervision of computer resources; and Marion Hinson for preparation of the manuscript.

This work was supported by a fellowship from the Deutsche Forschungsgemeinschaft (JG), an Established Investigator Award (RPT) and grant-in-aid No. 85-1282 (MGC) from the American Heart Association with funds contributed in part by their South Carolina affiliate, and by NHLBI grants No. 28936/37704 (RPT) and 29566 (MGC).

References

1. Back, H., Stumpf, W.E., Ando, E., Nokihara, K., Forssmann, W.G. (1986) Immuno-cytochemical evidence for CDD/ANP-like peptides in strands of myoendocrine cells associated with the ventricular conduction system of the rat heart. Anat Embryol 175: 223-226

2. Bompiani, G.D., Rouiller, C.H., Hatt, P.Y. (1959) Le tissu de conduction du coeur chez le rat. Etude au microscope electronique. Archives des Maladies du Coeur et des Vaisseaux 52: 269-292

3. Canale, E.D., Campbell, G.R., Smolich, J.J., Campbell, J.H. (1986) Cardiac Muscle. Springer, Berlin Heidelberg New York Tokyo, pp 185-186

4. Cantin, M., Ding, J., Thibault, G., Gutkowska, J., Salmi, L., Garcia, R., Genest, J. (1987) Immunoreactive atrial natriuretic factor is present in both atria and ventricles. Mol Cell Endocrinol 52: 105-113

5. Cantin, M., Gutkowska, J., Thibault, G., Milne, R.W., Ledoux, S., Min Li, S., Chapeau, C., Garcia, R., Hamet, P., Genest, J. (1984) Immunocytochemical localization of atrial natriuretic factor in the heart and salivary glands. Histochem 80: 113-127

6. Challice, C.E., Viragh, S. (1973) The embryonic development of the mammalian heart. In: Challice C.E. and Viragh, S. (eds.) Ultrastructure of the Mammalian Heart. Academic Press, New York, London, pp 91-126

7. Currie, M.G., Geller, D.M., Cole, B.R., Siegel, N.R., Fok, K.F., Adams, S.P., Eubanks, S.R., Galluppi, G.R., Needleman, P. (1984) Purification and sequence analysis of bioactive atrial peptides (atriopeptins). Science 223: 67-69

8. Currie, M.G., Newmann, W.H. (1986) Evidence for a-1 adrenergic regulation of atriopeptin release from the isolated rat heart. Biochem Biophys Res Comm 1: 94-100

9. Forsgren, S. (1984) The differentiation of the Purkinje fibres in the mammalian heart; comparisons with the ordinary myocytes. In: Legato, M.J. (ed) The Developing Heart. Martins Nijhoff Publ. Co., Boston, pp 47-67

10. Forssmann, W.G., Hock, D., Kirchheim, F., Metz, J., Mutt, V., Reinecke, M. (1984) Cardiac hormones: Morphological and functional aspects. Clin and Exper Theory and Practice, A6 (10-11), 1873-1878

11. Flynn, T.H., DeBold, M.L., DeBold, A.J. (1983) The amino acid sequence of an atrial peptide with potent diuretic and natriuretic properties. Biochem Biophys Res Communn 117: 359-865

12. Jamieson, J.D., Palade, G.E. (1964) Specific granules in atrial muscle cells. J Cell Biol 23: 151-172

13. Kisch, B. (1956) Electron microscopy of atrium heart. I. Guinea pig. Exp Med Surg 14: 99-112

14. Leak, L.V., Burke, J.F. (1964) The ultrastructure of human embryonic myocardium. Anat Rec 149: 623-650

15. Lemanski, L.F., Fitts, E.P., Marx, B.S. (1975) Fine structure of the heart in the Japanese Medaka *Oryzais latipes*. J Ultrastruct Res 53: 37-65

16. Lindroth, J., Starr, C., Thompson, R., Abutaleb, A. (1986) A digital cluster analysis system to locate labelled cell regions in autoradiographic images. In: Orphanoudakis, S.C. (ed) Proceedings XII Northeast Bioengineering Conference pp 87-90

17. Mall, F.P. (1912) On the development of the human heart. Amer J Anat 13: 249-298

18. Manasek, F.J. (1969) The appearance of granules in the Golgi complex of embryonic cardiac myocytes. J Cell Biol 43: 605-609

19. Mason, T.E., Phifer, R.F., Spicer, S.S., Swallow, R.A., Dreskin, R.B. (1969) An immunoglobulin-enzyme bridge method for localizing tissue antigens. J. Histochem Cytochem 17: 563-569

20. Metz, J., Mutt, V., Forssmann, W.G. (1984) Immunohistochemical localization of cardiodilatin in myoendocrine cells in the cardiac atria. Anat Embryol 170: 123-127

21. Nanot, J., LeDouarin, G. (1970) Etude au microscope electronique du myocarde du foetus de souris. C R Soc Biol 164: 890-893

22. Patten, B.J. (1948) Embryology of the Pig. McGraw-Hill, New York

23. Sissman, N.J. (1970) Developmental landmarks in cardiac morphogenesis: comparative chronology. Am J Cardiol 25: 141-148

24. Saper, C.B., Standaert, D.G., Currie, M.G., Schwartz, D., Geller, D.M:, Needleman, P. (1985) Atriopeptin-immunoreactive neurons in the brain: presence in cardiovascular regulatory areas. Science 227: 1047-1049

25. Sternberger, L.A. (1979) Immunocytochemistry, 2nd edition. Wiley, New York

26. Thompson, R.P., Simson, J.A., Currie, M.G. (1986) Atriopeptin distribution in the developing rat heart. Anatomy and Embryology 175: 227-233

27. Thompson, R.P., Wong, M., Fitzharris, T. (1983) A computer graphic study of cardiac truncal septation. Anat Rec 206: 207-214

28. Toshimori, H., Toshimori, K., Oura, C., Matsuo, H. (1987a) Immunohisto- chemical study of atrial natriuretic polopeptides in the embryonic, foetal and neonatal rat heart. Cell Tissue Res 248: 627-633

29. Toshimori, H., Toshimori, K., Oura, C., Matsuo, J. (1987b) Immunohistochemistry and immunocytochemistry of atrial natriuretic polypeptide in porcine heart. Histochem 86: 595-601
30. Truex, R.C., Smythe, M.Q. (1965) Comparative morphology of the cardiac conduction tissue in animals. Annals NY Acad Sci 127: 19-33
31. Wei, Y., Rodi, C.P., Day, M.L., Wiegand, R.C., Needleman, L.D., Cole, B.R., Needleman, P. (1987) Developmental changes in the rat atriopeptin hormonal system. J Clin Invest 79: 1325-1329
32. Wong, Y.M., Thompson, R., Cobb, L., Fitzharris, T. (1983) Computer reconstrucion of serial sections. Computers in Biomed Res 16: 580-586

Color codes.

Green = atrial walls (lines), or ANP-positive cells (dots); Red/Gold = left ventricular lumen (lines), or ANP cells (dots); Blue = right ventricular lumen (lines), or ANP cells (dots); Magenta = exterior of the ventricles (Bar equals 500 microns along horizontal axis)

Plate 1. Cardiac ANP distribution in the neonatal rat (one day old, anterior view). Note numerous ANP cells throughout the atria. In the left ventricle, ANP-positive cells were found along the septal wall and within the base of the anterior papillary muscle (highlighted in gold). A few ANP-positive cells were found at scattered subendocardial sites within the right ventricle. (cf. Fig. 3)

Plate 2. Cardiac ANP distribution in an 18mm pig embryo (cardiac frontal plane, color codes as in Plate 1). Note ANP-positive cells along myocardial strands within atrial appendages and lesser foci across atrial wall. Left ventricular ANP-positive cells were clustered along the septal wall and scattered across the mural myocardium at the base of the papillary muscles. In the right venticle, clusters of immunoreactive cells were seen along the developing right bundle branch with a few faintly positive cells at the crest of the muscular interventricular septum. (cf. Fig. 4)

Plate 3. Cardiac ANP distribution in a 17mm pig embryo (a= right lateral view; b= left superior oblique view, color codes as in Plate 1). Strands of ANP-positive cells were seen along the posterior wall of the incompletely divided atria. In the right ventricle (a), ANP-positive cells were restricted to the primitive moderator band just below the right ventricular infundibulum. In the left ventricle, as viewed from near the embryonic left shoulder with left atrium and primitive mitral valve removed (b), ANP-positive cells were distributed along a ring of myocardial trabeculae coursing from the septal wall (at top) around the interior of the left ventricle in the primordial papillary muscles. (cf. Fig. 5)

Morphological review: immunohistochemistry and ultrastructure of the endocrine heart

W. G. Forssmann

Department of Anatomy and Cell Biology, University of Heidelberg, FRG

Introduction

According to early light and electron microscopic investigations, two main myocardial cell types, which were functionally characterized as working myocardial fibers and cells of the conductive system, could be distinguished in the heart. In addition to these two cell lines a third myocardial cell type was first recognized about 30 years ago. In 1956, Kisch described a special form of myocardial cell with dense inclusions, which were further mentioned by Poche (128), characterized by Bompani et al. (18), and Jamieson and Palade (84), and analyzed by several other authors in the 1960s and 1970s. In some of these publications, the secretory nature of these specific granules was postulated and then strongly supported by Jamieson and Palade (84). Most investigators (10,31,32,81, 121,147,160) expected the granules to contain catecholamines, as initially implied by works of Shore et al. (143), Östlund et al. (125), Bloom et al. (15) and Palade (126). It was, however, already evident from fluorescence microscopy on cardiac nerves that these granules contain the bulk of atrial catecholamines (28,48). In a number of studies published in the early 1970s the nature of these specific endocrine-like cells was again investigated by ultrastructural, histochemical, and autoradiographic methods (11, 96,104, 121,151,138) with respect to the possibility of producing a protein substance (9,21,23,33,34,37,75,80,81,82,83,161). However, in 1976, a crucial discovery was made by Marie et al. (103) who observed that the atrial granules were influenced by changes in the sodium-electrolyte balance, which was further elaborated on by DeBold (30). Consequently, on the basis of these results, DeBold and coworkers (35) studied the biological effects of crude atrial extracts, which were found to produce a strong diuresis in rats, and thus introduced a diuretic test for atrial extracts. Also, when studying the biological activities of atrial peptides, other groups found a relaxant effect on vascular smooth muscle. This vasorelaxant effect exerted by an unknown atrial substance was first described by Deth and coworkers (38), and probably independently by others (26, 56,69). The diuretic and vasorelaxant hormones from the heart could thus be isolated by following the bioactivity, the vasorelaxant or diuretic effect, during the isolation procedure. In our laboratory, we isolated a bioactive substance from porcine atria using the differential relaxation of precontracted smooth muscle strips of the aortic, renal arterial and mesenteric arterial walls as a biotest (56), which was followed by different chromatography steps. In this work, in collaboration with Mutt from Stockholm, we reported the first successful isolation of "*cardiodilatin*", a new peptide hormone of the right atrium which was shown to contain a completely unknown amino acid sequence when compared to proteins and hormonal polypeptides published so far at that time. During the course of this work, other groups were also active in this field and isolated cardiac hormones, which were then published under various names. In 1983, Flynn et al. described *"cardionatrin"* from rat which contained 28 amino acids and, in 1984, Kangawa and Matsuo isolated a homologous *"alpha–ANP"* (alpha atrial natriuretic polypeptide) from human atria. Many additional publications during the same year on the isolation of various polypeptide fragments of this family document the strong interest in this field (49). After studies on the sequence of cDNA and the genetics of cardiac hormones (70,122,113,141,115) had been carried out, various results of different authors could be explained. It was found that the hormone precursor with a length of 151 amino acids is cleaved to form a signal peptide of 25 amino acids and the cardiodilatin-1-126, which we had isolated (56,50).

Methodological basis of the morphological studies

The morphological studies for this review were carried out with several mammalian species, using perfusion-fixed hearts for light microscopical immunohistochemistry, electron microscopy, and tracer studies. Detailed methodological descriptions are published elsewhere (46,59). Cardiodilatin was analyzed by immunocytochemical methods using antibodies raised in our laboratory. The hydrophilic regions of the molecule were synthetized and various oligopeptide fragments were used for immunization. The antibodies raised against the N-terminal (pos.1-7 and pos. 8-24) and C-terminal (pos. 99-126, pos. 117-126) regions of the prohormone molecule were particularly successful in their application. The immunogold method was used, as mentioned in more detail in an earlier publication of our study group (102). The specific antibodies against various synthetic CDD-fragments (119) resulted, in general, in the same immunohistochemical detection.

General structural organization of the endocrine heart and immunohistochemistry of the myoendocrine cell

As shown in several previous publications the immunoreactivity of the cardiodilatin-like substance in the mammalian atrial appendage is seen both with the fluorescence and peroxidase-anti-peroxidase methods (Plate 4, in the front of the book). Cardiodilatin

Plate 4: Fig. 1. Light microscopy of endocrine cells in the heart atrium; using immunocytochemical methods the myoendocrine cells of the right atrium are stained by the PAP (peroxidase-anti-peroxidase) technique applying anti-CDD antibodies. Note the CDD-IR (cardiodilatin-immunoreactivity) is found mostly in the perinuclear areas. (x 560)
Fig. 2. Single strands of CDD-IR cells are also found in the ventricle, then confined to the conductive system. Here some septal Purkinje-fibers of the porcine heart are seen stained by an anti-CDD-(1-7) antibody. (x 560)
Fig. 3. CDD-IR in atrial myoendocrine cells of the porcine heart in transverse section stained with the immunofluorescence method using a CDD-99-126 specific antibody. Note all cells are strongly fluorescent in the cytoplasmic regions around the nuclei. (x 560)
Fig. 4. Similar demonstration of CDD-IR of transsected atrial myoendocrine cells as shown in the fotoraph *(left)*, but stained when the PAP-method is applied using anti-CDD-99-126 antibodies. (x 560)

14

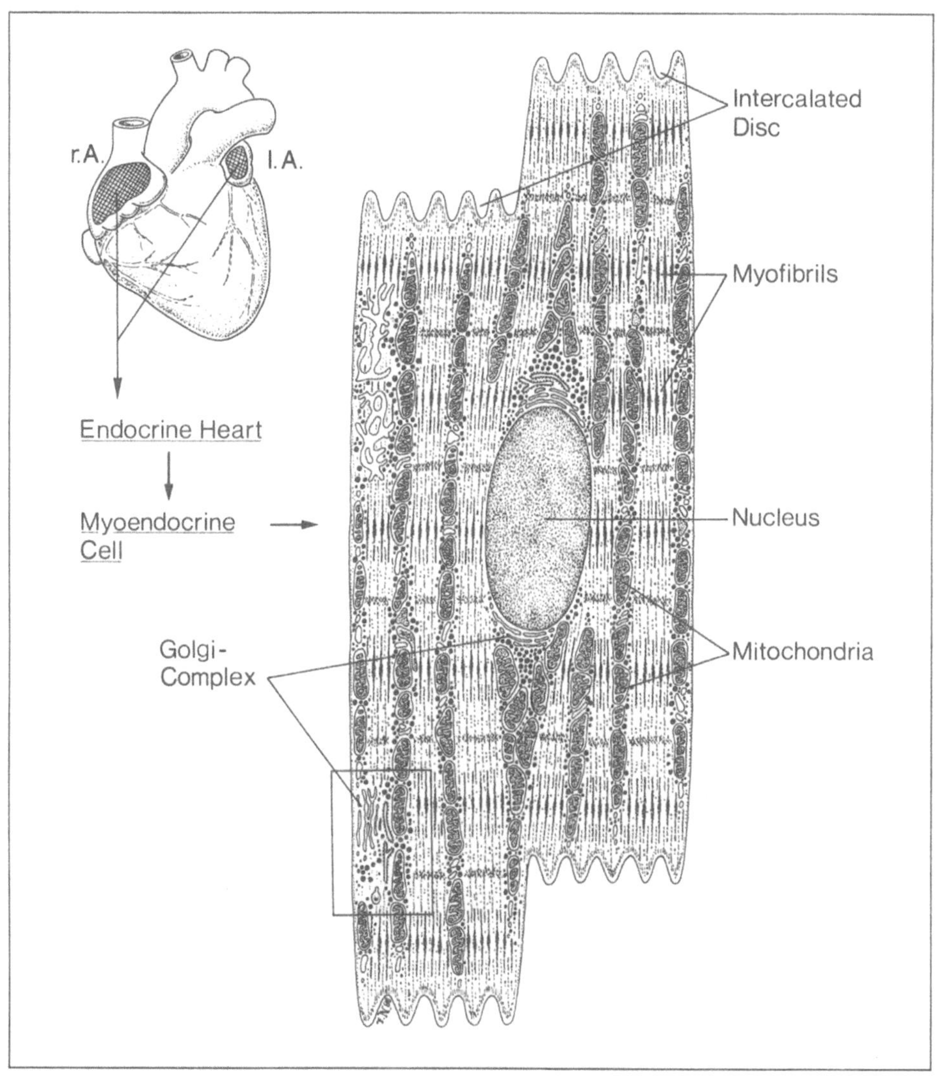

Fig. 5. Schematic view of the endocrine heart and its myoendocrine cells. The compact endocrine organ of the heart is located in the thin-walled, trabecular parts of the atria which form the atrial appendages normally well delimited from the other parts of the atrial walls. The rest of the heart contains a diffuse distribution of endocrine active cells, mainly confined to the conductive system. The myoendocrine cell exhibits a structure similar to the working myocardial cell however, additionally, like other polypeptide-producing endocrine cells, the rough endoplasmic reticulum, the Golgi apparatus, and the secretory granules are encountered as an endocrine synthetic apparatus. A magnification of the Golgi apparatus is seen in Fig. 19.

15

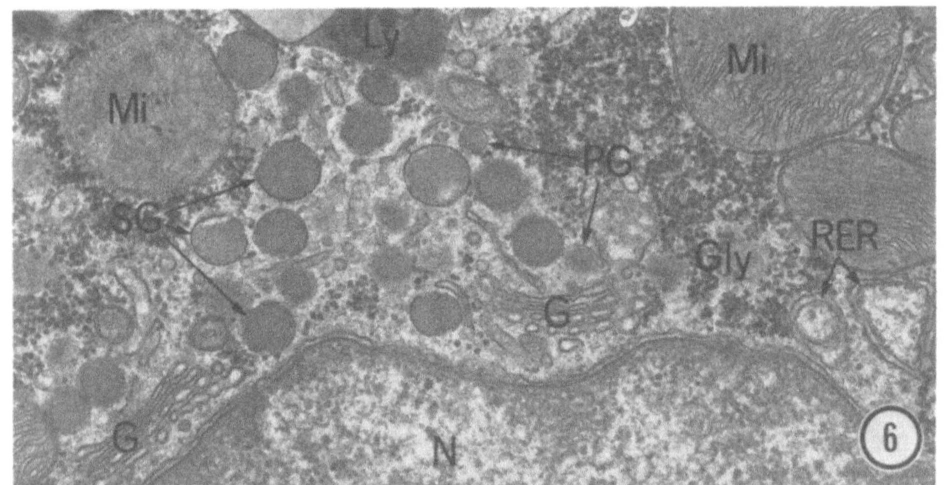

Mi
Ly
Mi
SG
PG
Gly
RER
G
G
N
6
myoendocrine 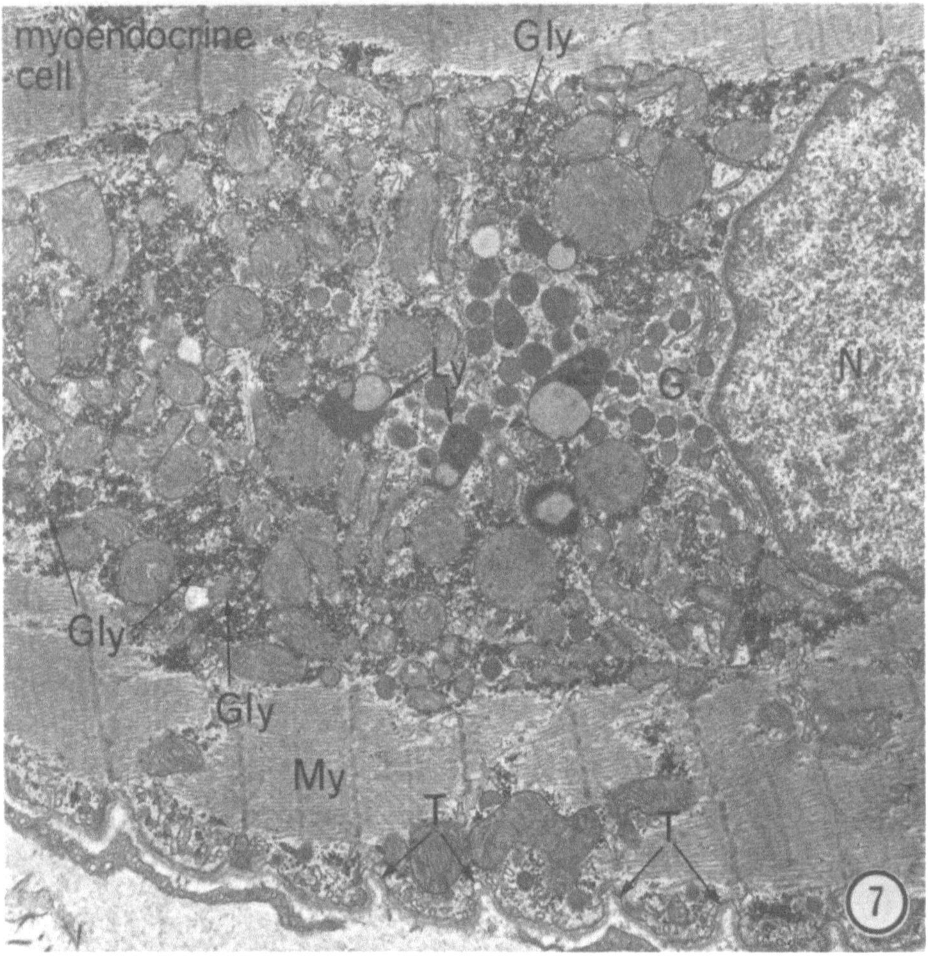
cell
Gly
Ly
G
N
Gly
Gly
My
T
T
Mi
7

Fig. 6. The central part of a myoendocrine cell is the perinuclear area, seen here by electron microscopy as a magnification of Fig. 7. The secretory apparatus is characterized by discrete cisternae of the rough endoplasmic reticulum (RER), the Golgi cisternae (G), and progranules (PG), and also the mature secretory granules (SG). Note the nucleus (N), mitochondria (Mi), lysosomes (Ly), and agglomerations of glycogen granules (Gly). (x 30 200)

Fig. 7. Atrial myoendocrine cell of human heart: note the poles of the nucleus (N) contain the heteromorphous area of the Golgi (G) depicted in higher magnification in Fig. 6. Lysosomes (Ly), mitochondria (Mi), and glycogen agglomerations (Gly) are also typical for this part of the cell. Furthermore, the muscular part of the cells are seen by its myofibrils (My) and invaginations of the T-system (T). (x 11 000)

immunoreactivity is detected in the perinuclear region where specific granules are concentrated in electron micrographs (Figs. 5-9,17,18). The use of antibodies against the C-terminal fragments results in a strong immunoreactivity in the perinuclear area of human atrial myoendocrine cells; by electron microscopy the accumulation of specific granules with immunoreactive label is confirmed (Fig. 10). The granules may vary in electron density, size and number, according to species; so far, precise morphometrical evalutions related to radioimmunoassayable cardiac hormone content are only available for the rat heart (76,77) although a number of morphometric studies have been published since the classical electron microscopical era (24,29,103). Since 1984 the immunohistochemical findings have been documented by a number of authors (1,22,50,55,102,110,150) showing the immunohistochemistry of the rat, cat, dog, tupaia, and human endocrine heart. It is generally accepted that the endocrine cells of the heart (Fig. 5), which appear macroscopically as right and left atrial appendages or auricles, are concentrated in thin-walled trabecular areas in the atrium. These atrial myoendocrine cells exhibit many general morphological features observed in polypeptide producing cells in other organs (47,53, 71). However, an extra-auricular diffuse distribution in the heart is now also well known since Back et al. (5) and Goebel et al. (67), independently described the distribution of CDD-immunoreactive conductive cells in the ventricles (Fig. 2), which was further confirmed by ultrastructural immunohistochemical studies (67,156). This localization may account for the detection of ventricular radioimmunoassayable cardiac hormones and the corresponding specific mRNA extractable from ventricular tissue in mammalian heart (65,72,116,158). It is under discussion whether these ventricular conductive cells alone, or also the ventricular working cells (which is probably the case in ontogenesis as Wei et al. stated in 1987 and in phylogenesis, see below) can express CDD/ANP, which seems possible according to the mentioned mRNA extraction studies.

The phylogenetic morphology of the endocrine heart is known from early electron microscopy (101,142,145,154) and recent immunohistochemistry. In summary, the same myoendocrine functions exist in all vertebrate hearts (132,133). However, CDD-immunoreactivity and the presence of specific granules are also found in all ventricular cells of some lower species (134,141). An interesting finding is that in the snail, a mollusk species, the presence of CDD-like immunoreactivity is confirmed to a specific type of atrial neurosecretory nerve cell while CDD production is not detectable in cardiac muscular cells of this species (114).

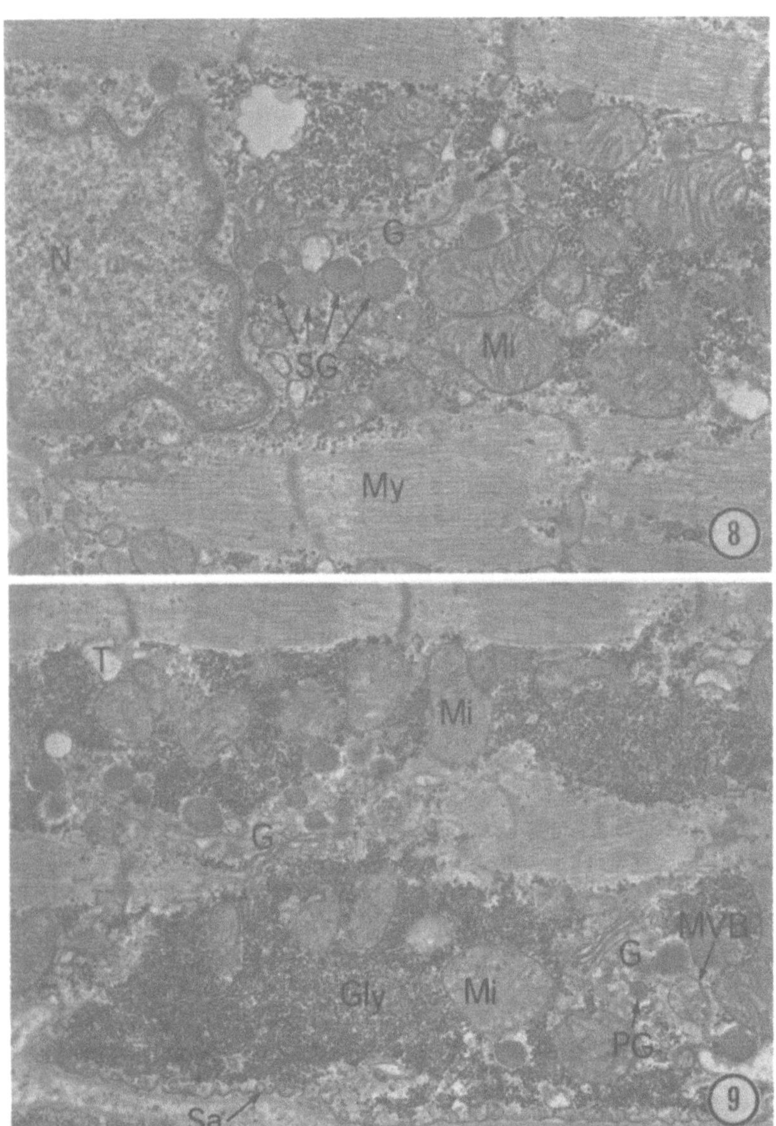

Figs. 8 and 9. Human atrial myoendocrine cells: the perinuclear and telenuclear location of the endocrine dictiosomes are observed. Figure 8 shows the average small Golgi (G) of a normotensive individuum where few secretory granules (SG) are observed around the cisternae which occasionally contain condensing progranules (double arrow). Note further the nucleus (N), myofibrils (My), and mitochondria (Mi). (x 22 800) Figure 9. Dictiosomes (G) close to the sarcolemma (Sa) and in the interfibrillar space. This is more frequent in stimulated atrial endocrine tissue. Note the progranules (PG) of the Golgi (G), the multivesicular bodies (MBV), mitochondria (Ma), T-tubules (T), glycogen agglomerations (Gly), and sarcolemma (Sa) which exhibits numerous micropinocytotic caveolae. (x 20 400)

18

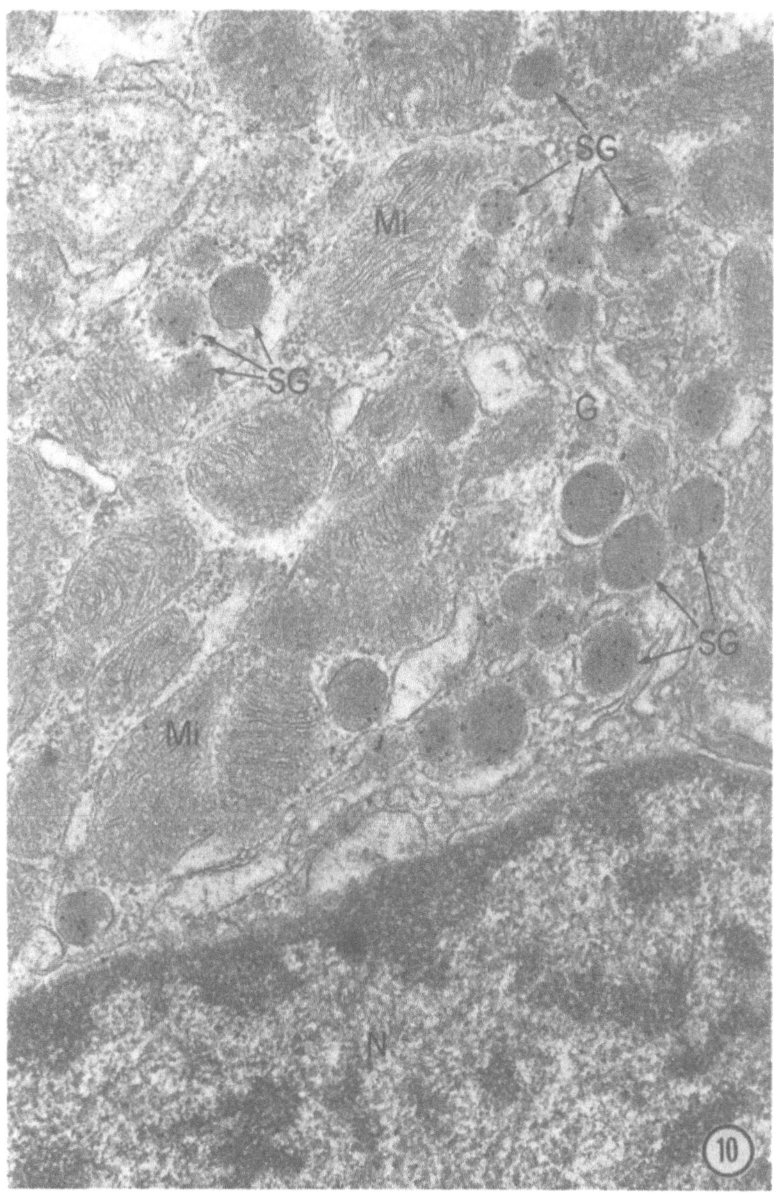

Fig. 10. Immunocytochemical proof of the storage of cardiodilatin in human atrial myoendocrine cells using the immunogold method with CDD-99-126 specific antibody. The specific secretory granules (SG) are labelled by the gold particles bound to the granule matrix after the anti-cardiodilatin-immunogold reaction (compare CDD-IR in the immuno-histochemical level of Figs. 1-4). The granules of the perinuclear region (N = nucleus) are mostly found around the Golgi-complex (G). Numerous mitochondria (Mi) are also seen. (x 37 700)

19

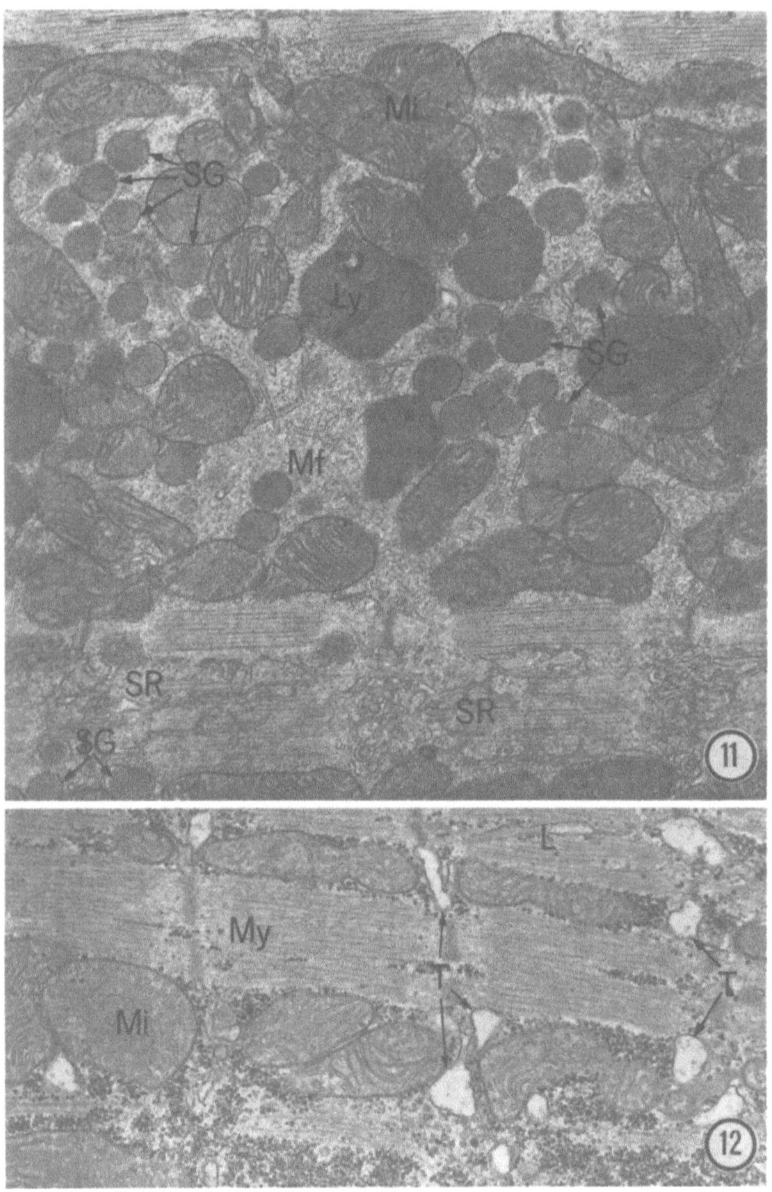

Fig. 11. Myoendocrine cells of other primates like small monkeys *(Tupaia belangeri)* exhibit similar morphological features as in human. Note the specific secretory granules (SG), mitochondria (Mi) lysosomes (Ly), and also some microfilaments and microtubules (Mf). Especially the extent of sarcoplasmic reticulum (SR = L-System) is developed in this species. (x 23 000)

Fig. 12. In larger mammalian species the T-system (T) of myoendocrine cells is found as in ventricular myocardium. Here a human atrial myoendocrine cell is shown, note also the longitudinal tubules (L), myofibrils (My) and mitochondria (Mi). (x 23 800)

20

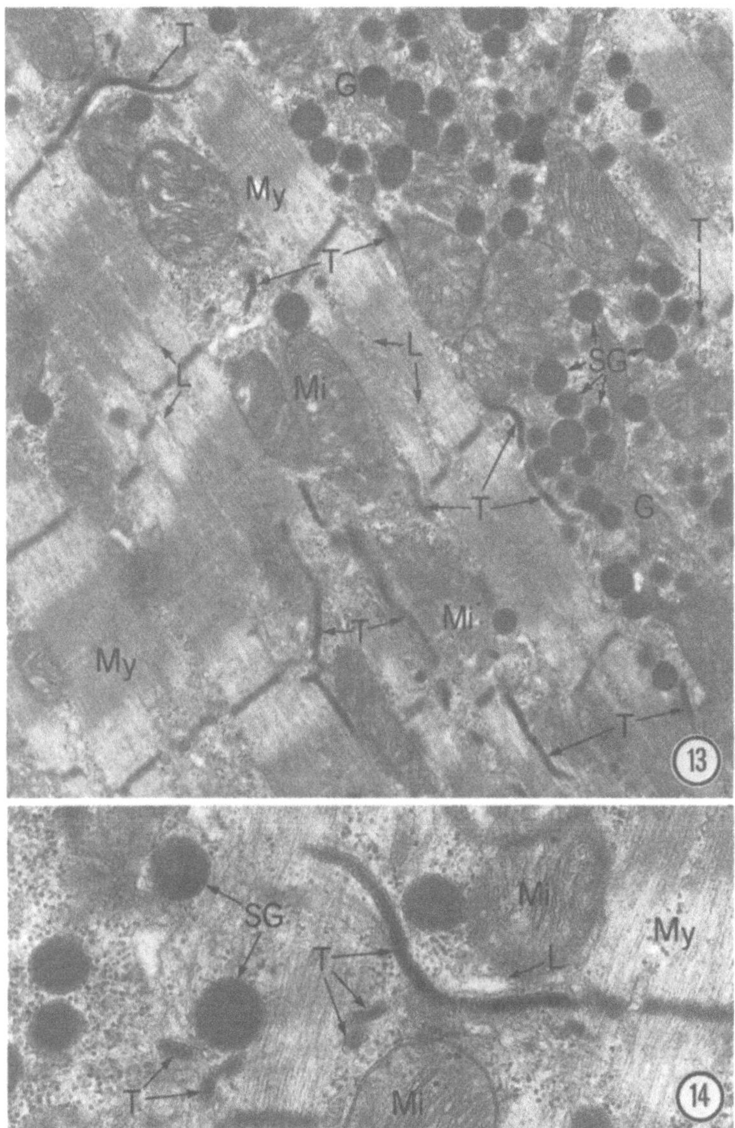

Fig. 13. A filigrane T-system (T) of many longitudinally running profiles is found in many rat atrial myoendocrine cells. These can only be demonstrated by HRP (horse raddish peroxidase) as a diffusion tracer. The intravenously injected peroxidase enters the tubules open to the interstitium and the staining reaction results in an opaque content of the tubules whereas the L-system (L) is not stained. Note also the Golgi-complexes (G) surrounded by specific secretory granules (SG), myofibrils (My) and mitochondria (Mi). (x17000)

Fig. 14. Higher magnification of T-tubules (T) from Fig.13 running close to secretory granules (SG) and adjacent to a cisterna of the sarcoplasmic reticulum (L-system = L). Note also mitochondria (Mi) and myofibrils (My). (x 37 400)

21

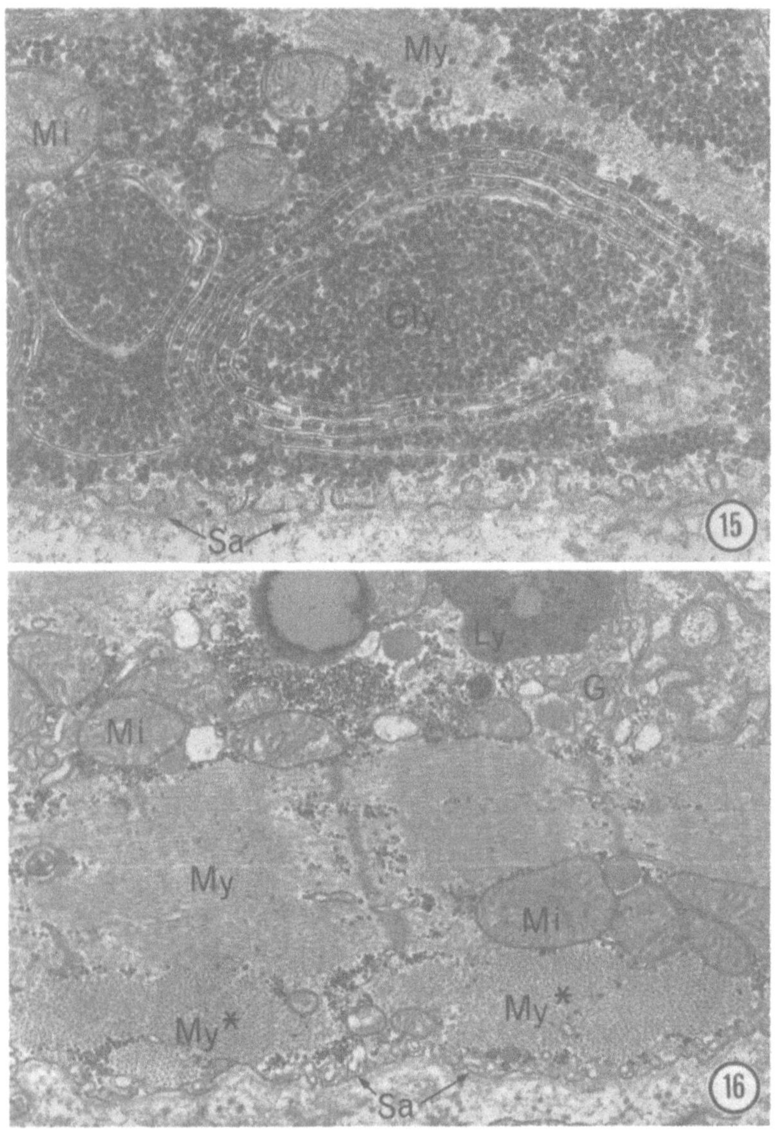

Fig. 15. Glycogen bodies of a myoendocrine cell exhibiting concentric lamellae of membranes. Glycogen (Gly), myofibrils (My), mitochondria (Mi), sarcolemma (Sa) which exhibit a high micropinocytotic activity. (x 44 500)

Fig. 16. Particular arrangement of myofibrils as so-called Ringbinden (My*) which run circular to the longitudinal axis of the cell and the normal myofibrils (My). Note mitochondria (Mi), Golgi-complex (G), lysosomes (Ly), and sarcolemma (Sa). (x 22 400)

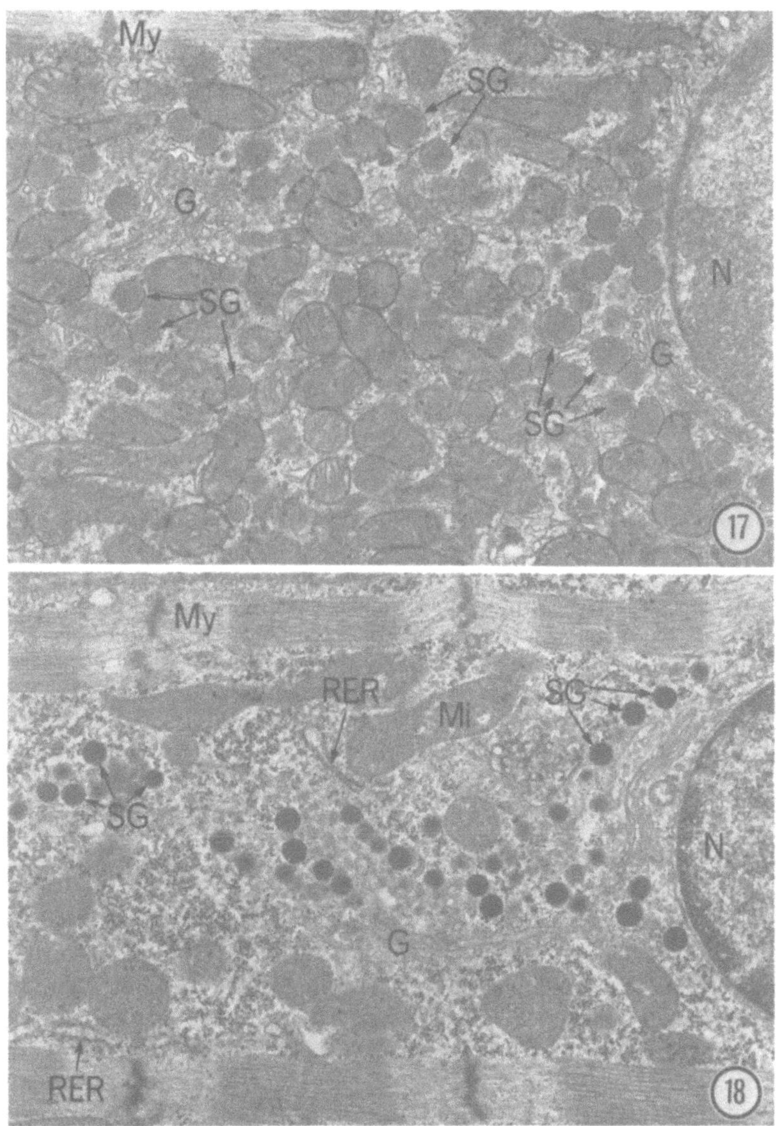

Figs.17 and 18. Comparison of the endocrine apparatus of atrial cells in two primate species (Fig. 17 = *Tupaia belangeri*; Fig. 18 = *Macacca mulatta*). Note the larger and pale secretory granules (SG) of *Tupaia* and the smaller dense granules in *Maccaca*. Golgi-complex (G), nucleus (N) rough endoplasmic reticulum (RER), myofibrils (My), and mitochondria (Mi). (x 17 700)

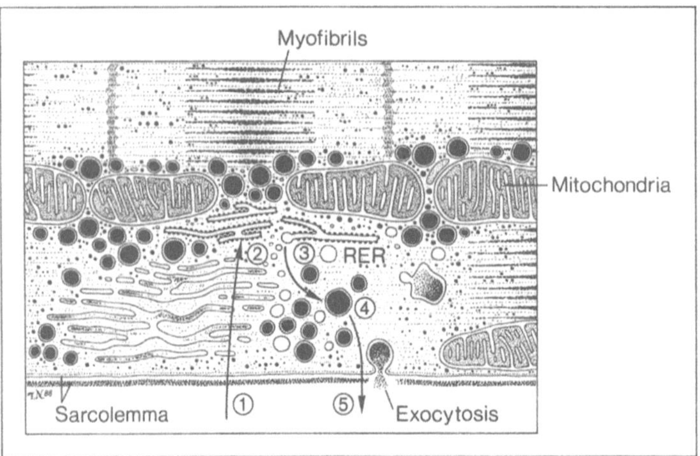

Fig. 19. Schematic drawing of the postulated secretory cycle of atrial myoendocrine cells (see magnification of the subsarcolemmal dictiosomes in Fig. 5): 1) uptake of amino acids through the sarcolemma and transport to the rough endoplasmic reticulum. 2) synthesis of the preprohormone which is the signal peptide elongated CDD-(1-126). 3) vesicular transport, cleavage of the signal peptide and condensation of the prohormone in pro-granules. 4) storage of the prohormone in specific secretory granules, 5) release of the secretory product by exocytosis and concomitant cleavage of the circulating CDD-(99-126).

Ultrastructure of the myoendocrine cell

Cellular Organization and Intercalated Disc: Atrial myoendocrine cells are generally organized as described for the ventricular myocardium (see also Fig. 5). Myoendocrine cells vary slightly in diameter, and the longitudinal continuity and branching of fibers is interrupted by dense transverse lines, the intercalated discs. These are cell borders across the full width of the fibers. A stepwise configuration is observed related to the repeating patterns of cross striations of myofibrils, the intercalated discs are always located at the level of the myofibril I-bands. The intercalated discs (Figs.21,22) guarantee a firm cohesion of the adjacent myocardial cells, transmitting mechanical tension and electrical signals along the fiber axis throughout the entire endocrine myocardium. The surface of intercalated discs is formed by grooves and pits of interdigitating cytoplasmic extensions delineated by the specialized sarcolemma. In this region of the surface membrane a wide range of specialization occurs, namely to guarantee surface-cell-to-cell cohesion by maculae adherentes and fasciae adherentes (17,62). The fasciae adherentes form a considerable part of the surface of the intercalated disc sarcolemma where the actin filaments of the myofibrils insert. Another part of the surface of the intercalated disc sarcolemma can be identified as two parallel dense lines separated by a small constant intercellular cleft of 15 to 20 nm. Maculae adherentes are only occasionally found in intercalated discs of myoendocrine cells. In the entire myocardium, maculae conducentes

24

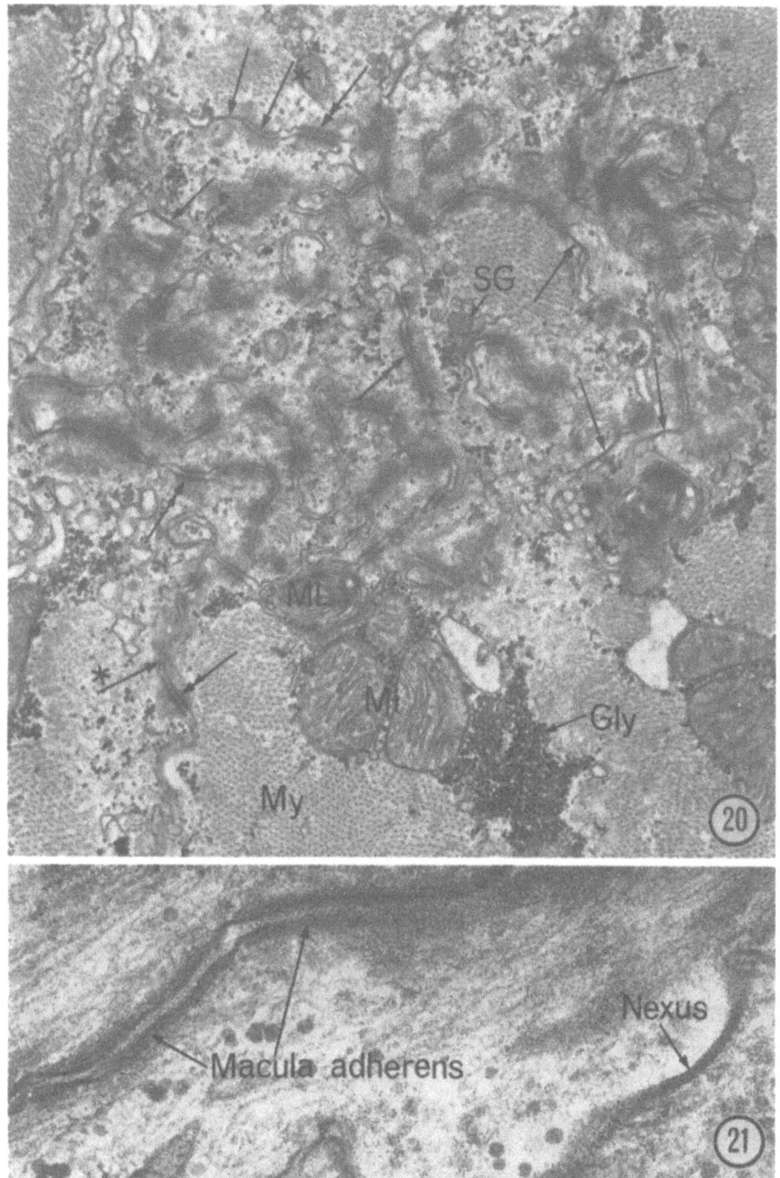

Fig. 20. Intercalated disc of myoendocrine cells; a transverse section depicts the strong interdigitation and many maculae conducens (nexus = arrows), fasciae adherentes (arrows with asterisks), and maculae adherentes (double arrows) are seen. Some lamellar, myelin-like bodies (Ml), mitochondria (Mi), glycogen (Gly), secretory granules (SG), and myofibrils (My). (x 25 000)
Fig. 21. Higher magnification of a region from an intercalated disc where the distinct characteristics of a maculae conducens (nexus) and a macula adherens are observed. (x 93 000)

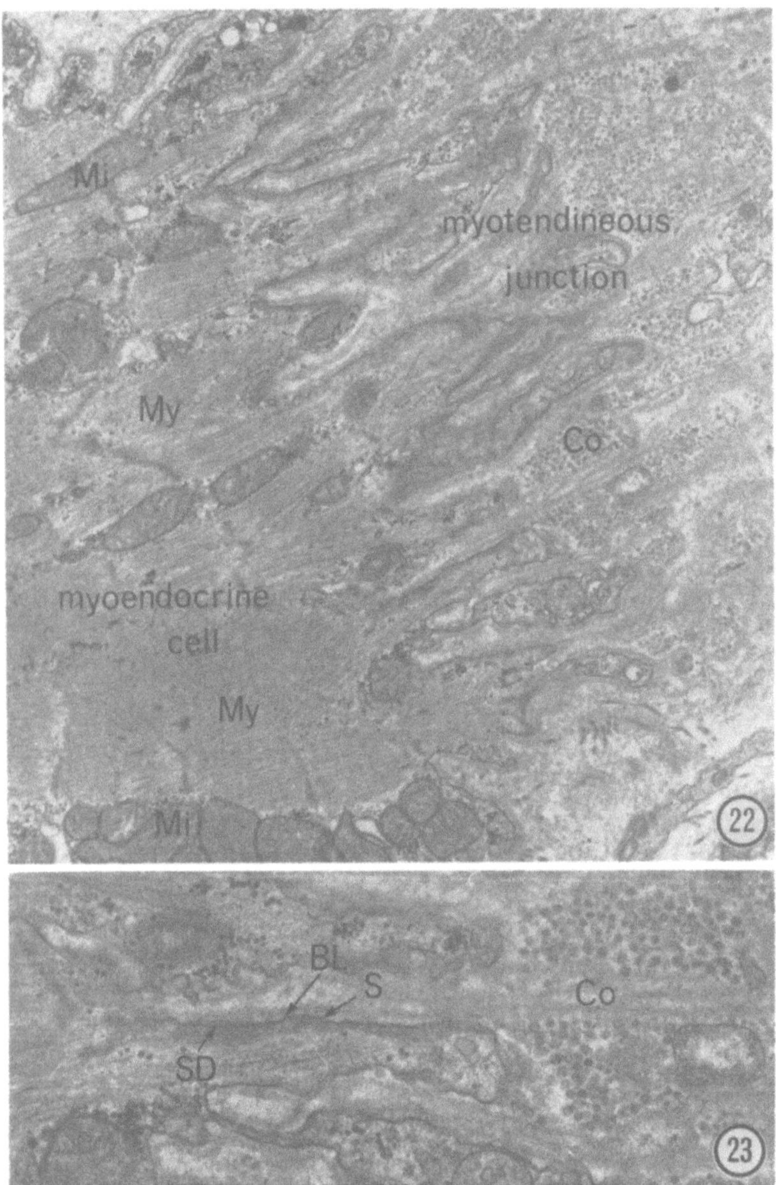

Fig. 22. Myotendineous junction of a myoendocrine cell close to the coronary sulcus directed towards the anulus fibrosus. A strong interdigitation of the cytoplasmic processus with the tendineous part of the interstitium is seen. Collagen (Co) in longitudinal and transverse section, myofibrils (My), and mitochondria (Mi). (x 15 000)

Fig. 23. Higher magnification of cytoplasmic processus at the myotendineous junction showing the collagen (Co) ending at the outer sarcolemma (BL = lamina basalis) which is separated from the inner sarcolemma (S = sarcolemmal cytoplasmic membrane). In this area many semidesmosomes (SD) are found. (x 35 000)

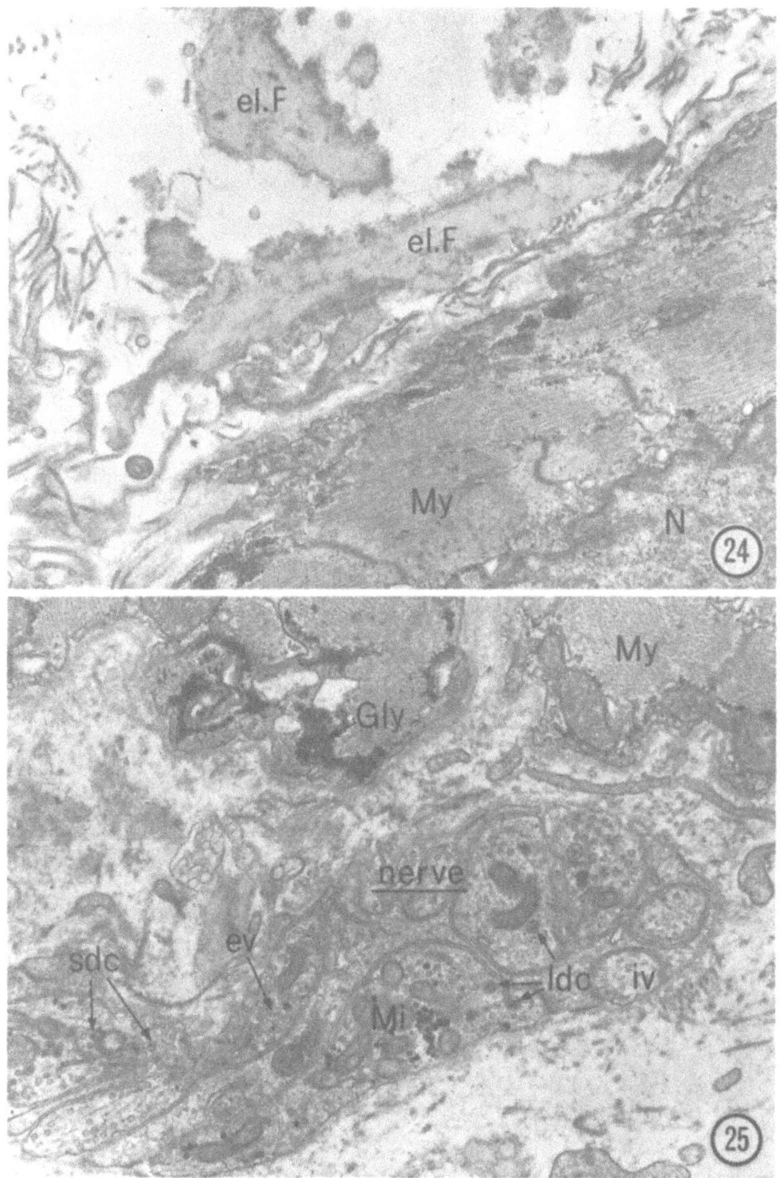

Fig. 24. Connective tissue of the endocrine heart. Note that many elastic fibers (el.F) are seen in regions with larger interstitium between the myoendocrine cells. Myofibrils (My) and cells nucleus (N) of the myoendocrine cell. (x 18 300)

Fig. 25. The strong intrinsic nerve plexus of the endocrine heart exhibits nerves with numerous varicosities containing empty vesicles (ev), small dense cored vesicles (sdc), and large dense cored vesicles (ldc). Note also mitochondria (Mi), the intervaricose axon (iv), as well as myofibrils (My), and glycogen (Gly) in the myoendocrine cell. (x 20 300)

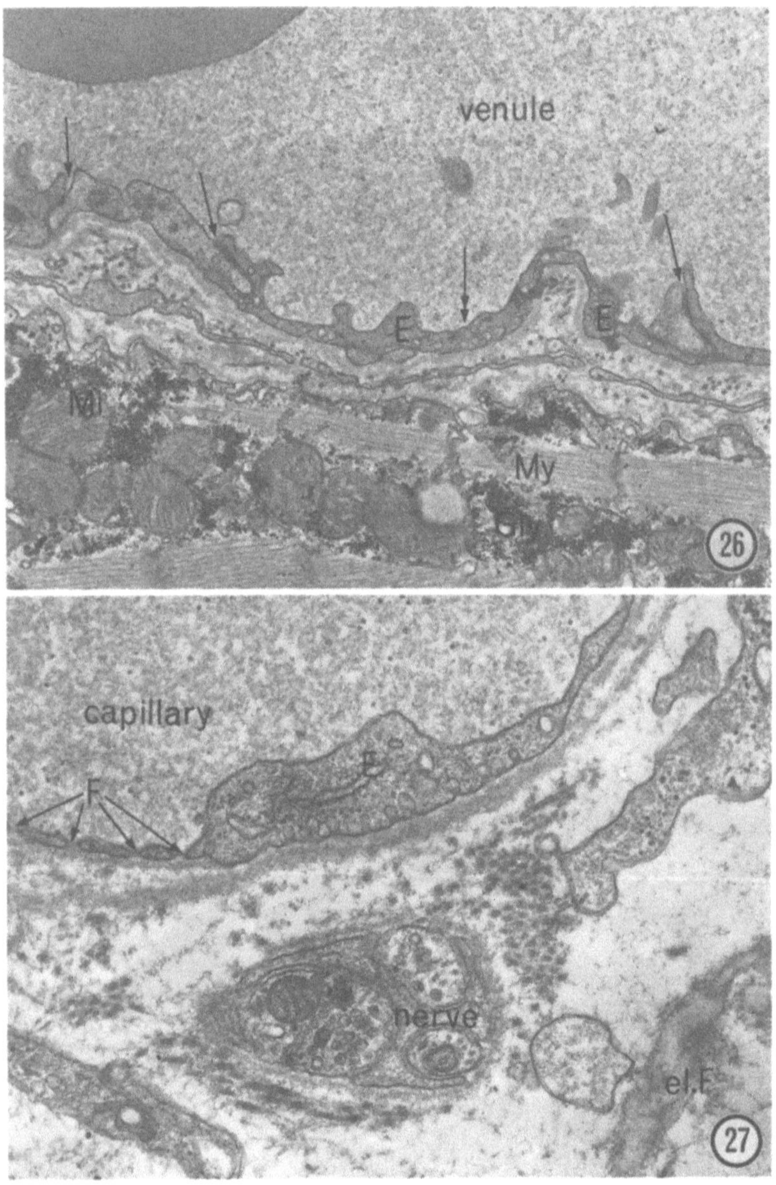

Figs. 26 and 27. Vascularization of the endocrine heart. Fig. 26 shows a typical post-capillary venule with a thin endothelium (E) containing the intercellular clefts (arrow) and few micropinocytotic vesicles (double arrow). Note the myofibrils (My), mitochondria (Mi), and glycogen (Gly) of the myoendocrine cell. (x 19 000)

Fig. 27 depicts a capillary with fenestrated (F) endothelium (E) of the endocrine heart surrounded by a nerve and loose connective tissue with elastic fibres (el.F). (x 31 100) (nexus or gap junctions) are regularly observed. At these patch-like regions the intercellular space is reduced to less than 2 nm, and low electrical resistance for the rapid spread of excitation and syncytial interconnection of myoendocrine cells is presumed. These intercellular connections may also occur in laterally opposed myoendocrine cells where the interstitial space is locally interrupted. Thus the endocrine heart exhibits all morphological features of cellular organization on the ventricular myocardium as described in many publications. Its cellular-syncytial network, therefore, represents the basis of a homogeneous morphofunctional endocrine unit in each auricular appendage.

Sarcolemma: The cytoplasmic membrane of atrial myoendocrine cells may be considered a special form of sarcolemma (Figs. 15,16). As in the ventricular myocardium the sarcolemma is formed by an outer layer, the equivalent of a basal lamina, and an inner layer, which is the cytoplasmic membrane proper. The latter is a membrane unit of two dark osmiophilic layers and a clear space between them. Membraneous particles, micropinocytotic vesicles, and sarcolemmal invaginations are observed in freeze fracture pictures.

Nucleus: One or several nuclei can be observed located centrally in the myoendocrine cells (Figs. 5,7,17,18). They are arranged in a longitudinal core without myofibrils. The oval-shaped nuclei exhibit little chromatin, a typical nuclear membrane with numerous nuclear pores. Myoendocrine cells fixed during relaxation have smooth surfaced nuclei while in contracted cells nuclear membrane lateral infoldings are regularly observed.

Myofibrils: The major compartment of the myoendocrine cell is built by myofibrils which are striated as in skeletal and ventricular myocardial muscles (Figs. 8,12,13). The thick myosin and thinner actin elementary filaments are well discerned and form the characteristic Z-, A- and I-bands containing sarcomeres. M-bands are also observed as a regular feature of myoendocrine cells. It has been shown that myoendocrine cells exhibit a myosin different from that in ventricular cells; these isomyosins can be immunologically differentiated (86). Transverse sections of myofibrils (Fig. 20) show the irregularly sized and irregularly shaped bundles of filaments surrounding the interfibrillar space and the sarcoplasmic reticulum.

Sarcoplasmic Reticulum and T-System: The sarcoplasmic reticulum in atrial myoendocrine cells has been described in detail (43,51) and a morphometric study has been published (19) (Figs. 5,11-14). The longitudinal network of cisternae along the interfibrillar space forms the distinct L-system, also called sarcoplasmic reticulum proper (Fig. 11). It is less developed than in working myocardial cells. The T-system runs predominantly from its apertures at the surface sarcolemma as continuous tubules of invaginated inner sarcolemma, mostly at the level of the I-bands. These transverse tubules frequently branch and give rise to longitudinal extensions. They also form a moderate number of diad-like contacts with the L-cisternae. Initially, the surface segments of the T-tubules may be lined by a basal lamina, the outer sarcolemma. It has been suggested that the density of the T-system is proportional to the fiber size when hypertrophied or species-dependent large diameter fibres are compared to small myoendocrine cells (79). A contact of secretory granules of myoendocrine cells with the T-system and a possible secretion by exocytosis into the tubular lumen is under discussion.

Mitochondria: The mitochondria (Figs.6-12), or sarcosomes (4) are located in longitudinal columns within the enlarged interfibrillar space. They are of elongated shape and rich in cristae. The sarcosomes contain a fine granulated matrix and some small dense mitochondrial granules. Some round or oval-shaped mitochondria are also found in the perinuclear regions where they are intermingled with the Golgi apparatus, the lysosomes, and the glycogen agglomerations.

Lysosomes: In myoendocrine cells lysosomes occur in the same different forms as observed in ventricular myocardial cells (Figs. 7,11,16). They are predominant structures in aging myocardium, such as lipofuscin granules or residual lysosomes. These heteromorphous cytoplasmic organellas are mainly encountered in the perinuclear zones. They were already observed in early publications on the atrial myocardium (128) and were related to atrial granules.

Peroxisomes: Peroxisomes have not been specifically analyzed in myoendocrine cells; however, it appears that they are identical or similar in structure and function as in ventricular myocardial cells (78).

Secretory apparatus (GERL) and secretory granules: The endocrine secretory apparatus in normal myoendocrine cells of human hearts seems not to be developed extensively; however, this may vary strongly according to functional states or species. The usual location of the Golgi complex is the perinuclear area (Figs 5-8,10,18). This means that the Golgi complex and its associated organelles are restricted to the enlarged sarcoplasmic region at the ends of the elongated nucleus, or when more than one nucleus occurs within a continuous column of nuclei-Golgi complexes. The scarce rough endoplasmic reticulum (Figs. 6,18) is intermingled with dictiosomes, glycogen granule plaques, lysosomes and mitochondria. The number of dictiosomes and their arrangement within a single perinuclear core may vary. However, the Golgi apparatus exhibits all usual components, especially the progranule formation within cisternae, the high number of vesicles and progranules. It also surrounds mature secretory granules indicating the activity and turnover of the endocrine secretory product in the Golgi apparatus. Mature secretory granules may migrate to the interfibrillar space and very close to the sarcolemma. This process seems to occur quickly as can be deduced from functional studies on stimulated myoendocrine cells (52). The cells release their products by exocytosis, which is very rarely observed (52,149), but can be deduced from the possible incidence of each exocytotic event (77). The formation of cardiac hormones within the atrial myoendocrine cells can be proven by immunogold-labeling. Segment-specific antibodies show that all granules contain the CDD-1-126-IR, this means that each granule possesses the antigen epitops of CDD-1-126, either in fragments or as the entire CDD-(1-126) molecule (119). According to extraction studies, the entire prohormone is most probably stored in the granules, (50,55); that has been confirmed by a number of authors (13,74,118,152,163).

The presence of the Golgi-complex and its associated structures which belong to the secretory unit is not restricted to the perinuclear zone. Especially in stimulated endocrine hearts, all interfibrillar regions may develop dictiosomes, such as the paranuclear (along the longitudinal side of the nucleus), the telenuclear (in the cytoplasmic interfibrillar space) and the subsarcolemmal cytoplasma (Figs. 5,9,16,19). This development may be stimulated (135) or inhibited (87).

Mature granules have been the subject of many papers, all indicating the spherical shape and rather constant size in different species; several morphometric features concerning the parameters of the granules have been evaluated (29,56,77,84,103). They are round, membrane bound and contain a homogeneous granulated matrix of variable density which depends on fixation and species.

Accessory apparatus of the endocrine heart

Vascularization: The microvasculature of the heart is generally described as a continuous system containing capillaries (Figs. 26,27). This means that the capillaries, which constitute a dense vascular plexus around the myofibrils, respresent an endothelium without fenestrations or clefts. The endocrine heart, however, contains some segments of the microvasculature with fenestrations which possibly function as additional pathways to circumvent the vesicular transport for peptides released from the myoendocrine cells to the systemic circulation. So far, these types of capillaries have only been described in connection with the microvasculature of the conductive system by Weihe and Kalmbach (195). This type of capillary is a general characteristic of peptide producing endocrine tissues.

Innervation: The general findings on cardiac innervation (48,127) also hold true for the endocrine heart. The endocrine heart shows a strong innervation with a dense plexus of catecholaminergic fibers (Figs. 25,27). Additionally they are peptidergic in nature - mostly VIP-IR, NPY-IR, SP-IR, and CGRP-IR nerves may be mentioned (52,131). The intrinsic nature of a large portion of the auricular nerves has been demonstrated by heart transplantations (25,52). The functional implication of these nerves concerning the secretory status of myoendocrine cells is not established.

Myotendineous junctions and connective tissue framework: The endocrine heart contains small interstitial tissue spaces which are entirely surrounded by the basal lamina of the myoendocrine cells, capillaries or nerves. Few interstitial cells are encountered, loose collagen fibers are seen, but in enlarged areas a framework of elastic tissue (93) is observed, which is particularly well developed in the atrial appendages (Fig. 24), indicating a possible mechanical function in this distensable part of the atrial wall. The border to the valve area and the connection to the ventricular tissue is formed by fibrillar rings of dense connective tissue into which the ending myoendocrine cells insert. Their typical myotendineous junctions (Figs. 22,23) with interdigitating cytoplasmic protrusions, half-desmosomes and inserted actin filaments are found as described earlier for skeletal and ventricular cardiac muscles (48,73,107).

Conclusions

The study of the ultrastructural morphology of the endocrine heart with the electron microscope has initiated one of the most remarkable steps in the field of cardiovascular and endocrinological research. As early as the late 1950s, several researchers, while looking for morphological differences between atrial and ventricular cells, and also when studying the cardiac conductive cells, observed a novel cell type containing granules. Kisch (90) merits discovering atrial granules for the first time in the guinea pig. Until the first extensive description of *specific atrial granules* by Jamieson and Palade (84) few reports were published dealing with the atrial myocardium and the conductive system muscle cells different from the ventricular myocardium according to physiological data (7,18,90-92,128,134,157). Only Bompiani et al. (18) paid more attention to these cells containing *"corps denses"* . Among the above mentioned Palade (126) was the first to postulate that the granules may be the intracellular storage sites for catecholamines because the number of granules to be reduced by reserpine treatment in rats. However, the same authors (84) later refuted this hypothesis, based on autoradiographic investigations of Wolfe et al. (160), as well as on their own histochemical studies. This assumption was also supported by comparative measurements of the catecholamine content in ventricles and atria of different species (16,125,143). However, it is now well known that these differences in catecholamine content are due to the density of catecholaminergic nerves in the different topographical regions of the heart (28,60). In an extensive description, Jamieson and Palade (84) wrote that the specific granules were *"presumably secretory in nature"*; their forecast has to be admired. In subsequent papers on atrial muscle, further details on the cytology (51,109), as well as extensive variabilities among species (8,123) were described. A remarkable series of publications on the nature of atrial cells with respect to their proteinaceous nature and metabolism of the granular product appeared when Cantin and his coworkers studied the atrial heart (21,23,82,83). However, the most prominent finding about atrial myocardial cells was published by Marie et al. (103), and later by DeBold (30), who discovered that the granular content can be related to a physiological function, namely the water-sodium homeostasis of the body. This event, indeed, may be considered the major stimulus for the isolation and characterization of the bioactive substance of the granules, which were achieved by specific biotests, such as diuresis (35) and vascular smooth muscle relaxation (26,38,56). Also, the separation of specific granules by ultracentrifugation (64), confirmed the diuretic activity of the granular content, which supported the intension to isolate the specific cardiac hormone.

After cardiac hormones had been characterized by biochemical methods and the amino acid sequence of this polypeptide hormone family was available (27,45,50,55,88) morphological studies on the myoendocrine cells and the atrial heart were more or less scarce compared to the compulsive number of publications including biochemical, physiological, genetic and clinical studies (20,54,94,95). Therefore, morphological studies were devoted to the functional changes of atrial cells after volume overload (136,137), because it was now evident that the release of atrial hormones, determined by radio-immunoassay, was significantly stimulated by volume overload (99,100). There was interest in finding evidence for the stimulated secretory cycle and particulary the demonstration of exocytosis, which has only been documented in two publications so far (52,149). Numerous experiments were carried out concerning the release mechanism and stimulation of the myoendocrine cells (2,3,6,12,14,36,39,40,97,106,146). Since the innervation may also play a role in the regulation of the atrial myoendocrine cell secretory cycle, several authors have demonstrated new data on atrial innervation (52,67).

It is now well known that atrial nerves constitute an extensive intrinsic plexus which persists after denervation of the heart (25,87).

Indeed, great attention has been paid to the immunohistochemistry and immunocytochemistry of atrial myoendocrine cells. The proof that atrial cells store cardiac hormones was the detection of the immunohistochemical reaction of cardiac hormones with specific antibodies, first and independently demonstrated by Cantin's group and ours (22,50,55,84). This method also helped later to reveal numerous extra-atrial locations of cardiac hormones (41,44,58,63,85,89,108,111,117,139,144,162). Ultrastructural immunocytochemistry clearly showed the specific storage site of cardiac hormones in the rat secretory granules (22) and human (102) heart. The method was also applied to prove the granular exocytosis (52), which is almost impossible to demonstrate by routine electron microscopy as derived from morphometric data calculated by Herbst et al. (76,77).

The purpose of this paper was to review the morphological features of the endocrine heart in the light of interdisciplinary matters. The morphology of the endocrine heart has been well described in many available publications, there is, however, certainly more morphology to be studied in relation to functional and pathological states of the endocrine heart. The vascularization, innervation, and connective tissue framework of the atrial appendages is far from being studied adequately. The few completed pathomorphological studies do not suffice to explain the basis of the morphological events during heart disease (42,98,105,112,120,136,137,149,153). This statement may stimulate further morphological work, which is necessary to establish the complete basis of our thinking on how the endocrine heart works.

Acknowledgement: The author wishes to thank B. Brühl and B. Herbold for their excellent technical help, as well as J. Sis, K. Grützner and G. Sürig for preparing the manuscript, and R. Nonnenmacher for his schematic drawings.

References

1. Ackermann, D.M., Edwards, B.S., Wold, L. E., Burnette, J.C. (1986) Atrial natriuretic peptide: localization in the human heart. JAMA 256: 1048
2. Agnoletti, G., Rodella, A., Ferrari, R., Harris, P. (1987) Release of atrial natriuretic peptide-like immunoreactive material during stretching of the rat atrium *in vitro*. J Mol Cell Cardiol 19: 217-220
3. Arjamaa, O., Vuolteenaho, O. (1985) Sodium ion stimulates the release of atrial natriuretic polypeptides (ANP) from rat atria. Biochem Biophys Res Comm 132: 375-381
4. Armiger, L.C., Benson, D.C. (1978) The fine structure of normal canine atrial myocardium, with particular reference to mitochondria. J Mol Cell Cardiol 10: 587-591
5. Back, H., Stumpf, W.E. , Ando, E., Nokihara, K., Forssmann,W.G. (1986) Immunocytochemical evidence for CDD/ANP-like peptides in strands of myoendocrine cells associated with the ventricular conduction system of the rat heart. Anat Embryol 175: 223-226
6. Baertschi, A.J., Hausmaninger, C. , Walsch, R.S., Mentzer Jr., R.M., Wyatt, D.A., Pence, R.A. (1986) Hypoxia-induced release of atrial natriuretic factor (ANF) from the isolated rat and rabbit heart. Biochem Biophys Res Comm 140: 427-433
7. Battig, C.G., Low, F.N. (1961) The ultrastructure of human cardiac muscle and its associated tissue space. Am J Anat 108: 349
8. Bencosme, S.A., Berger, J.M. (1971) Specific granules in mammalian and non-mammalian vertebrate cardiocytes. In: E Bajusz, G., Jasmin, S. (eds) Methods and Achievements in Experimental Pathology, vol. 5. Karger, Basel, pp 173-213
9. Berger, J.M., Bencosme, S.A. (1971) Fine structural cytochemistry of granules in atrial cardiocytes. J Mol Cell Cardiol 3: 111-120

10. Berger, J.M., Sosa-Lucero, J.C., de la Iglesia, F.A., Lumb, G., Bencosme, S.A. (1972) Relationship of atrial catecholamines to cytochemistry and fine structure of atrial specific granules. In: Bajusz, E., Rona, G. (eds) Recent Advances in studies on cardiac structure and metabolism. Vol. 1, Myocardiology. University Park Press, Baltimore, pp 340 - 350

11. Berlinguet, J.C., Huet, M., Cantin, M. (1976) Autotransplantation hétérotopique de l'oreillette chez le rat. Effet sur les granules spécifiques. Pathol Biol (Paris) 24: 175-182

12. Bilder, G.E., Schofield, T.L., Blaine, E.H. (1986) Release of atrial natriuretic factor. Effects of repetitive stretch and temperature. Am J Physiol 251: F817-F821

13. Bloch, K.D., Scott, J.A., Zisfein, J.B., Fallon, J.T., Margolies, M.N., Seidmann, C.E., Matsueda, G.R., Homcy, C.J., Graham, R.M., Seidman, J.G. (1985) Biosynthesis and secretion of proatrial natriuretic factor by cultured rat cardiocytes. Science 230: 1168-1171

14. Bloch, K.D., Seidmann, J.G., Naftilan, J.D., Fallon, J.T., Seidman, C.E. (1986) Neonatal atria and ventricles secrete atrial natriuretic factor via tissue-specific secretory pathways. Cell 47: 695-702

15. Bloom, G.D. (1962) The fine structure of cyclostome cardiac muscle cells. Z. Zellforsch 57: 213-239

16. Bloom, G., Östlund, E., v. Euler, U.S., Lishajko, F., Ritzén, M., Adams-Ray, J. (1961) Studies on catecholamine-containing granules of specific cells in cyclostome hearts. Acta Physiol Scan 53 (suppl 185): 1-34

17. Bloom, W., Fawcett, D.W. (1975) A textbook of Histology, 10th edition. Cardiac muscle. Saunders, Philadelphia-London-Toronto, pp 315-332

18. Bompiani, G.D., Rouiller, C.H., Hatt, P.Y. (1959) Le tissu de conduction du coeur chez le rat. Étude au microscope électronique. Arch Mal Coeur 52: 1257-1274

19. Bossen, E.H., Sommer, J.R., Waugh, R.A. (1981) Comparative stereology of mouse atria. Tissue Cell 13: 71-77

20. Brenner, B.M., Laragh, J.H. (eds.) (1987) American Society of Hypertension Symposium Series, vol. 1: Biologically active atrial peptides. Raven Press, New York

21. Cantin, M., Benchimol, S., Castonguay, Y., Berlinguet, J.C., Huet, M. (1975) Ultrastructural cytochemistry of atrial muscle cells. V. Characterization of specific granules in the human left atrium. J Ultrastructural Res 52: 179-192

22. Cantin, M., Gutkowska, J., Thibault, G., Milne, R.W., Ledoux, S., MinLi, S., Chapeau, C. (1984) Immunocytochemical localization of atrial natriuretic factor in the heart and salivary glands. Histochemistry 80: 113-127

23. Cantin, M., Timm-Kennedy, M., El-Khatib, E., Huet, M., Yunge, L. (1979) Ultrastructural cytochemistry of atrial muscle cells. VI. Comparative study of specific granules in right and left atrium of various animal species. Anat Rec 193: 55-96

24. Chang, W.W.L., Bencosme, S.A. (1969) Quantitative electron microscopic analysis of the specific granule population of rat atrium. Can J Physiol Pharmacol 47: 483-485

25. Cuevas, P., Golitsin, A., Gonzalez, A.M., Forssmann, W.G. (1987) Ultrastructural evidence of cardiodilatin in rat transplanted heart. Transplantation Proc 19: 3789-3791

26. Currrie, M.G., Geller, D.M., Cole, B.R., Boylan, J.G., YuSheng, W., Holmberg, S.W., Needleman, P. (1983) Bioactive cardiac substances: Potent vasorelaxant activity in mammalian atria. Science 221: 71-73

27. Currie, M.G., Geller, D.M., Cole, B.R., Siegel, N.R., Fok, K.F., Adams, S.P., Eubanks, S.R., Gallupi, G.R., Needleman, P. (1984) Purification and sequence analysis of bioactive atrial peptides (atriopeptins). Science 223: 67-69

28. Dahlström, A., Fuxe, K., Mya-Tu, M., Zetterström, B.E.M. (1965) Observations on adrenergic innervation of dog heart. Am J Physiol 209: 689-692

29. DeBold, A.J. (1978) Morphometric assessment of granulation in rat atrial cardiocytes: effect of age. J Mol Cell Cardiol. 10: 717-724

30. DeBold, A.J. (1979) Heart atria granularity effects of changes in water-electrolyte balance. Proc Soc Exp Biol Med 161: 508-511

31. DeBold, A.J., Bencosme, S.A. (1973a) Studies on the relationship between the catecholamine distribution in the atrium and the specific granules present in atrial muscle cells. 1. Isolation of a purified specific granule subfraction. Cardiovascular Res 7: 351-363

32. DeBold, A.J., Bencosme, S.A. (1973b) Studies on the relationship between the catecholamine distribution in the atrium and the specific granules present in atrial muscle cells. 2. Studies on the sedimentation pattern of atrial noradrenaline and adrenaline. Cardiovascular Res 7: 364-369

33. DeBold, A.J., Bencosme, S.A. (1975a) Autoradiographic analysis of label distribution in mammalian atrial and ventricular cardiocytes after exposure to tritiated leucine. In: Roy, P.E., Harris, P. (eds) Recent advances in studies on cardiac structure and metabolism, vol.8, The cardiac sarcoplasm. pp 129-138

34. DeBold, A.J., Bencosme, S.A. (1975b) Selective light microscopic demonstration of the specific granulation of the rat atrial myocardium by lead-hematoxylin- tartrazine. Stain Technol 50: 203-205

35. De Bold, A.J., Borenstein, H.B., Veress, A.T., Sonnenberg, H. (1981) A rapid and potent natriuretic response to intravenous injection of atrial extracts in rats. Life Sci 28: 89-94

36. DeBold, A.J., DeBold, M.L., Sarda, I.R. (1986) Functional-morphological studies on *in vitro* cardionatrin release. J Hypertension 4 (suppl): S3-S7

37. DeBold, A.J., Raymond, J.J., Bencosme, S.A. (1978) Atrial specific granules of the rat heart light microscopic staining and histochemical reactions. J Histochem Cytochem 26: 1094-1102

38. Deth, R.C., Wong, K., Fukozawa, S., Rocco, R., Smart, J.L., Lynch, C.J., Awad, R. (1982) Inhibition of rat aorta contractile response by natriuresis-inducing extract of rat atrium. Fed Proc 41: 983

39. Dietz, J.R. (1984) Release of natriuretic factor from rat heart-lung preparation by atrial distension. Am J Physiol 247: R1093-1096

40. Dietz, J.R. (1987) Control of atrial natriuretic factor release from a rat heart-lung preparation. Am J Physiol 252: R498-R502

41. Feller, S., Meyer, M., Hock, D., Forssmann, W.G. (1988) Extraauriculäre Lokalisation von Cardiodilatin. Acta Anat (Basel) 132: 80

42. Fengolio J.J. Jr., Pham, T.D., Hordof, A., Edie, R.N., Wit, A.L. (1979) Right atrial ultrastructure in congenital heart disease. II. Atrial septal defect: effects of volume overload Am J Cardiol 43: 820-827

43. Ferguson, D.G., Leeson, T.S. (1983)Postnatal development of sarcolemmal invaginations in right atrial myocardium of the rat. Acta Anat (Basel) 117: 289-302

44. Flügge, G., Inagami, T., Fuchs, E. (1987) Atrial natriuretic peptide detected by immunocytochemistry in peripheral organs of Tupaia belangeri. Histochemistry 86: 479-483

45. Flynn, T.G., DeBold, M.L., DeBold, A.J. (1983) The amino acid sequence of an atrial peptide with potent diuretic and natriuretic properties. Biochem Biophys Res Comm 117: 859-865

46. Forssmann, W.G. (1969) A method for *in vivo* diffusion tracer studies combining perfusion fixation with intravenous tracer injection. Histochemie 20: 277-286

47. Forssmann, W.G. (1970) Ultrastructure of hormone-producing cells of the upper gastrointestinal tract. In: Creutzfeld (ed) Origin, chemistry, physiology and pathophysiology of gastrointestinal hormones. Schattauer, Stuttgart pp 31-70

48. Forssmann, W.G. (1982) Morphologie des Skelettmuskels und des Muskel-Sehnen-überganges. In: Groher W., Noack, W., (Hrsg) Sportliche Belastungsfähigkeit des Haltungs- und Bewegungsapparates (Symposium Berlin 1981). Thieme, Stuttgart-New York S. 1-18

49. Forssmann, W.G. (1986) Cardiac hormones. I. Review on the morphology, biochemistry and molecular biology of the endocrine heart. Eur J Clin Invest 16: 439-451

50. Forssmann, W.G., Birr, C., Carlquist, M., Christmann, M., Finke, R., Henschen, A., Hock, D., Kirchheim, H., Kreye, V., Lottspeich, F., Metz, J., Mutt, V., Reinecke, M. (1984) The auricular myocardiocytes of the heart constitute an endocrine organ. Characterization of a porcine cardiac peptide hormone, Cardiodilatin-126. Cell Tissue Res 238: 425-430

51. Forssmann, W.G., Girardier, L. (1970) A study of the T-system in rat heart. J Cell Biol 44: 1-19

52. Forssmann, W.G., Greenberg, J., Schulz-Knappe, P., Rippegather, G. (1988) Morphological changes induced by stimulation and inhibition of atrial myoendocrine cells. Eur J Clin Invest (in press)

53. Forssmann, W.G., Grube, D. (1985) Die disseminierten endokrinen Zellen. In: Fleischhauer, K. (ed) Benninghoff. Makroskopische und Mikroskopische Anatomie des Menschen, 2. Band, 13./14. Aufl.. Urban & Schwarzenberg, München-Wien-Baltimore, S 604-612

54. Forssmann, W.G., Heidland, A., Lang, R.E. (eds.) (1986) I. Workshop on cardiac hormones "Potential role in health and disease." Klin Wochenschr. 64 (suppl. VI)

55. Forssmann, W.G., Hock, D., Kirchheim, F., Metz, J., Mutt, V., Reinecke, M. (1984) Cardiac hormones: morphological and functional aspects. Clin Exp Hypertension A6: 1873-1878

56. Forssmann, W.G., Hock, D., Lottspeich, F., Henschen, A., Kreye, V., Christmann, M., Reinecke, M., Metz, J., Carlquist, M., Mutt, V. (1983) The right auricle of the heart is an endocrine organ: Cardiodilatin as a peptide hormone candidate. Anat Embryol 168: 307-313

57. Forssmann, W.G., Lang, R.E., Aoki, A., Reinecke, M., Rippegather, G., Hock, D. (1987) Cardiodilatin as a neuropeptide (Cardiac polypeptide hormones are also neuropeptides). Exp Brain Res 16: 43-50

58. Forssmann, W.G., Mutt, V. (1985) Cardiodilatin-immunoreactive neurons in the hypothalamus of Tupaia. Anat Embryol 172: 1-5

59. Forssmann, W.G., Pickel, V., Reinecke, M., Hock, D., Metz, J. (1981) Immunohistochemistry and immunocytochemistry of nervous tissue. In: Heym, Ch, Forssmann, W.G. (eds) Techniques in Neuroanatomical Research. Springer, Berlin Heidelberg New York, pp 171 - 205

60. Forssmann, W.G., Reinecke, M., Weihe, E. (1982) Cardiac innervation. In: Bloom, S.R., Polak, J.M., Lindenlaub, E. (eds.) Systemic role of regulatory peptides, Symposium Oosterbeek Netherlands 1982. FK Schattauer, Stuttgart-New York pp 329-349

61. Forssmann, W.G., Triepel, J., Daffner, C., Heym, Ch.,Cuevas, P., Noble, M.I.M., Yanaihara, N. (1987) Vasoactive intestinal peptide in the heart. Ann New York Acad Sci 527: 405-420

62. Forssmann, W.G., Weihe, E. (1981) Histologie und Ultrastruktur des Herzens. In: Krayenbühl, H.P., Kübler, W. (Hrsg) Kardiologie in Klinik und Praxis. Band I Anatomie, Physiologie, Diagnostik, Allgemeine Pathophysiologie. Thieme, Stuttgart S. 4.1 - 4.13

63. Fujio, N., Ohashi, M., Nawata, H., Kato, K., Tateishi, J., Matsuo, H., Ibayashi, H. (1987) Unique distribution of natriuretic hormones in dog brain. Reg Pept 18: 131-137

64. Garcia, R., Cantin, M., Thibault, G., Ong, H., Genest, J. (1982) Relationship of specific granules to the natriuretic and diuretic activity of rat atria. Experientia (Basel) 38: 1071- 1073

65. Gardner, D.G., Deschepper, C.F., Ganong, W.F., Hane, S., Fiddes, J., Baxter, J.D., Lewicki, J. (1986) Extra-atrial expression of the gene for atrial natriuretic factor. Proc Natl Acad Sci USA 83: 6697-6701

66. Göbel, J., Back, H., Forssmann, K., Daffner, C., Stumpf, W.E., Forssmann, W.G. (1988) Evidence for the existence of cardiac hormones in the conductive system of the mammalian heart. In: Forssmann, W.G., Scheuermann, D.W., Alt, J. (eds) Functional Morphology of the Endocrine Heart. Steinkopff, Darmstadt pp 43 - 49

67. Göbel, J., Metz, J., Forssmann, W.G. (1986) Korrelation zwischen myoendokrinen Zellen und peptiderger Innervation im Herz. Verh Anat Ges 80: 551-553

68. Gorza, L., Sartore, S., Schiaffino, S. (1982) Myosin types and fibre types in cardiac muscle. II Atrial myocardium. J Cell Biol 95: 838-845

69. Grammer, R.T., Fukumi, H., Inagami, T., Misono, K.S. (1983) Rat atrial natriuretic factor: purification and vasorelaxant activity. Biochem Biophys Res Comm 116: 696-703

70. Greenberg, B.D., Bencen, G.H., Seilhamer, J.J., Lewicki, J.A., Fiddes, J.C. (1984) Nucleotide sequence of the gene encoding human atrial natriuretic factor precursor. Nature 312: 656-658

71. Grube, D., Forssmann, W.G. (1979) Morphology and function of the entero-endocrine cells. Horm Metab Res 11: 589-606

72. Hamid, Q., Wharton, J., Terenghi, G., Hassall, C.J.S., Aimi, J., Taylor, K.M., Nakazato, H., Dixon, J.E., Burnstock, G., Polak, J.M. (1987) Localization of atrial natriuretic peptide mRNA and immunoreactivity in the rat heart and human atrial appendage. Proc Natl Acad Sci USA 84: 6760-6764

73. Hanak, H., Böck, P. (1971) Die Feinstruktor des Muskel-Sehnenverbindung von Skelett- und Herzmuskel. J Ultrastruct Res 36: 68-85

74. Hassall, C.J.S., Wharton, J., Gulbenkian, S., Anderson, J.V., Frater, J., Bailey, D.J., Merighi, A., Bloom, S.R., Polak, J.M., Burnstock, G. (1988) Ventricular and atrial myocytes of newborn rats synthesise and secrete atrial natriuretic peptide in culture: light- and electron-microscopical localization and chromatographic examination of stored and secreted molecular forms. Cell Tissue Res 251: 161-169

75. Heine, H. (1979) Funktionelle Morphologie und Ultrazytochemie der spezifischen Granula der Vorhofmuskelzellen bei Säugetieren. Acta Anat (Basel) 105: 86-93

76. Herbst, W.M., Mall, G., Weers, J., Mattfeldt, T., Lang, R.E., Forssmann, W.G. (1987) Quantitive Morphologie der atrialen sekretorischen Granula (AG). Z Kardiologie 76: Suppl. 1: 44

77. Herbst, W.M., Mall, G., Weers, J., Mattfeldt, T., Lang, R.E., Forssmann, W.G. (1988) Combined stereological and biochemical study of the endocrine heart. In: Forssmann, W.G., Scheuermann, D.W., Alt, J. (eds) Functional Morphology of the Endocrine Heart. Steinkopff, Darmstadt, pp 75 - 82

78. Herzog, V.,Fahimi, H.D. (1976) Identification of peroxisomes (microbodies) in mouse myocardium. J Mol Cell Cardiol 8: 271-281

79. Hock, D., Fey, E., Brühl, B., Forssmann, W.G. (1988) Das endokrine Herz beim Rind. Verh Anat Ges 82

80. Huet, M., Benchimol, S., Berlinguet, J.C., Castonguay, Y., Cantin, M. (1974) Cytochemie ultrastructurale des cardiocytes de l'oreillette humaine. IV. Digestion des granules spécifiques par les protéases. J Microscopie 21: 147-158

81. Huet, M., Benchimol, S., Castonguay, Y., Cantin, M. (1974) Ultrastructural cytochemistry of atrial muscle cells. III. Reactivity of specific granules in man. Histochemistry 41: 87-105

82. Huet, M., Cantin, M. (1974) Ultrastructural cytochemistry of atrial muscle cells. I. Characterization of the carbohydrate content of specific granules. Lab Invest 30: 514-524

83. Huet, M., Cantin, M. (1974) Ultrastructural cytochemistry of atrial muscle cells. II. Characterization of protein content of specific granules. Lab Invest 3o: 525-532

84. Jamieson, J.D., Palade, G.E. (1964) Specific granules in atrial muscle cells. J Cell Biol 123: 151-172

85. Jirikowski, G.F., Back, H., Forssmann, W.G., Stumpf, W.E. (1986) Coexistence of atrial natriuretic factor (ANF) and oxytocin in neurons of the rat hypothalamus. Neuropeptides 8: 243-249

86. Jockusch, H., Füchtbauer, E.M., Füchtbauer, A., Leger, J.J., Leger, J., Maldonado, C.A., Forssmann, W.G. (1986) Long-term expression of isomyosins and myo-endocrine functions in ectopic grafts of atrial tissue. Proc Natl Acad Sci USA 83: 7325-7329

87. Kaczmarczyk, G., Forssmann, W.G., Noble, M.I.M., Gaul, E., Schnoy, N. (1987) Atrial granules in the dog heart after cardiac denervation. Cell Tissue Res 249: 701-705

88. Kangawa, K., Matsuo, H. (1984) Purification and complete amino acid sequence of alpha-human atrial natriuretic polypeptide (alpha-hANP). Biochem Biophys Res Comm 118: 131-139

89. Kawata, M., Nakao, K., Moril, N., Kiso, Y., Yamashita, H., Imura, H., Sano, Y. (1985) Atrial natriuretic polypeptide: topographical distribution in the rat brain by radioimmunoassay and immunohistochemistry. Neuroscience 16: 521-546

90. Kisch, B. (1956) Electron microscopy of the atrium of the heart. I. Guinea pig. Exp Med Surg 14: 99-112

91. Kisch, B. (1959) Electron microscopy investigations of the heart of cattle. I. The atrium of the heart of cows. Exp Med Surg 17: 247-261

92. Kisch, B. (1963) A significant electron microscopic difference between the atria and the ventricules of the mammalian heart. Exp Med Surg 21: 193-221

93. Klein, W., Böck, P. (1983) Elastica-positive material in the atrial endocardium. Light and electron microscopic identification. Acta Anat (Basel) 116: 106-113

94. Kramer, H.J. (ed.) (1987) Natriuretic hormones in hypertension. Official satellite symposium to the 11th meeting of the International Society of Hypertension, Heidelberg, 7th - 9th September, 1986. Klin Wochenschr 65: (suppl VIII)

95. Kreye, V.A.W., Bussmann, W.D. (Hrsg) (1987) ANP-Atriales natriuretisches Peptid und das kardiovaskuläre System. Steinkopff, Darmstadt

96. Kuhn, H., Richards, J.G., Tranzer, J.P. (1975) The nature of rat "specific heart granules" with regard to catecholamines: an investigation by ultrastructural cytochemistry. J Ultrastructure Res 50: 159-166

97. Lachance, D., Garcia, R., Gutkowska, J., Cantin, M., Thibault, G. (1986) Mechanisms of release of atrial natriuretic factor. I. Effect of several agonists and steroids on its release by atrial minces. Biochem Biophys Res Comm 135: 1090-1098

98. Lambertenghi-Deliliers, G., Zanon, P.L., Pozzoli, E.F., Bellini, O., Praga, C. (1978) Ultrastructural alterations of atrial myocardium induced by adriamycin in chronically treated animals. Tumori 64: 15-24

99. Lang, R.E., Thölken, H., Ganten, D., Luft, F.C., Ruskoaho, H., Unger, Th. (1985) Atrial natriuretic factor - a circulating hormone stimulated by volume loading. Nature 314: 264-266

100. Ledsome, J.R., Willson, N., Courneya, C.A., Rankin, A.J. (1985) Release of atrial natriuretic peptide by atrial distension. Can J Physiol Pharmacol 63: 739-742

101. Leknes, I.L. (1980) Ultrastructure of atrial endocardium and myocardium in three species of gadidae (Teleostei). Cell Tissue Res 219: 1 - 10

102. Maldonado, C.A., Saggau, W., Forssmann, W.G. (1986) Cardiodilatin- immunoreactivity in specific atrial granules of human heart revealed by the immunogold stain. Anat Embryol 173: 295-298

103. Marie, J.P., Guillemot, H., Hatt, P.Y. (1976) Le degré de granulation des cardiocytes auriculaires. Étude planimétrique au cours de différents apports d' eau et de sodium chez le rat. Pathol Biol 24: 549-554

104. Martinez-Palomo, A., Bencosme, S.A. (1986) Electron microscopic observations on myocardial specific granules and residual bodies in vertebrates. Anat Rec 154: 473

105. Mary-Rabine, L., Albert, A., Pham, T.D., Hordof, A., Fenoglio, J.J., Malm, J.R., Rosen, M.R. (1983) The relationship of human atrial cellular electrophysiology to clinical function and ultrastructure. Circulation Res 52: 188-199

106. Matsubara, H., Nishikawa, M., Umeda, Y., Taniguchi, T., Iwasaka, T., Kurimoto, T., Yamane, Y., Inada, M. (1987) The role of atrial pressure in secreting atrial natriuretic polypeptides. Am Heart J 113: 1457-1462

107. Matter, A., Forssmann, W.G. (1968) Muskelsehnenverbindungen. Verh Anat Ges 62: 73-81

108. McKenzie, J.C., Tanaka, I., Misono, K.S., Inagami, T. (1985) Immunocytochemical localization of atrial natriuretic factor in the kidney, adrenal medulla, pituitary and atrium of rat. J Histochem Cytochem 33: 828-832

109. McNutt, N.S., Fawcett, D.W. (1969) The ultrastructure of the cat myocardium. II Atrial muscle. J Cell Biol 42: 46-67

110. Metz, J., Mutt, V., Forssmann, W.G. (1984) Immunohistochemical localization of cardiodilatin in myoendocrine cells of the cardiac atria. Anat Embryol 170: 123-127

111. Meyer, M., Feller, S., Gagelmann, M., Hock, D., Nokihara, K., Forssmann, W.G. (1988) Lokalisation von Cardiodilatin (CDD) im Nebennierenmark. Verh Anat Ges (in press)

112. Mitsui, H., Kawamura, K., Ishizawa, K., Omae, M., Kawai, C. (1976) Ultrastructural features of diseased human atrial muscle cells. Recent Adv Stud Cardiac Struct Metab 12: 299-313

113. Nakayama, K., Ohkubo, H., Hirose, T., Inayama, S., Nakanishi, S. (1984) mRNA sequence for human cardiodilatin-atrial natriuretic factor precursor and regulation of precursor mRNA in rat atria. Nature 310: 699-701

114. Nehls, M., Reinecke, M., Lang, R.E., Forssmann W.G. (1985) Biochemical and immunological evidence for a cardiodilatin-like substance in the snail neurocardiac axis. Proc Natl Acad Sci USA 82: 7762-7766

115. Nemer, M., Chamberland, M., Sirois, D., Argentin, S., Drouin, J., Dixon, R.A.F., Zivin, R.A., Condra, J.H., (1984) Gene structure of human cardiac hormone precursor, pronatriodilatin. Nature 312: 654-656

116. Nemer, M., Lavigne, J.P., Droouin, J., Thibault, G., Gannon, M., Antakly, T. (1986) Expression of atrial natriuretic factor in heart ventricular tissue. Peptides 7: 1147-1152

117. Netchitailo, P., Feuilloley, M., Pelletier, G., Leboulenger, F., Cantin, M., Gutkowska, J., Vaudry, H. (1987) Atrial natriuretic factor-like immunoreactivity in the central nervous system of the frog. Neuroscience 22: 341-359

118. Nokihara, K., Ando, E., Forssmann, W.G. (1987) Isolation of cardiodilatin from porcine atria using immunoaffinity chromatography and production of monoclonal antibody. In: Miyazawa, T. (ed) Peptide Chemistry. Protein Research Foundation, Osaka, pp 7 - 10,

119. Nokihara, K., Forssmann, W.G. (1988) Immunohistochemical characterization of molecular forms of cardiac hormones using segment-specific antibodies against cardiodilatin. Kyoto Symposium on ANP, 1988. (in press)

120. Nomura, T., Mizukawa, K., Otsuka, N. (1987) Myocardial lesions and the alteration of atrial granularity in dystrophic mice. Arch Histol Jap 50: 335-346

121. Oberpriller, J.O., Oberpriller, J.C (1977) Modified hamster atrial cardiac muscle cells isolated in the anterior chamber of the eye. J Mol Cell Cardiol 9: 1013-1017

122. Oikawa, S., Imai, M., Ueno, A., Tanaka, S., Noguchi, T., Nakazato, H., Kangawa, K., Fukuda, A., Matsuo, H. (1984) Cloning and sequence analysis of cDNA encoding a precursor for human atrial natriuretic polypeptide. Nature 309: 724-726

123. Okamoto, H. (1969) An electron microscopic study of the specific granules in the atrial muscle cell upon the administration of agents affecting autonomic nerves. Arch Histol Jap 30: 467-478

124. Otsuka, N., Okamoto, H., Tomisawa, M. (1969) Electron and fluorescence microscopic study of the specific granules in rat atrial muscle cells. Arch Histol Jap 30: 367-374

125. Östlund, E., Bloom, G., Adams-Ray, J., Ritzen, M., Siegman, M., Nordenstam, H., Lishajko, F., von Euler, U.S. (1960) Storage and release of catecholamines, and the occurence of a specific submicroscopic granulation in hearts of cyclostomes. Nature 188: 324-325

126. Palade, G.E. (1961) Secretory granules in the atrial myocardium. Anat Rec 139: 262

127. Papka, R.E. (1978) Development of innvervation to the atrial myocardium of the rabbit. Cell Tissue Res 194: 219-236

128. Poche, R. (1958) Submikroskopische Beiträge zur Pathologie der Herzmuskelzelle bei Phosphorvergiftung, Hypertrophie, Atrophie und Kaliummangel. Virchows Arch 331: 165-248

129. Reinecke, M., Betzler, D., Forssmann, W.G. (1987) Immunocytochemistry of cardiac polypeptide hormones (Cardiodilatin/atrial natriuretic polypeptide) in brain and hearts of Myxine glutinosa (Cyclostomata). Histochemistry 86: 233-239

130. Reinecke, M., Betzler, D., Forssmann, W.G., Thorndyke, M., Askensten, U., Falkmer, S. (1987) Electronmicroscopical, immunohistochemical, immunocyto-chemical and biological evidence for the occurence of cardiac hormones (ANP/CDD) in chondrichthyes Histochemistry 87: 531-538

131. Reinecke, M., Forssmann, W.G. (1984) Regulatory peptides (SP, NT, VIP, PHI, ENK) of autonomic nerves in the guinea pig heart. J Hypertension A6 (10/11): 1867-1871

132. Reinecke, M., Forssmann, W.G. (1985) Untersuchungen zur Phylogenese cardialer Hormone. Acta Anat (Basel) 125: 286

133. Reinecke, M., Nehls, M., Forssmann, W.G. (1985) Phylogenetic aspects of cardiac hormones as revealed by immunocytochemistry, electron microscopy and bioassay. Peptides 6: (suppl 3): 321-331

134. Rhodin, J.A.G. , Missier, P., Reid, L.C. (1961) The structure of the specialized impulse-conducting system of the steer heart. Circulation 24: 349

135. Riegger, G.A.J., Elsner, D., Kromer, E.P., Daffner, C., Forssmann, W.G., Muders,F., Pascher E.W., Kochsiek,K. (1988) Atrial natriuretic peptide in congestive heart failure in the dog: plasma levels, cyclic guanosine monophosphate, ultrastructure of atrial myoendocrine cells, and hemodynamic, hormonal and renal effects. Circulation 77: 398-406

136. Rippegather, G., Maldonado, C.A., Schulz-Knappe, P., Hock, D., Lang, R.E., Forssmann, W.G. (1987) Morphologie der myoendokrinen Zellen des Rattenherzens bei akuten Veränderungen des Blutvolumens. Verh Anat Ges 81: 109-113

137. Rippegather, G., Saggau, W., Forssmann, W.G. (1987) Pathomorphologie des endo-krinen Herzens. In: Kreye, V.A.W., Bussmann, W.D. (eds) ANP - Atriales Natriure-tisches Peptid und das kardiovaskuläre System. Steinkopff, Darmstadt pp 169-178

138. Saetersdal, T., Jodalen, H., Lie, R., Rotevatn, S., Engedal, H., Myklebust, R. (1979) Effects of isoproterenol on the dense core and perigranular membrane of atrial specific granules. Cell Tissue Res 199: 213-224

139. Saper, C.B., Standaert, D.G., Currie, M.G., Schwartz, D., Geller, D.M., Needleman, P. (1985) Atriopeptin-immunoreactive neurons in the brain: presence in cardio-vascular regulatory areas. Science 227: 1047-1049

140. Schipp, R., Beyerle-v. Wehren, A. (1979) Zur funktionellen Bedeutung der osmiophilen Granula in Herzorganen niedriger Vertebraten. Eine vergleichend elektronenmikroskopische und pharmakologische Untersuchung am isolierten Herzen des Frosches und anderer poikilothermer Vertebraten. Z Zellforsch 108: 243-267

141. Seidmann, C.E., Bloch, K.D., Klein, K.A., Smith, J.A., Seidman, J.G. (1984) Nucleotide sequences of the human and mouse atrial natriuretic factor genes. Science 226: 1206-1209

142. Shibata, Y., Yamamoto, T. (1976) Fine structure and cytochemistry of specific granules in the lamprey atrium. Cell Tissue Res 172: 487-501

143. Shore, P.A., Cohn, V.H., Highman, B., Maling, H.M. (1958) Distribution of norepinephrine in the heart. Nature 181: 848

144. Skofitsch, G., Jocobowitz, D.M., Eskay, R.L., Zamir, N. (1985) Distribution of atrial natriuretic factor-like immunoreactive neurons in the rat brain. Neuroscience 16: 917-948

145. Sommer, J.R., Johnson, E.A. (1969) Cardiac muscle. A comparative ultrastructural study with special reference to frog and chicken hearts. Z Zellforsch 98: 437-468

146. Sonnenberg, H., Veress, A.T. (1984) Cellular mechanism of release of atrial natri-uretic factor. Biochem Biophys Res Comm 124: 443-449

147. Sosa-Lucero, J.C., De La Iglesia, F.A., Lumb, G., Berger, J.M., Bencosme, S. (1969) Subcellular distribution of catecholamines and specific granules in rat heart. Laboratory Investigation 21: 19-26

148. Späth, J., Rippegather, G., Lang, R.E., Schulz-Knappe, P., Saggau, W., Forssmann, W.G. (1986) Pathomorphological study of the endocrine heart in various cardiovascular diseases. Klin Wochenschr 64: (suppl VI): 83-88

149. Sugawara, M. (1987) Ultrastructural evidence for exocytotic release of atrial natriuretic polypeptide in the rat atrial muscle with the use of the tannic acid method. Biomedical Res 8: 9-17

150. Tang, J., Fei, H., Xie, C.W., Suen, M.Z., Han, J.S., Webber, R.J., Chang, D., Chang, J.K. (1984) Characterization and localization of atriopeptin in rat atrium. Peptides 5: 1173-1177

151. Theron, J.J., Biagio, R., Meyer, A.C., Boekkoois (1978) Ultrastructural observations on the maturation and secretion of granules in atrial myocardium. J Mol Cell Cardiol 10: 567-572

152. Thibault, G., Garcia, R., Gutkowska, J., Bilodeau, J., Lazure, C., Seidah, N.G., Chretien, M., Genest, J., Cantin, M. (1987) The propeptide Asnl-Tyr126 is the storage form of rat atrial natriuretic factor. Biochem J 241: 256-272

153. Thiedemann, K.U., Ferrrans, V.J. (1977) Left atrial ultrastructure in mitral valvular disease. Am J Pathol 89: 575-604

154. Tomisawa, M. (1969) Atrial specific granules in various mammals. Arch Histol Jap 30: 449-465

155. Toshimori, H., Nakazato, M., Toshimori, K., Asai, J., Matsukura, S., Oura, C., Matsuo, H. (1988) Distribution of atrial natriuretic polypeptide (ANP)-containing cells in the rat heart and pulmonary vein. Cell Tissue Res 251: 541-546

156. Toshimori, H., Toshimori, K., Oura, C., Matsuo, H. (1987) Immunohistochemistry and immunocytochemistry of atrial natriuretic polypeptide in porcine heart. Histochemistry 86: 595-601

157. Viragh, S., Porte, A. (1960) Le noeud de Keith et Flack et les différents fibres auriculaires du coeur de rat. Études en microscopic optique et électronique. Compt Rend Acad Sc 251: 2086

158. Wei, Y., Rodi, C.P., Day, M.L., Wiegand, R.C., Needleman, L.D., Cole, B.R., Needleman, P. (1987) Developmental changes in the rat atriopeptin hormonal system. J Clin Invest 79: 1325-1329

159. Weihe, E., Kalmbach, P. (1978) Ultrastructure of capillaries in the conductive system of the heart in various mammals. J Mol Cell Cardiol 192: 77-78

160. Wolfe, D.E., Axelrod, J., Potter, L.T., Richardson, K.C. (1962) Localization of nor-epinephrine in adrenergic axons by light- and electron-microscopic autoradiography. In: Breese S.S. (ed) Electron Microscopy. Fifth International Congress of Electron Microscopy, vol. 2. Academic Press, Philadelphia pp L11-L12

161. Yunge, L., Benchimol, S., Cantin, M. (1979) Ultrastructural cytochemistry of atrial muscle cells. VII. Radioautographic study of synthesis and migration of glycoproteins. J Mol Cell Cardiol 11: 375-388

162. Zamir, N., Skofitsch, G., Eskay, R.L., Jacobowitz, D.M. (1985) Distribution of immunoreactive atrial natriuretic peptides in the central nervous system of the rat. Brain Res 365: 105-111

163. Zisfein, J.B., Sylvestre, D., Homcy, C.J., Graham, R.M. (1987) Analysis of atrial natriuretic factor biosynthesis and secretion in adult and neonatal rat atrial cardiocytes. Life Sci 41: 1953-1959

Evidence for the existence of cardiac hormones in the conductive system of the mammalian heart

J. Göbel, H. Back, K. Forssmann, C. Daffner , [1]W.E.Stumpf, W.G. Forssmann

Department of Anatomy and Cell Biology, University of Heidelberg, Heidelberg, FRG,
[1]University of North Carolina, Chapel Hill, North Carolina, USA

Introduction

In recent years, the identity of the endocrine heart has been established by means of isolation of polypeptide hormones from atria of several species. Antibodies are currently available to perform studies using specific immunohistochemistry and radioimmuno-assays in order to investigate these new polypeptides (see reviews 3,4). It soon became evident that the same family of polypeptides was also present in various extra-auricular sites, particularly the central nervous system (6) as well as in the pituitary, the adrenal medulla, the peripheral autonomic nervous system, and the kidney. In addition to these sites, several research groups detected cardiac hormones in the ventricles of lower vertebrates (12). Furthermore, RIA and mRNA results showed that CDD/ANP poly-peptides occur in mammalian ventricles. Initial studies demonstrated that atrial peptides are also found in ventricular myoendocrine cells (1,8,13). It has become apparent that the immunoreactive material of cardiac hormones in the ventricle may be confined to the conductive system. We investigated this subject in greater detail by systematically ana-lyzing the heart for the detection of extra-auricular CDD/ANP immunoreactivity using light and electron microscopic immunocytochemistry, RIA, and HPLC combined with RIA and biotests of topographically defined areas of the ventricular tissue.

Our study of CDD/ANP localization in the conductive system included the following main points:

1) Are the myoendocrine cells of the atrial appendages identical to those of the ven-tricles?

2) How are the extra-auricular and ventricular myoendocrine cells structurally composed?

3) What is the exact distribution pattern of the extra-auricular myoendocrine cells?

4) What is the functional implication of the extra-auricular CDD/ANP sites?

Preliminary results of this study were presented at the "Verhandlungen der Anato-mischen Gesellschaft 1987" in Leipzig (7).

Material and methods

Porcine hearts were obtained from a slaughterhouse (we are grateful to Mr. Adler, Mannheim-Viernheim, for his assistance). The hearts were immediately excised and the tissue immediately processed for morphological and biochemical studies. For light and electron microscopy the hearts were perfused retrogradedly entering the coronary vessels by cannulating the ascending aorta. The morphological methods are described elsewhere (see the morphological review of this edition, 5). For biochemistry, the atrial and ventricular tissue was dissected and immediately frozen in liquid nitrogen. Specific topographical regions of atrial and ventricular regions were separately extracted and analyzed by RIA, biotests, and HPLC, combined with the latter methods. Detailed descriptions of the biochemical analysis are also found in the biochemical review of this edition (9).

Morphological results

For each porcine heart perfused for immunohistochemistry, an identical control localization of CDD-IR using the immunofluorescence and immunoperoxidase technique as previously established was ascertained prior to the ventricular localization (10). Thus CDD may be produced and stored in the ventricular segments of the His-bundle, the Purkinje-fibers and the moderator band. In earlier light microscopical investigations of mammalian hearts (8), a ventricular localization of CDD-IR was established when ventricles were used as controls for our immunohistochemical studies of atrial tissue. In these original preparations of porcine ventricles we observed a subendocardial localization of CDD/ANP immunoreactive strands topographically located at sites in the ventricular conductive system. In porcine heart, in contrast to smaller mammalian species, a discernable conductive system provides a simple discrimination of the conductive fibers from the myocardial working cells. In the atrio-ventricular conductive system at the initial segment where the His-bundle initially arises from the AV-node, cells with myoendocrine function are detected (Fig. 1, see plate 5 in front of the book). This is also confirmed by electron micrographs which show the typical endocrine synthetic apparatus in both longitudinally and transversely sectioned cells of the atrial region of the His-bundle. The abundance of specific secretory granules indicating the endocrine function is striking. The His-bundle fibers in this region exhibit a developed Golgi apparatus. This differentiated secretory apparatus in the cells of the atrial segment of the His-bundle in the rat heart was originally recognized by Bompani, Rouiller and Hatt in 1959, who showed several of these conductive cells to be particularly rich in specific granules.

To date no results are available concerning ventricular sites of CDD/ANP-IR or other morphology depicting endocrine active cells in the ventricle. Subsequent to this study others (14) published similar results. Furthermore, careful investigation by electron microscopy revealed large bundles of CDD-IR fibers in the ventricular septum. From ultrastructural micrographs it can be deduced that the fiber bundles belonging to the

Fig. 1. Light microscopic immunohistochemistry of a CDD-immunoreactive septal branch of the His-bundle. Note the immunoreactive spots confined to the conductive system whereas the ventricular myocardium shows no reaction. (x 500)

Fig. 2. Immunoreactive strand of fibers in the ventricular septum belonging to branches of the His-bundle. Note the immunoreactive spots confined to the cells of the Purkinje fiber type. (x 500)

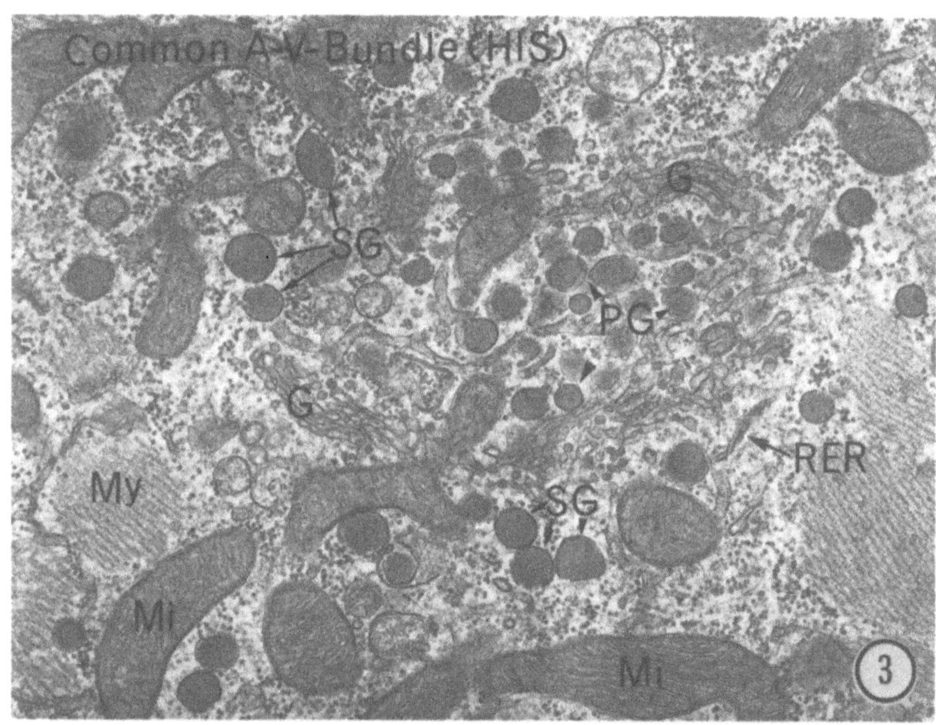

Common A-V-Bundle (HIS)
G
SG
PG
G
RER
My
SG
Mi
Mi
3

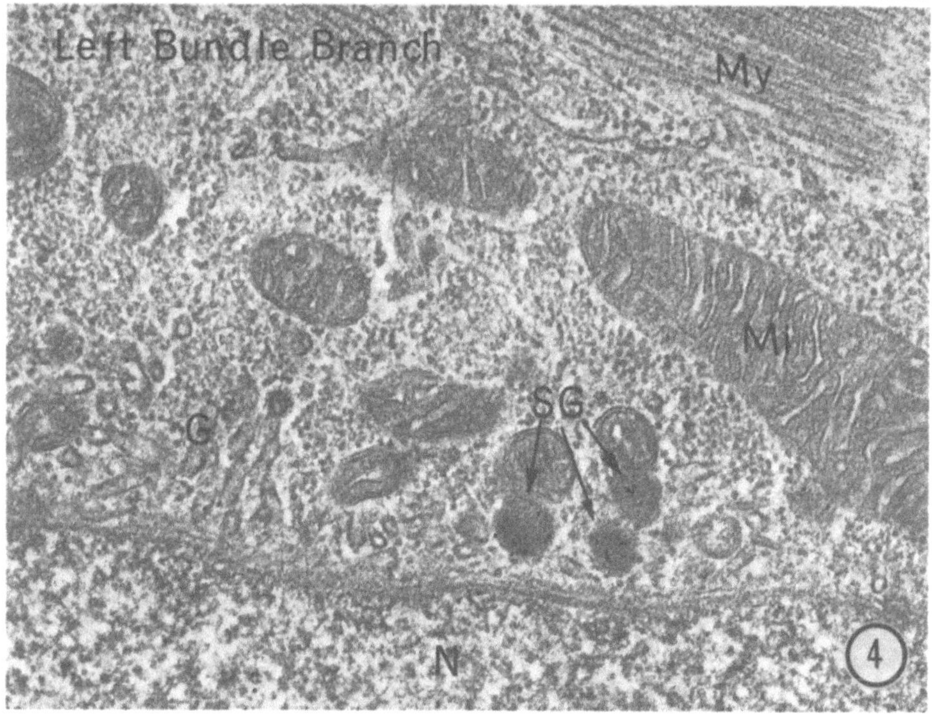

Left Bundle Branch
My
Mi
G
SG
N
4

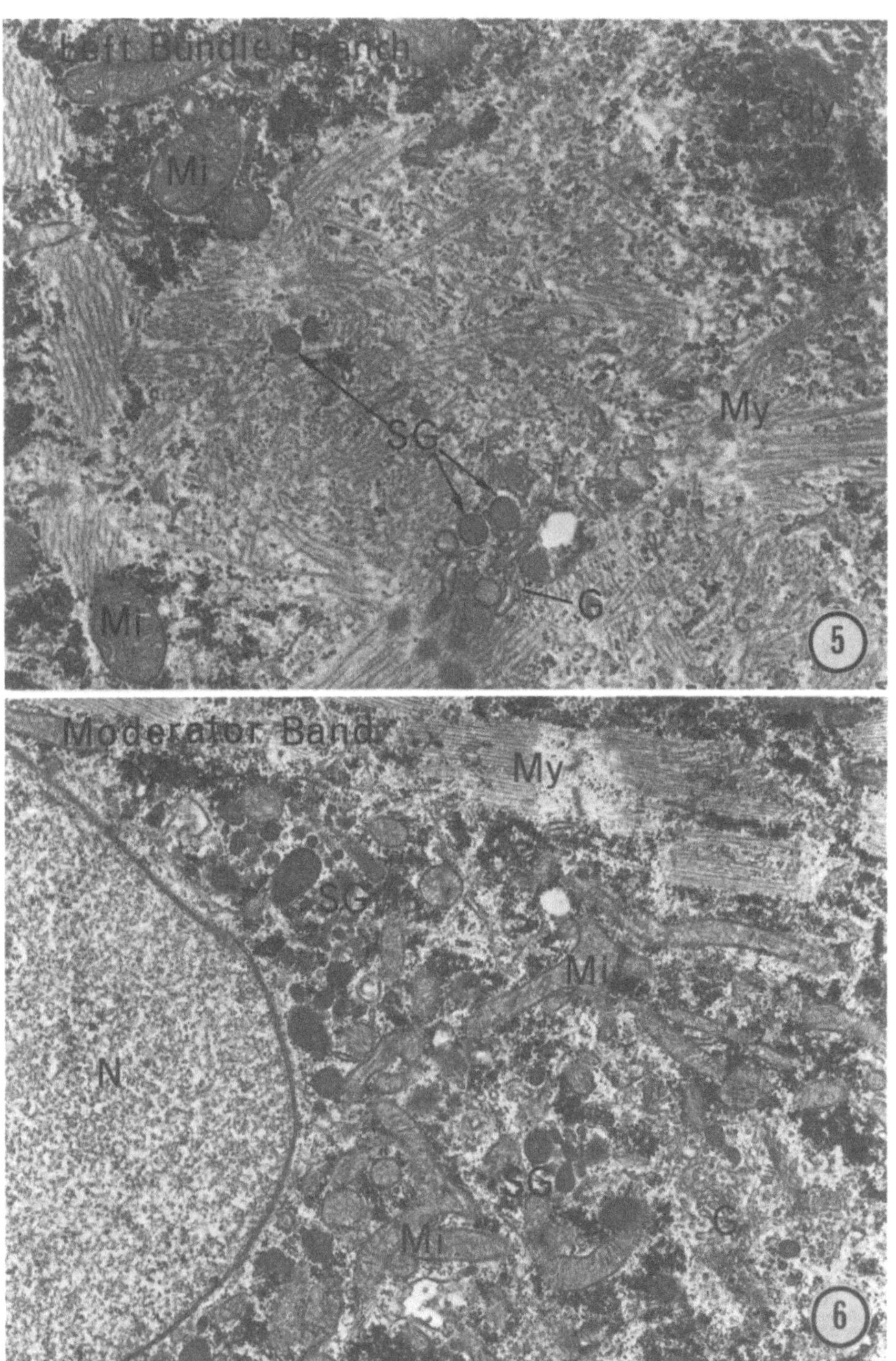

Left Bundle Branch
Mi
Gly
My
SG
G
5
Moderator Band
My
SG
Mi
N
SG
C
Mi
6

Fig. 3. Electron micrograph of a myoendocrine cell from the common AV-bundle of His from the porcine heart. This cell type is characterized by a high amount of secretion granules (SG), and a developed Golgi apparatus (G) is found in the initial segment of theatrioventricular conductive system. Note the high number of specific granules and also progranules (PG) as well as the rough endoplasmic reticulum (RER) which indicates a pronounced secretory endocrine activity. Note also the myofibrils (My) and mitochondria (Mi). (x 26 000)

Fig. 4. Immunohistochemical proof of the cardiodilatin-immunoreactivity in a conductive cell of the left bundle branch from the porcine ventricular septum. Note that the secretory granules (SG) are specifically labelled by anti-cardiodilatin antibodies characterized by numerous gold particles superimposed upon the secretory granules (SG). Note also the Golgi apparatus (G), the cell nucleus (N), mitochondria (Mi), and myofibrils (My).
(x 40 000)

Fig. 5. Electron micrograph of a cell from the conductive system in the ventricular septum of porcine heart. A conductive cell of the left bundle branch with its characteristic array of (My), glycogen condensations (Gly), and mitochondria (Mi) are seen. In a circumscribed region, a Golgi apparatus (G) with typical specific secretory granules is seen (SG).
(x 22 000)

Fig. 6 . A conductive cell of the moderator band from a porcine heart showing the perinuclear area. A developed Golgi apparatus (G) surrounded by progranules (arrows) and secretory granules (SG) is shown. Note the typical large nucleus (N), myofibrils (My), and mitochondria (Mi). (x 14 400)

ventricular conductive system contain a peptide-producing secretory apparatus. Figure 1 shows a section of one of these ventricular conductive fibers where isolated areas of CDD/ANP-IR are observed by light microscopic immunocytochemistry. Bundles of fibers or single strands of the CDD-IR are clearly deliniated from the surrounding ventricular myocardium. All fibres exhibiting the immunoreactive spots clearly correspond to cells of the Purkinje-fiber type (Fig. 2, see plate 5 in front of the book). Immunocytochemistry reveals that the CDD/ANP-IR may not only be located adjacent to the cell nuclei, but it is predominantly within the large regions of the sarcoplasma which are free from myofibrils.

Immunohistochemical proof that the Purkinje-fibers possess a secretory apparatus producing cardiodilatin was established by the immunogold method which demonstrated the binding of specific CDD/ANP antibodies to the secretory granules (Fig. 4). Systematic ultrastructural analysis shows the discrete Golgi apparatus of large conductive cells surrounded by a few small secretory granules of the progranular type, also some mature specific secretory granules detectable in lower amounts. These cells exhibit ultrastructural features corresponding to the ventricular conductive system such as the large oval shaped cell nucleus, the richness of glycogen, and rudimentary myofibrils (Figs. 5 and 6). The low background agrees with a specific localization of CDD/ANP within the subcellular storage site. Identical morphological results were obtained in all fibers of the ventricular conductive system including the moderator band (Fig. 6).

Biochemical results

These immunocytochemical results compelled us to prove the physiological and biochemical presence of CDD/ANP in the ventricular conductive system. Extractions of ventricular tissue were carried out in order to demonstrate the vasorelaxant activity similar to that found in atrial extracts. A highly potent vasorelaxant activity is found on aortic smooth muscle strips. Also, the biochemical analysis of the CDD concentration using RIA and HPLC revealed the presence of several molecular forms of cardiodilatin within the different septal regions. The main molecular forms of CDD in septal extracts are the prohormone CDD-126 and CDD-88 . Extremely small amounts of CDD-28, the hormone circulating in the blood, can also be detected. The distribution of radioimmuno-assayable cardiodilatin in the right and left ventricle with particular reference to the septum was established by extraction and radioimmunoassay in topographical exact zones of the tissue. It was found that the septal myocardium exhibits a prominent cardiodilatin-immunoreactivity which is concomitant of the presence of the conductive fibers we observed.

Discussion and conclusions

Our investigation showed that the porcine heart contains various types of endocrine active myocardial cells and that:
1) Typical myoendocrine cells occur in the atrial appendages; 2) The atrio-ventricular conductive system exhibits many small endocrine cells belonging to the initial segment of the His-bundle; and, 3) The ventricular conductive system characterized by large glycogen-containing cells contains circumscribed cytoplasmic areas which show an endocrine-secretory apparatus. In ventricular segments of the His-bundle, CDD may be produced and stored in the Purkinje-fibers and the moderator band.

In addition to the compact endocrine organ of mammalian atrial appendages where the atrial wall is entirely composed of *myoendocrine cells,* it is now clearly shown that cells of myoendocrine function also occur in the atrial and ventricular conductive system. All criteria substantiating an intracellular synthetic apparatus related to this endocrine function are established by the combined morphological-biochemical analysis presented here. This confirms the early investigation of (2), who described the presence of secretory granules in the cells of the rat His-bundle, as well as the results recently published by (14). Furthermore, it is necessary to emphasize that only (11) have claimed that cells of the conductive system exhibit a markedly developed synthetic apparatus for proteins. In addition, we have now revealed specific granules to be clearly present in all segments of the ventricular conductive cells such as the His-bundle, Purkinje-fibers and the moderator band.

Thus, mammalians exhibit typical myoendocrine cells not only within atrial regions morphologically defined as appendages, but also in other locations of the atria and ventricles, which together constitute a diffuse endocrine system. One remaining question concerns the functional significance of CDD production within the ventricles. The amount of ventricular cardiodilatin compared to that of atrial appendages is negligible and hardly contributes to the total concentration of the systemic circulation. In contrast to the compact atrial endocrine organ which is responsible for the systemic function of the endocrine heart, the ventricular production sites may only be involved in local endocrine regulation which is, at present, not certain.

References

1. Back, H., Stumpf, W.E., Ando, E., Nokihara, K., Forssmann, W.G. (1986) Immuno-cytochemical evidence for CDD-ANP-like peptides in strands of myoendocrine cells associated with the ventricular conduction system of the rat heart. Anat Embryol 175: 223-226
2. Bompiani, G.D., Rouiller, Ch., Hatt, P.Y. (1959) Le tissue de conduction du coeur chez le rat. Étude au microscope électronique. I. Le tronc commun du faisceau de His et les cellules claires de l'oreillette droite. Arch Mal Coeur 52: 1257-1274
3. Cantin, M., Genest, J. (1985) The heart and the atrial natriuretic factor. Endocr Rev 6: 107-127
4. Forssmann, W.G. (1986) Cardiac hormones. I. Review on the morphology .bio-chemistry and molecular biology of the endocrine heart. Eur J Clin Invest 16: 439-451
5. Forssmann, W.G. (1988) Morphological review: immunohistochemistry and ultra-structure of the endocrine heart. In: Forssmann, W.G., Scheuermann, D.W., Alt, J. (eds) Functional Morphology of the Endocrine Heart. Steinkopff, Darmstadt, pp 13-42
6. Forssmann, W.G., Lang, R.E., Aoki, A., Reinecke, M., Rippegather, G., Hock, D. (1987) Cardiodilatin as a Neuropeptide (Cardiac Polypeptide Hormones are also Neuropeptides). Exp Brain Res 16: 43-50
7. Göbel, J., Daffner, C., Rippegather, G., Forssmann, K., Hock, D., Forssmann, W.G. (1987). Das endokrine Herz: Verteilung der CDD-IR myoendokrinen Zellen im Vorhof und im Ventrikel des Schweineherzens. Verh Anat Ges (in press)
8. Göbel, J., Metz, J., Forssmann, W.G. (1986) Korrelation zwischen myoendo-krinen Zellen und peptiderger Innervation im Herz. Verh Anat Ges 80: 551-553
9. Hock, D. (1988) Biochemistry of cardiac hormones. In: Forssmann, W.G., Scheuermann, D.W., Alt, J. (eds) Functional morphology of the endocrine heart. Steinkopff, Darmstadt, pp 93 - 101
10. Maldonado, C.A., Saggau, W., Forssmann, W.G. (1986) Cardiodilatin-immuno-reactivity in specific atrial granules of human heart revealed by the immunogold stain. Anat Embryol 173: 295-298
11. Mochet, M., Moravec, J., Guillemot, H., Hatt, P.Y. (1975) The ultrastructure of rat conductive tissue; an electron microscopic study of the atrioventricular node and the bundle of His. J Mol Cell Cardiol 7: 879-889
12. Reinecke, M., Nehls, M., Forssmann, W.G. (1985) Phylogenetic aspects of cardiac hormones as revealed by immunocytochemistry, electronmicroscopy, and bioassay. Peptides 6 (suppl 3): 321-331
13. Thompson, R.P., Simson, J.A.V., Currie, M.G. (1986) Atriopeptin distribution in the developing rat heart. Anat Embryol 175: 227-233
14. Toshimori, H., Toschimori, K., Oura, C., Matsuo, J. (1987) Immunohisto-chemistry and immunohistochemistry of atrial natriuretic polypeptide in porcine heart. Histochemistry 86: 595-601

Catecholamine-containing granules in cardiac chromaffin cells as compared with atrial granules

D. W. Scheuermann

Institute of Histology and Microscopic Anatomy, University of Antwerp, Antwerp, Belgium

Summary

In the atrial wall of the heart of various animals, atrial-specific granules can be observed in contractile myocardial cells. In addition, cells without myofibrils may contain secretory granules. In higher vertebrates, these non-contractile granule-containing cells appear mostly singly and less frequently in groups; in some lower vertebrates, e.g., dipnoan fish, numerous non-contractile granule-containing cells form a thick subendothelial layer in the auricle, occasionally showing single cells interposed among atrial myocardial working fibers. In *Protopterus and Lepidosiren*, both types of granule-containing cells have a well-developed Golgi complex and seem innervated by cholinergic nerve terminals. In conventional electron microscopy, the atrial muscular "specific" granules and those from the granule-containing cells which lack myofibrils bear a close resemblance in size, shape, and electron density. Cytochemical investigation of the auricular wall shows that the granules of the myocytes are immunoreactive for α-human ANF and are not chromaffin, unlike the granules of the non-contractile cells which show a positive chromaffin reaction without ANF immunoreactivity. Microspectrofluorimetric analysis demonstrates that the chromaffin cells mainly contain dopamine. It thus appears that the heart is an endocrine gland for at least two hormones. Since adrenergic agonists appear to induce the release of ANF, it is suggested that the role of both cardiac hormones may be interrelated and that they participate in the maintenance of cardiovascular homeostasis by affecting vascular resistance, kidney function, and myocardial contractility.

Introduction

For several decades, the presence of a population of dense granules within mammalian atrial cardiocytes has been studied by a wide range of authors. As morphological and experimental studies have reported, atrial-specific granules closely resemble the secretory granules of various endocrine organs (11). The occurrence of large amounts of catecholamines in some granule fractions isolated from hearts of cyclostome (4) has led to the suggestion that the atrial-specific granules may have a catecholamine-storage function. This is consistent with the finding that the average catecholamine amount in the atria of the mammalian heart was found to be significantly higher than in the ventricles (34) as well as with the result after reserpine (known to deplete catecholamines elsewhere) administration when the number of atrial-specific granules in rats decreased (21,23,24).

Material and methods

Four adult specimens of the African lungfish *Protopterus annectens* and two adult specimens of the South-American lungfish *Lepidosiren paradoxa* were used. The animals were anesthetized with an intraperitoneal injection of pentobarbital sodium. For electron microscopy, the hearts of two speciments of *Protopterus* and one specimen of *Lepidosiren* were fixed by vascular perfusion with 1.25% glutaraldehyde (GA) buffered at pH 7.4 by 0.15 M sodium cacodylate for 10 min. After removal of the heart, small blocks of tissue from the auricle were immersed for 4h in 4% GA buffered at pH 7.4 by 0.15 M sodium cacodylate and postfixed in a 2% osmium tetroxide solution. For the chromaffin reaction at the ultrastructural level, other portions of the auricular tissue remained in 4% GA for one month and were further treated for 4h with a solution of 4% GA containing 2.5% potassium dichromate and 2% sodium sulphate in a 0.2 M acetate buffer (pH 4.1). After fixation by either method, the tissues were rinsed with distilled water, dehydrated in a graded acetone series and embedded in Durcupan. Thick (1μm) sections were cut for light microscopy and stained with toluidine blue/basic fuchsine. Thin sections were counterstained with uranyl acetate followed by lead citrate. Granules in semi-thin and thin sections of tissues treated with the GA dichromate technique were studied without on-grid staining and analyzed with a Kevex energy-dispersive x-ray microanalyser to detect the incorporated chromium at the site of the reaction product.

For fluorescence microscopy, the auricle of the heart of two specimens of *Protopterus* and one specimen of *Lepidosiren* was freeze-dried; tissue blocks were exposed to paraformaldehyde powder in air, at a constant relative humidity of 60%, then embedded in filtered paraffin under vacuum, serially sectioned at ± 10μm, and mounted in fluorescence-free paraffin oil. Microspectrofluorimetry was performed using a micro-spectrofluorimeter, improved in our laboratory with the instrumental configuration and differentiation mode of biogenic amines as reported earlier (36).

For immunocytochemical localization of ANF, tissue blocks of the auricle of *Protopterus*, fixed by vascular perfusion with 1.25% GA and stored in 4% GA for one month, were thin-scctioned; sections were collected on nickel grids and first treated with 10 % hydrogen peroxide for 5 min. Following a short rinse in distilled water and after blocking the non-specific binding with normal goat serum, they were incubated in the primary anti-α-human ANF (1-28) serum (UCB Bio-products); diluted 1:100 to 1:1000 in 0.05 M Tris Buffered Saline (TBS), 0.1% bovine serum albumin (BSA), 0.01% sodium azide for 16h at 4°C. After rinsing in 0.05 M TBS, 0.1% BSA, 0.01% sodium azide, the incubation in a solution containing goat immunogold to rabbit immuno-globulin (Auroprobe EM GAR/G10, Janssen Life Sciences; diluted 1:20 in the same dilution medium as for the primary antiserum) was carried out at room temperature for 60 min. Sections were first washed with 0.05 M TBS, 0.1% BSA, 0.01% sodium azide and then with distilled water. After drying, they were stained with a low concentration of uranyl acetate (0.5% instead of 2%), for 30 sec only, and with lead citrate, for 5 min only. These immunocytochemical results were compared with those of the immunocyto-

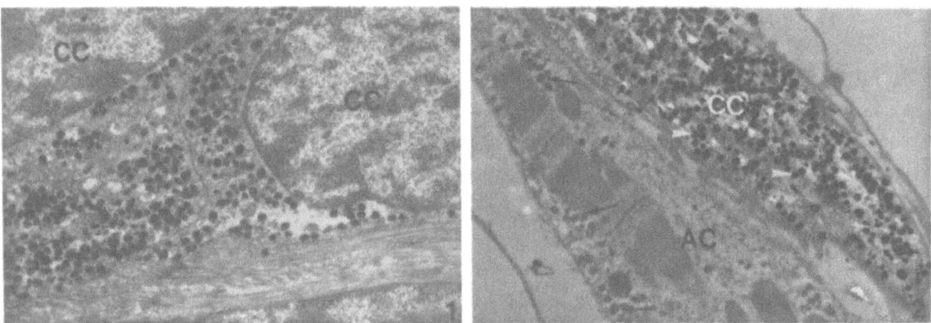

Fig. 1. Granule-containing cells (CC) in the wall of the auricle of the *Protopterus* heart. Around some dense cores, an electron-lucent halo can be observed. (x9 600).

Fig. 2. Chromaffin cell (CC) adjacent to an auricular cardiocyte (AC). Note the difference in electron densit between atrial granules (arrows) and catecholamine-containing granules (arrowheads). (x7 350).

chemical localization of ANF in the atrium of the cat and the rat, under the same conditions. For controls, the primary antiserum was replaced by normal rabbit antiserum and by primary antiserum after absorption with synthetic α-human ANF.

Results and discussion

In electron micrographs from the auricle of the investigated dipnoan fish, the chromaffin cells are distributed singly or in groups, located either on the surface of heart muscle bundles, i.e., close to the endocardium and pericardium, or in between contractile cardiocytes (Fig. 1, 2). The most conspicuous feature of these cells is the presence of numerous spherical or more irregular electron-dense granules with a diameter ranging from 60-300 nm. A limiting smooth membrane encloses the osmiophilic content entirely. Usually there is a narrow clear halo interposed between the central dense core and the membrane. In larger vesicles the dense core may be small, eccentrically located, and surrounded by a wide electron-lucent space. These granules possess the same fine structure as the neurosecretory granules found as a normal component in adreno-medullary cells. As in the latter, ends of the lamellae of the well-developed Golgi complex are frequently dilated and multiple parallel cisternae of rough endoplasmic reticulum are observed. Some pictures suggest that the granule content is liberated from the cell by exocytosis. A discontinuous sheath of satellite cell cytoplasm, covering clusters of granule-containing cells, may be observed. Usually, single cells within a cluster are not serarated by satellite cell processes. These cells are often tightly associated and occasionally connected to each other by attachment plaques. Granule-containing cells can be observed close to the endothelium of the auricular blood cavity and to capillaries. Nerve terminals are found closely adjacent to the granule-containing cells. These fibers, which contain both microfibrils and tubules , are

54

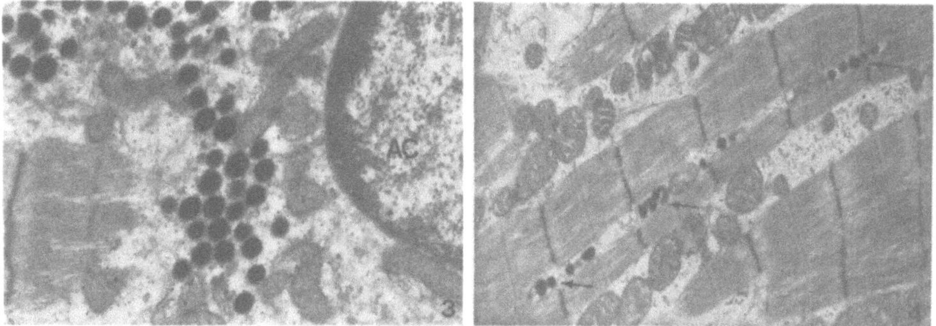

Fig. 3. Atrial granules lying in the perinuclear region of a cardiocyte (AC). (x20 600)

Fig. 4. Atrial granules (arrows) situated between the myofibrils of a cardiocyte. (x11 520)

nearly always surrounded by Schwann cell cytoplasm. Within the axon terminal, and mitochondria, a few dense-cored vesicles (about 100 nm in diameter) and numerous clear vesicles (about 55 nm in diameter) can be found. At the zone of contact between the nerve terminal and the granule-containing cell, asymmetrical membrane specializations near an accumulation of small translucent vesicles, as found in cholinergic synapses, can be observed.

The atrial-specific granules, as observed in cardiocytes of the auricle of *Protopterus* and *Lepidosiren*, occur both in the perinuclear sarcoplasm (Fig. 3), and in the interfibrillar and subsarcolemmal regions (Fig.4). They are mostly spherical with a diameter ranging from 100 to 400 nm, and surrounded by a smooth limiting membrane. Likewise, the opaque content of the granules is separated from the limiting membrane by a thin, light band, about 20 nm wide. These granules show a difference in fine structure from those of chromaffin cells in that they are generally larger in size and less dense in electron opacity after postfixation with osmium tetroxide and counterstaining with lead salts only (Fig. 2). If no specific techniques are used, however, the atrial-specific granules often closely resemble the chromaffin granules.

Investigation of the two kinds of granule-containing cells in the auricle of *Protopterus* and *Lepidosiren* was performed after initial fixation with glutaraldehyde and subsequent oxidation with dichromate. Only primary amines produce an amine-glutaraldehyde complex (in the case of dopamine and noradrenaline an insoluble isoquinoline; in the case of 5-hydroxytrytamine a beta-carboline derivative) reacting with salts of chromium. The amount of chromium present in the granules, detectable by x-ray microanalysis, entails the assessment that the investigated granules in cells which lack myofibrils contain a primary biogenic monoamine, without allowing a further differentiation between them. Conversely, granules of the contractile cardiocytes show no positive chromaffin reaction.

Identification of the biogenic amine in the granule-containing cells can be achieved after treatment with formaldehyde vapor, since the latter produces a ring-closing condensation reaction with the biogenic amine, yielding heterocyclic compounds. Subsequent dehydrogenation of the latter results in specific fluorophores which can be

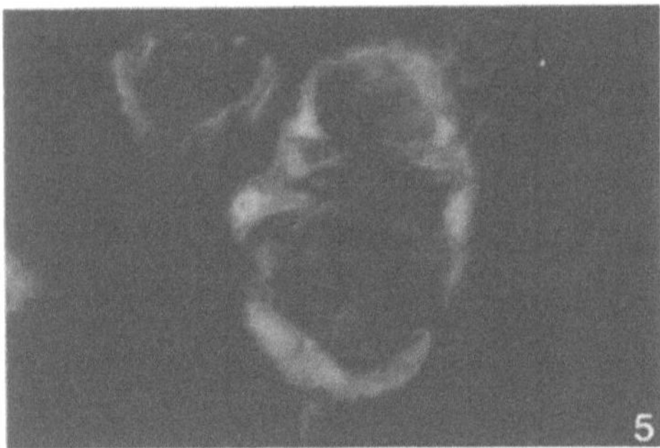

Fig. 5. Fluorescent chromaffin cells in the auricular wall. (x65)

characterized microspectrofluorimetrically (3). Hence, the auricular wall of the heart of dipnoan fish discloses cells showing a bright, blue-white, formaldehyde-induced fluorescence with an emission maximum at 480 nm (Fig. 5). After excessive hydrochloric acid treatment, the maximum excitation peak at 420 nm, obtained at neutral pH, reveals a bathochromic shift to 380 nm with additional peaks at 320-330 nm and 260 nm, yielding peak ratios 380/320-330 of 1.030 and 320-330/260 of 1.523, characteristic for the dopamine in its non-quinoidal form (27,36). The fading curves of the acidified fluorescent cells show a slow photodecomposition rate of about 30% after 5 min irradiation at the most effective wavelength, characteristic for dopamine. The fluorescent cells and the chromaffin cells in the heart of the investigated Dipnoi occur in similar parts of the auricle. It can therefore be assumed that these cells are one and the same. Hence, microspectrofluorimetrically, these chromaffin, formaldehyde-induced fluorescent cells mainly contain dopamine. The auricular cardiocytes, which display no chromaffin reaction, show no formaldehyde-induced fluorescence either. They do, however, react with antibodies against ANF.

Thin sections of the auricle of *Protopterus* have been incubated in the primary antiserum directed against synthetic α-human ANF (dilution 1:1000) and further processed for immunocytochemistry using the immuno-gold technique. ANF immunoreactivity is accumulated in the auricular cardiocytes and appears consistently associated with atrial-specific granules (Figs. 6, 7). The distribution patterns of ANF-like immunoreactivity in atrial cardiocytes of the cat and the rat are found to be identical to that of the auricle of *Protopterus*. From this it can be assumed that the auricle of *Protopterus* contains a peptide which, immunologically, appears related to α-human ANF.

Adreno-medullary cells of the rat have been shown to exhibit immunoreactivity for ANF (12,19). The adrenal medulla of mammals is homologous with the chromaffin tissue of the auricle of the heart of lungfish. Nevertheless, with the immuno-gold technique no significant label of chromaffin granules can be found, while in the same section a specific immunoreaction is obtained for the atrial-specific granules. The possibility remains that the sensitivity of the immuno-gold technique used is too weak to demonstrate potential small amounts of ANF in the chromaffin cells of the investigated animals.

56

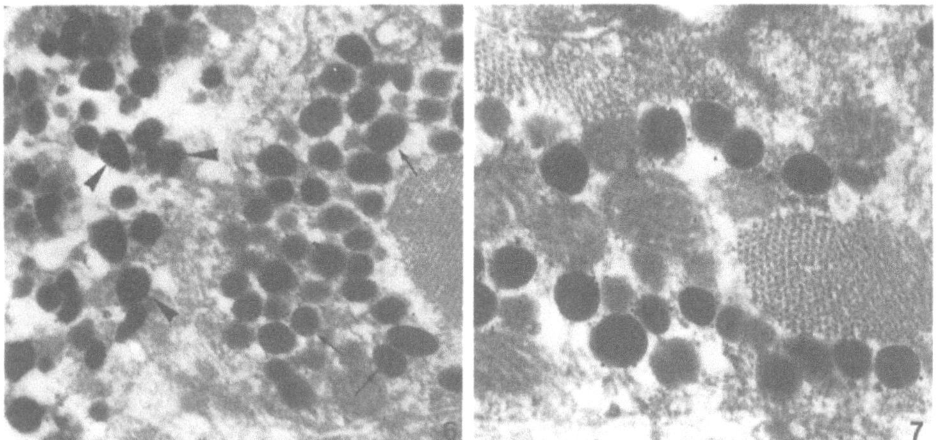

Fig. 6. Atrial granules (arrows) showing immunoreactivity for α–human ANF. In the catecholamine-containing granules (arrowheads) of the adjacent chromaffin cell, no immunoreactivity could be found. (x40 000)

Fig. 7. Higher magnification of atrial granules in the auricle, showing immunoreactivity for α-human ANF. (x60 000)

The auricular chromaffin cells of the investigated dipnoan fish are innervated and often located close to capillaries. It is therefore possible that, under neural control, they liberate catecholamines into the blood circulation . Since serial sections have shown that the chromaffin cells are not confined to the cardiac ganglia, it is assumed that their function is not directly related to that of ganglion cells (15). Atrial catecholamine-containing cells have been demonstrated in all animal species investigated so far. Although there are obviously large species variations in the number of atrial-catecholamine-containing cells, the described results relating to the comparison of the two kinds of granule-containing cells in the auricle of *Protopterus* probably also apply to other species.

On the basis of similarities between, the innervated clusters of dopamine-containing cells in the auricle of the heart of various animals examined so far and the carotid and aortic bodies, the possibility of a chemoreceptor function must be considered. From this it might be presumed that catecholamines are secreted in response to substances present in the wall of the atrium. Indeed, it has been reported that catecholamines may increase the number of atrial-specific granules (1,21,23,37) and that release of ANF is responsive to adrenergic-agonist stimulation (5,26,33,35).

Since the mechanism responsible for the release and action of ANF is not clearly defined, future work should be aimed at investigating the possible regulatory interaction between the two kinds of endocrine cells described in this work and colocalized side-by-side in the atrial wall of the vertebrate heart.

References

1. Artemian, N.A., Potapova, V.B., Shperling, I.D. (1980) Functional significance of specific atrial granules. J Mol Cell Cardiol 12: 10
2. Berger, J.M., Bencosme, S.A. (1971) Fine structural cytochemistry of granules in atrial cardiocytes. J Mol Cell Cardiol 3: 111-120
3. Björklund, A., Ehinger, B., Falck, B. (1968) Method for differentiating dopamine from noradrenaline in tissue sections by microspectrofluorometry. J Histochem Cytochem 16: 263-270
4. Bloom, G., Östlund, E., von Euler, U.S., Lishajko, F., Ritzén, M., Adams-Ray, J. (1961) Studies on catecholamine-containing granules of specific cells in cyclostome hearts. Acta Physiol Scand (suppl) 185: 1-34
5. Brown, A., Imada, T., Takayanagi, B., Inagami, T. (1986) Beta-adrenergic modulation of the release of atrial natriuretic factor from rat cardiac atria in vitro. Fed Proc 45: 526
6. Chang, W.L.W., Bencosme, S.A. (1968) Selective staining of secretory granules of adrenal medullary cells by silver methenamine: a light and electron microscope study. Can J Physiol Pharmacol 46: 745-747
7. DeBold, A.J., Bencosme, S.A. (1973) Studies on the relationship between the catecholamine distribution in the atrium and the specific granules present in atrial muscle cells. studies on the sedimentation pattern of atrial noradrenaline and adrenaline. Cardiovasc Res 7: 364-369
8. DeBold, A.J., Raymond, J.J., Bencosme, S.A. (1978) Atrial specific granules of the rat heart light microscopic staining and histochemical reactions. J Histochem Cytochem 26: 1094-1102
9. Ehinger, B., Falck, B., Persson, H., Sporrong, B. (1968) Adrenergic and cholinesterase-containing neurons of the heart. Histochemie 16: 197-205
10. Ellison, J.P., Hibbs, R.G. (1974) Catecholamine-containing cells of the guinea pig heart: an ultrastructural study. J Mol Cell Cardiol 6: 17-26
11. Forssmann, W.G. (1986) Cardiac hormones. 1. Review on morphology, biochemistry and molecular biology of the endocrine heart. Eur J Clin Invest 16: 439-451
12. Fuchs, E., Shigematsu, K., Saavedra, J.M. (1986) Binding sites of atrial natriuretic peptide in tree shrew adrenal gland. Peptides 7: 873-876
13. Giacomini, E. (1906) Sulle capsule surrenali e sul simpatico dei Dipnoi. Ricerche in Protopterus annectens. Atti R Accad Lincei 15: 394-398
14. Holmes, W. (1950) The adrenal homologues in the lungfish Protopterus. Proc Roy Soc London 137: 549-562
15. Jacobowitz, D. (1967) Histochemical studies of the relationship of chromaffin cells and adrenergic nerve fibres to the cardiac ganglia of several species. J Pharmacol Exp Ther 158: 227-240
16. Jamieson, J.D., Palade, G.E. (1964) Specific granules in atrial muscle cells. J Cell Biol 23: 151-172
17. Malor, R., Taylor, S., Chesher, G.B., Griffin, C.J. (1974) The intramural ganglia and chromaffin cells in guinea pig atria: an ultrastructural study. Cardiovasc Res. 8: 731-744
18. Martinez-Palomo, A., Bencosme, S.A. (1966) Electron microscopic observations on myocardial specific granules and residual bodies in vertebrates. Anat Rec 154: 473

19. McKenzie, J.C., Tanaka, I., Misono, K.S., Inagami, T. (1985)   Immunocytochemical localization of atrial natriuretic factor in the kidney, adrenal medulla, pituitary and atrium of rat. J Histochem Cytochem 33: 828-832

20. Nielsen, K.C., Owman, C. (1968) Difference in cardiac adrenergic innervation between hibernators and non-hibernating mammals. Acta Physiol Scand 316 (suppl): 1-30

21. Okamoto, H. (1969)   An electron microscopic study of the specific granules in the atrial muscle cell upon the administration of agents affecting autonomic nerves. Arch Histol Jap 30: 467-478

22. Östlund, E., Bloom, G., Adams-Ray, J., Ritzén, M., Siegman, M., Nordenstam, H., Lishajko, F., von Euler, U.S. (1969)   Storage and release of catecholamines and the occurrence of a specific submicroscopic granulation in hearts of cyclostomes. Nature 188: 324-325

23. Otsuka, N., Okamoto, H., Tomisawa, M. (1969) Electron and fluorescence microscopic study of specific granules in rat atrial muscle cells. Arch Histol Jap 30: 367-374

24. Palade, G.E. (1961) Secretory granules in the atrial myocardium. Anat Rec 139:262

25. Papka, R.E. (1974)   A study of catecholamine-containing cells in the hearts of fetal and postnatal rabbits by fluorescence and electron microscopy . Cell Tissue Res 154: 471-484

26. Rankin, A.J., Wilson, N., Ledsome, J.R. (1987) Influence of isoproterenol on plasma immunoreactive atrial natriuretic peptide and plasma vasopressin in the anesthetized rabbit. Pflüg Arch 408: 124-128

27. Reinhold, C., Hartwig, H.-G. (1982)   Progress in microfluorometric identification of monoamine fluorophores. Brain Res Bull 9: 97-105

28. Scheuermann, D.W. (1974) Cytochemistry of granules in atrial cardiocytes. Acta Morphol Neerl-Scand 13: 107-109

29. Scheuermann, D.W. (1978) Über den Feinbau des Myokards von Protopterus annectens. Verh Anat Ges 72: 297-307

30. Scheuermann, D.W. (1979) A study of chromaffin cells in the heart of Protopterus. Proc 5th Eur Anat Congr: 362

31. Scheuermann, D.W. (1980) A study of chromaffin cells in the heart of Protopterus. Folia Morphol 28: 93-98

32. Scheuermann, D.W., Stilman, C. (1982) Dopamine-containing cells in the heart of Lepidosiren paradoxa. Biol Cell 45: 143

33. Seiden, D. (1979) Specific granules of the rat atrial muscle cell. Anat Rec 194: 587-602

34. Shore, P.A., Cohn, V.H. jr., Highman, B., Maling, H.M. (1958)   Distribution of norepinephrine in the heart. Nature 181: 848-849

35. Sonnenberg, H., Veress, A.T. (1984) Cellular mechanism of release of atrial natriuretic factor. Biochem Biophys Res Commun 124: 443-449

36. Stilman, C., Scheuermann, D.W. (1987) Computer-operated microspectrofluorimetry to identify formaldehyde-induced fluorophores of biogenic monoamines and precursor substances in models and tissue sections. Histochemistry 86: 255-267

37. Tomisawa, M. (1969) Atrial specific granules in various mammals. Arch Histol Jap 30: 449-465

Seasonal changes of the endocrine heart

A. Aoki, C. A. Maldonado, W.G. Forssmann

Center of Electron Microscopy, Cordoba National University, Cordoba, Argentina and Department of Anatomy and Cell Biology, University of Heidelberg, Heidelberg, FRG

Summary

The endocrine heart of the toad *Bufo arenarum Hensen* exhibits a seasonal secretory activity and constitutes an excellent experimental model for the study of the synthesis and secretion of cardiac hormones and the role played by the cell population in the regulation of hormonal secretion.

In winter, concurrent with the lowest hormonal production, there are fewer myocardiocytes differentiated as myoendocrine cells, which contrasts with the summer heart where the majority of atrial cells are engaged in hormonal secretion to meet an increased requirement of cardiac hormones. Morphometric analysis also reveals a significant diminution of both cell number and surface density of myocardiocytes in winter. A number of involuting myocardiocytes are always found in winter but this appears as a slow and progressive cell renewal process. The small number of secretory granules per cell during the summer is interpreted as an accelerated synthesis and intracellular processing of hormonal peptides, bypassing the packaging step in the Golgi complex.

The discovery of cardiac hormones has produced a remarkable impact on life sciences as judged by the volume of investigations performed in recent years which has few precedents in basic and applied research (see 1, 2, for reviews).

These cardiac hormones or family of peptides with hormonal activity are found in a wide range of species with little or no molecular changes. They are synthesized and secreted predominantly in the atrium by specific myocardiocytes identified as myoendocrine cells and are easily recognized in the electron microscope by the presence of typical secretory granules. These are characterized by a round profile, a rather uniform size and a dense content which is stained specifically for cardiodilatin with immunocytochemical techniques (3,5,6).

A systematic ultrastructural analysis of the atrium reveals a functional heterogeneity of myoendocrine cells. Some cells exhibit a structural organization compatible with the secretory cycle of highly stimulated cells which coexist with cells in quiescent states. These characteristics are similar to those described in other endocrine cell systems like the lactotrophs of the pituitary gland (7,8). As in the lactotrophs, this morphological heterogeneity is greatly reduced when the secretory activitiy of endocrine cells is submitted to prolonged stimulation or inhibition. However, the heart cannot be subjected experimentally to chronic stimulation or inhibition of its endocrine secretion without introducing profound hemodynamic changes in the whole organism. Since the analysis of the secretory cycle of cardiac hormone secretion seems to be critical for the interpretation of the synthesis and cell processing of atrial peptides, we tested these situations in the heart of animals whose reproductive metabolic activities are seasonal.

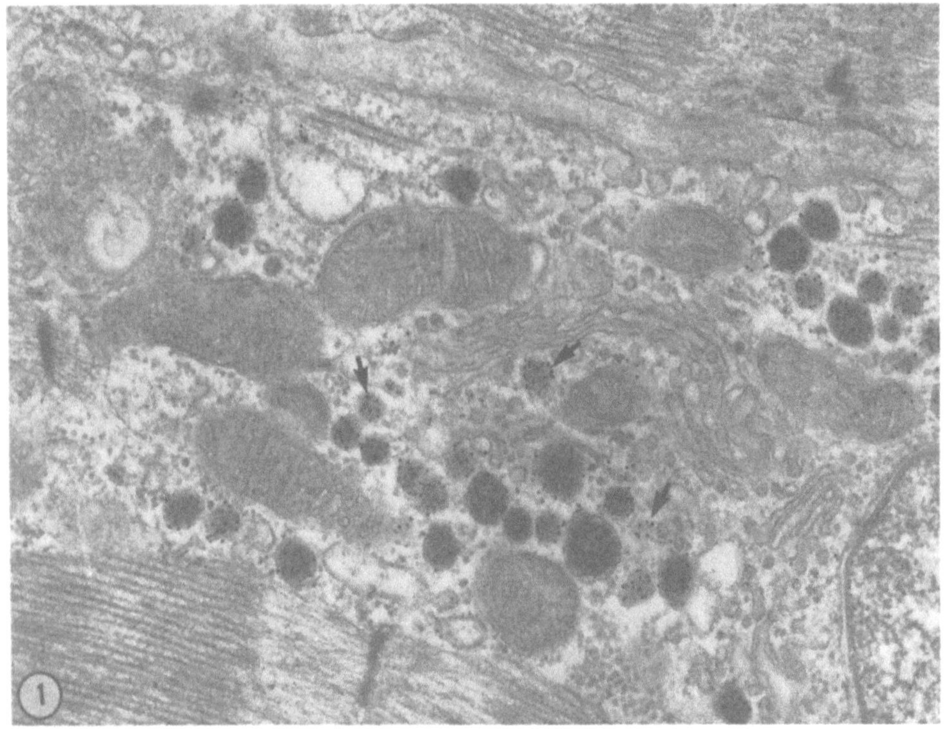

Fig.1. Electron microscope immunocytochemistry of a myoendocrine cell of a toad *Bufo arenarum Hensen*, after thin sections of atrial tissue embedded in L.R. White were incubated with antiserum against cardiodilatin-126 (dilution 1:6000), followed by protein A-colloidal gold complex. A strong labelling of secretory granules and progranules (arrows) is seen in contrast with the lack of immunoreaction at the Golgi cisternae. (x46 000)

For this purpose we studied several autochthonous species of Argentina and chose an amphibian, the *Bufo arenarum Hensen*, which has been extensively studied by South American biologists because of its many peculiarities and their documentation by the late Professor Houssay.

The *Bufo arenarum* is a common toad which hybernates from March or April through August or September, depending on climatic conditions. In this period it remains buried in deep holes without access to food or water. Its heart beat is reduced to a minimum, which contrasts with the high activity developed for foraging and reproduction in the spring and summer.

The frog's myoendocrine cells exhibit similar characteristics of cells described in other species and their secretory granules show a strong labelling when stained immunocyto-chemically. Figure 1 illustrates the tagging of secretory granules with colloidal gold-Protein A in acrylic-embedded atrial tissue after incubation with antiserum against synthetic fragments of the C-terminal of cardiodilatin-126 (5). In addition to mature secretory granules, some immature granules and vesicles associated with the "trans" face

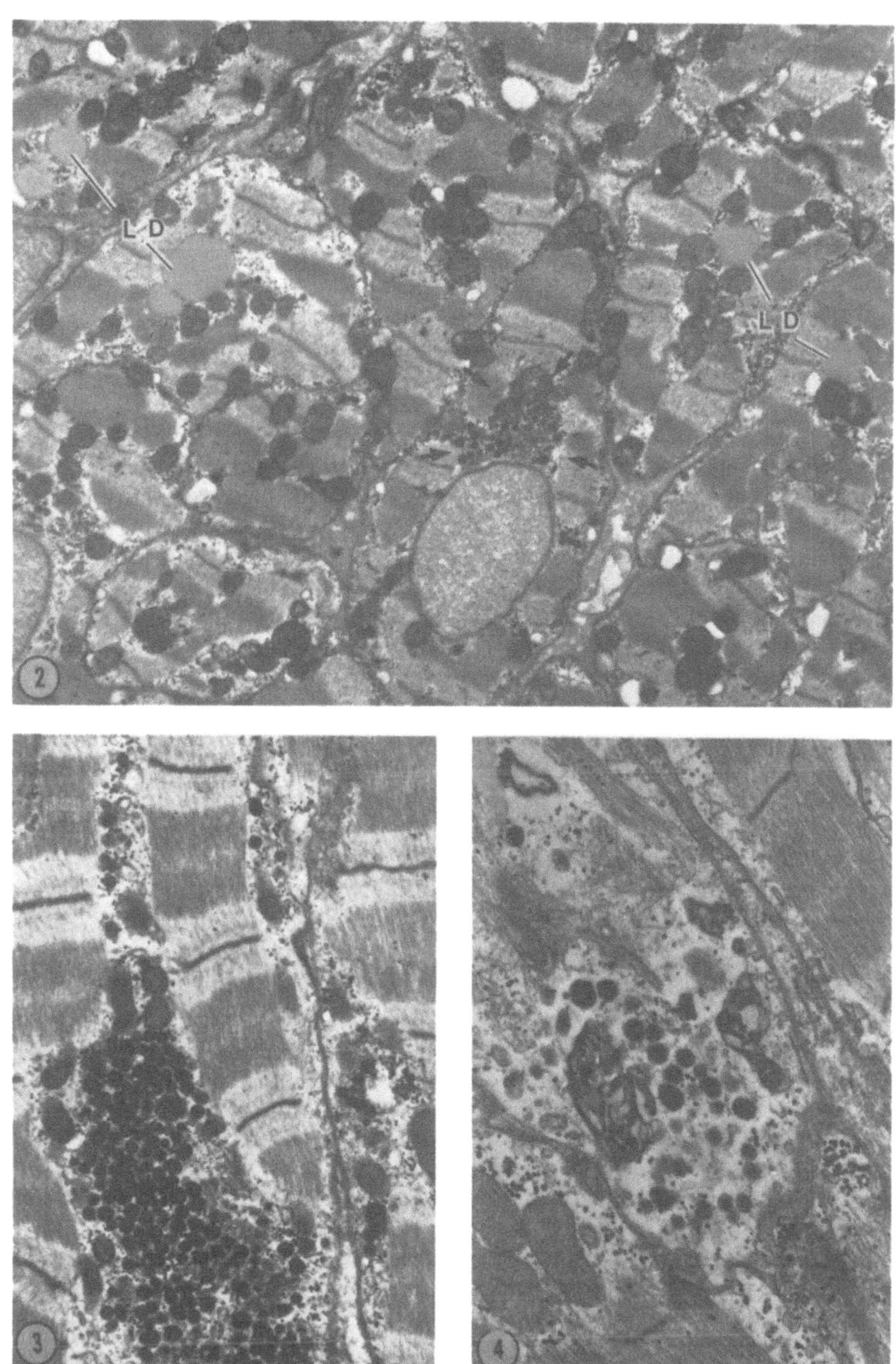

L D
L D
2
3
4

Fig. 2. Low magnification survey electron micrograph of the atrium of winter toad. Only few myocardiocytes containing typical secretory granules of cardiac endocrine cells are detected. The secretory granules occur in clusters usually in the perinuclear areas of the cytoplasm (arrows). (x10 000)

Fig. 3. In winter the atrium exhibits fewer myoendocrine cells.These are easily recognized by thepresence of conspicuous aggregates of secretory granules as it is seen in this electron micrograph. Secretory granules are often intermingled with polymorphic mitochondria and large stores of glycogen particles. (x15 000)

Fig. 4. Many myoendocrine cells in winter hearts undergo an involutive process characterized by the presence of a disorganized cytoplan containing anssortment of degenerated organelles. Quite often these changes are accompanied by the formation of lipid inclusions. (x46 000)

of the Golgi complex are also labelled with gold particles. However, neither the cisternae of this organelle nor the rough endoplasmic reticulum were stained with this technique. The ultrastructure of the atrium in winter, at the lowest period of activity, reveals only few myoendocrine cells. These cells are only recognized in sections through certain areas of the cytoplasm where typical secretory granules are found. At low magnification electron micrographs, the presence of cardiodilatin containing granules allows the identification of myoendocrine cells and their intracellular distribution (Fig. 2). In the majority of myoendocrine cells, the secretory granules are usually found clustered in a paranuclear region and associated with smooth endoplasmic reticulum intermingled with dense polymorphic mitochondria and large amounts of glycogen particles (Fig. 3). The finding of lipid droplets is not infrequent. No involuting myocardiocytes are found but quite frequently the presence of myoendocrine cells with a disorganized cytoplasm containing degenerated organelles is observed, which seems to appear concurrent with the cessation of the secretory activity (Fig. 4).

In summer the majority of atrial myocardiocytes exhibit secretory granules, indicating that the the cells are capable of producing cardiodilatin (Fig. 5). Quite frequently, myocardiocytes contain just a few isolated granules, an observation that can be interpreted as either a rapid turnover of secretory granules or a hormone secretion bypassing the packaging step of cardiodilatin into secretory granules.

In other cells, the secretory granules are sparse with no particular distribution, but they are more often located in the proximity of the sarcoplasmic reticulum or associated with mitochondria which are more numerous, but smaller than those found in winter (Fig. 6). In specimens of both summer and winter the secretory granules are rarely seen in the process of exocytosis. Sometimes several granules may fuse before the release of their hormonal content into the extracellular compartment as was seen in hyperstimulated lactotrophs (Fig. 7).

Stereological analysis reveals that in the summer heart, the atrium exhibits almost a 30% increase in the cell count, indicating a moderate hyperplasia of atrial myocardiocytes (Fig. 8).

These data closely correlate with the surface density of secretory cells when the results of the winter and summer hearts are compared. In summer all the myocardiocytes contain

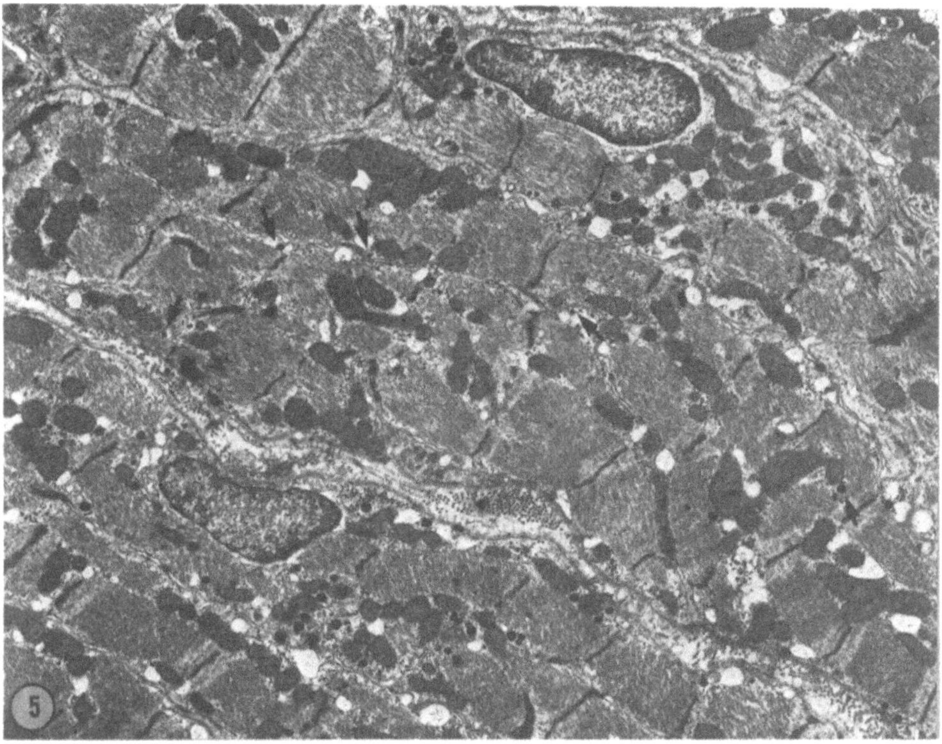

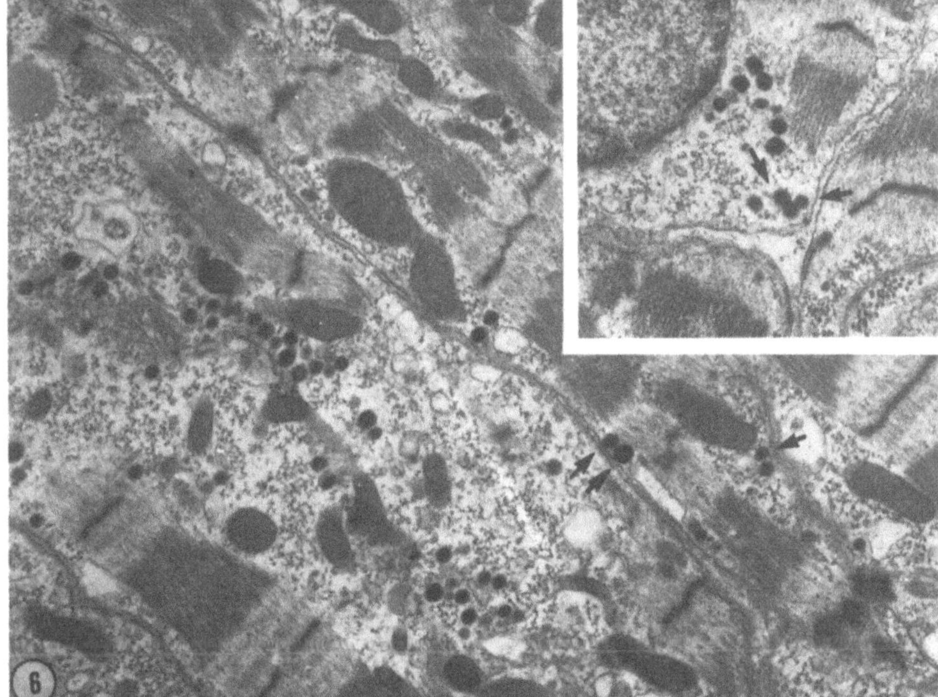

Fig 5. Survey electron micrograph of a section of atrial tissue processed in summer. The majority of myocardiocytes present a variable number of secretory granules. Some cells contain just a few isolated secretory granules usually located at the cell periphery. (x10 000)

Fig.6. Electron micrograph of two adjacent myocardiocytes. Many secretory granules are found either in proximity to, or in contact with the plasmalema. Exocytotic profiles are rarely seen in electron micrographs but the observation of secretory granules attached to cell membranes (arrows) is not infrequent. In highly stimulated cells secretory granules may fuse before their content is discharged into the intracellular compartment (inset). (x15 000; inset x16 600)

secretory granules when sections through the cell nucleus are analysed. The absolute number of these secretory granules is almost 2.5 fold that are found in the winter period. However, the number of secretory granules per myoendocrine cell appears to be smaller in summer but this information has not yet been assessed stereologically.

Ultrastructural and morphological observations led us to conclude that the secretory activity of the atrial myoendocrine cells is intimately associated with the metabolic activity of the organism, increasing both their number and secretory function concurrent with the enhancement in hormonal requirements.

During hybernation the number of cells engaged in cardiodilatin production diminishes significantly and their secretory activity is restricted to a minimum. In spite of this remarkable decrease in myoendocine cells no degenerated cells were observed in this period, as is the rule for endocrine cells after interruption of the stimulatory factors (4). Perhaps the involution of organelles involved in the synthesis and processing of cardiac peptides are equivalent in the heart to those in endocrine organs where the cells can be replaced without introducing changes; this might impair the structural and functional integrity of an organ like the heart. And the significant diminution in the number of myocardiocytes found in winter months may be the result of a progressive process of cell renewal occuring over a long period of time in spring and autumn.

The stereological analysis of the atrium in summer, at the peak of metabolic and reproductive activities, reveals that the majority of the myocardiocytes are engaged in the synthesis and secretion of cardiac hormones. This observation suggests that all atrial cells and not a specialized cell sustain the genetic information to produce hormonal peptides, and that this capacity might differentiate according to the levels of hormonal requirements of the organism.

The study of species with defined seasonal activity is well suited for studying different aspects of the metabolic activity of cardiac hormones and the role of cell populations in regulating hormonal secretion, and it provides new data for the interpretation of the synthesis and the intracellular processing of hormones which otherwise cannot be assessed.

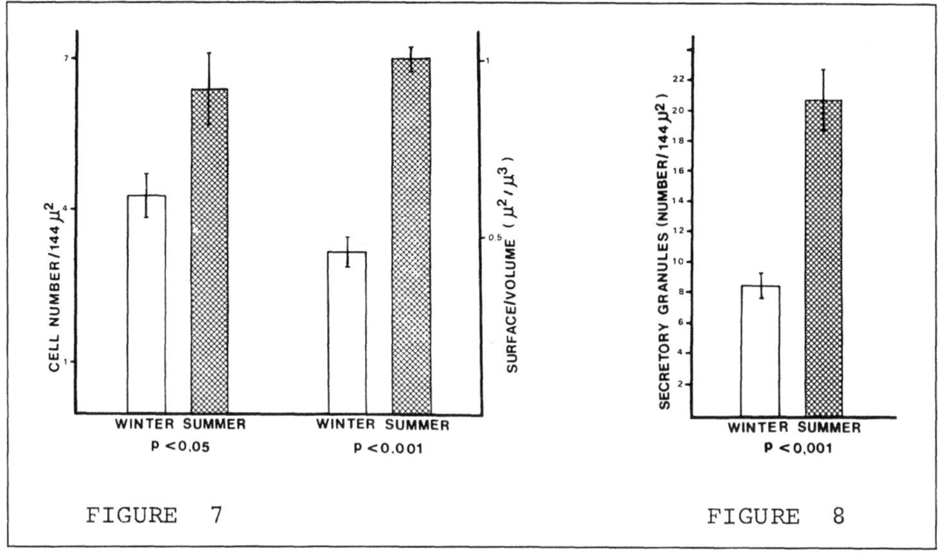

Fig. 7. Graphic representation of the morphometric data obtained from atrial of six summer and winter toads. A total number of 240 electron micrographs were taken for each point and processed statistically applying the Student's *t*-Test. To calculate the cell number, all nuclear sectioned cells were counted. The surface density was determined applying the method of Weibel. Significant differences were found with both procedures.

Fig. 8. Total numbers of secretory granules were counted in atria of both winter and summer toads. A significant difference was established between these two seasons. In winter the number of myocardiocytes with secretory granules decreased considerably.

Acknowledgements: This investigation was performed under the Scientific Cooperation Programme Deutscher Akademischer Austauschdienst (DAAD), FRG, and the Consejo Nacional de Investigaciones Cientificas y Tecnicas (CONICET), Argentina. (A.A. and C.A. M. are career members of CONICET.)

References

1. Cantin, M., Genest, J. (1985) The heart and the atrial natriuretic factor. Endocr Rev 6: 107-127

2. Forssmann, W.G. (1986) Cardiac hormones: I Review on the morphology, biochemistry and molecular biology of the endocrine heart. Eur J Clin Invest 16: 439-451

3. Forssmann, W.G., Hock, D., Lottspeich, F., Henschen, A., Kreye, V., Christmann, F., Reinecke, M., Metz, J., Mutt, V. (1983) The right auricle of the heart is an endocrine organ. Cardiodilatin as a peptide hormone candidate. Anat Ambryol 168: 307-313

4. Haggi, E.S., Torres, A.I., Maldonado, C.A., Aoki, A. (1986) Regression of redundant lactotrophs in rat pituitary gland after cessation of lactation. J Endocrinol 111: 367-373

5. Maldonado, C.A., Saggau, W., Forssmann, W.G. (1986) Cardiodilatin-immuno-reactivity in specific atrial granules of human heart revealed by the immunogold stain. Anat Embryol 173: 295-298

6. Metz, J. , Mutt , V., Forssmann, W.G. (1984) Immunohistochemical localization of cardiodilatin in myoendocrine cells of the cardiac atria. Anat Embryol 170: 123-127

7. Torres, A.I., Aoki, A. (1985) Subcellular compartmentation of prolactin in ratlactotrophs. J Endocrinol 105: 219-225

8. Torres, A.I., Aoki, A. (1987) Release of big and small molecular forms of prolactin: dependence upon dynamic state of the lactotroph. J Endocrinol 111: 367-373

Storage and secretion of atrial natriuretic peptide (ANP) by cultured atrial and ventricular cells from the neonatal rat

T. Brand[1], H. Jockusch[1], G. Rippegather[2], W. G. Forssmann[2]

[1]Developmental Biology Unit, University of Bielefeld, Bielefeld, FRG, [2]Department of Anatomy and Cell Biology, University of Heidelberg, Heidelberg, FRG

Introduction

In the heart of the neonatal rat, myoendocrine ANP producing cells are located in the right and left atria and in the conduction tissue mainly of the left ventricle (2,10,11). When minced atrial and ventricular tissues from neonatal rats were transplanted to ectopic sites and analyzed after several weeks, strong ANP immunoreactivity was found only in atrial grafts (6). In cell culture atrial cells were ANP-immunoreactive but ventricular cells were not (3).

In this study we asked whether atrial and ventricular cells in monolayer cultures retain their region-specific differences in storage and production of ANP, and whether it is possible to find ventricular cells in cultures which stain for ANP. For this purpose cells from defined regions of the neonatal rat heart were cultured for various periods.

Material and methods

Cell cultures were set up as outlined in Fig.1. Twenty-five neonatal rat (Wistar) hearts were collected in HEPES buffered saline-glucose (HBS-glucose; 154 mM NaCl, 5.6 mM KCl, 2.2 mM $CaCl_2$,0.12 mM $MgCl_2$ 2.5 mM glucose, 5.0 mM HEPES, pH 7.3) (1). Tissues were minced and dissociated into single cells by rotating 2 x 30 min at 100 rpm in 2mg/ml collagcnase, 2 mg/ml hyaluronidase, 1 mg/ml protease (B. subtilis), and 0.5 mg/ml DNase in HBS-glucose. Cells were incubated for three hours in MEM with 20% horse serum (HS) on TC dishes (7). Non-attached cells were seeded at a density of 5 x 10^5 cells per collagen coated 35mm ø dish in 1ml medium (MEM, 5% fetal calf serum FCS, 5% horse serum HS, 4 mM glutamine, penicillin-streptomycin). The plating efficiency was about 60% and independent of the regional origin of the cells. After three days of culture, immunocytochemical staining for myosin heavy chains (4) showed that 50% of the cells were myocytes. The cells were cultured for up to 136 h.

Immunochemical staining for ANP was done using a PAP-method (5) on acetone fixed cultured cells. The polyclonal rabbit antibody anti-CDD 016 had been raised against a synthetic 28 residue peptide (positions 99-126 of human CDD) coupled to thyreoglobulin. The ANP-antiserum is reactive to C-terminus of CDD (residues 118-126).

Proportions of ANP-positive cells were determined as follows: 40 randomly focused fields of 2.14 mm^2 were counted for immunoreactive cells. Counts of cultures from different cardiac regions were compared to the mean total count of immunoreactive cells from the right atrium (average 160 cells) which was taken as 100%.

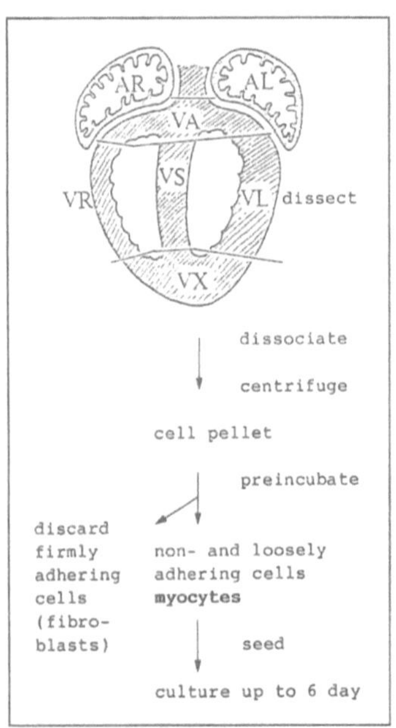

Fig. 1. Flow chart for setting up heart muscle cultures: AR, right atrium; AL, left atrium; VA, atrio-ventricular region; VR, right ventricle; VS, ventricular septum; VL, left ventricle; and VX, apex

For measurements of stored and secreted ANP, cells were fed with fresh medium. After one hour, the medium was harvested, made 0.5 mM in phenylmethylsulfonylfluoride, and frozen at -70°C. Cells were lysed according to Bloch et al. (3) and the lysate was frozen. Radioimmunoassay (RIAs) of cellular and secreted ANP were done essentially as outlined by Lang et al. (8). The amount of ANP was expressed as ng ANP/mg protein. The rate of secretion was expressed as ng ANP/mg protein/h. Protein was determined by the Lowry method.

Results

After three days in culture, atrial specific granules were found by electron microscopy in atrial but not in ventricular cells (Fig. 2). After 136 h, atrial myocytes and a subpopulation of cells from the right ventricular showed strong ANP immunoreactivity (Fig. 3). Most ventricle cells were unreactive or only weakly reactive. Proportions of immunoreactive cells were lower in cultures from right (33%) and left (13%) ventricle than in those from right atria (100% by definition). Immunopositive cells in cultures from right atria showed a granular staining, whereas a large proportion of the cells from the ventricular walls showed a diffuse staining pattern (Fig. 4). These differences between cells from atrial and ventricular walls were retained after six days in culture.

70

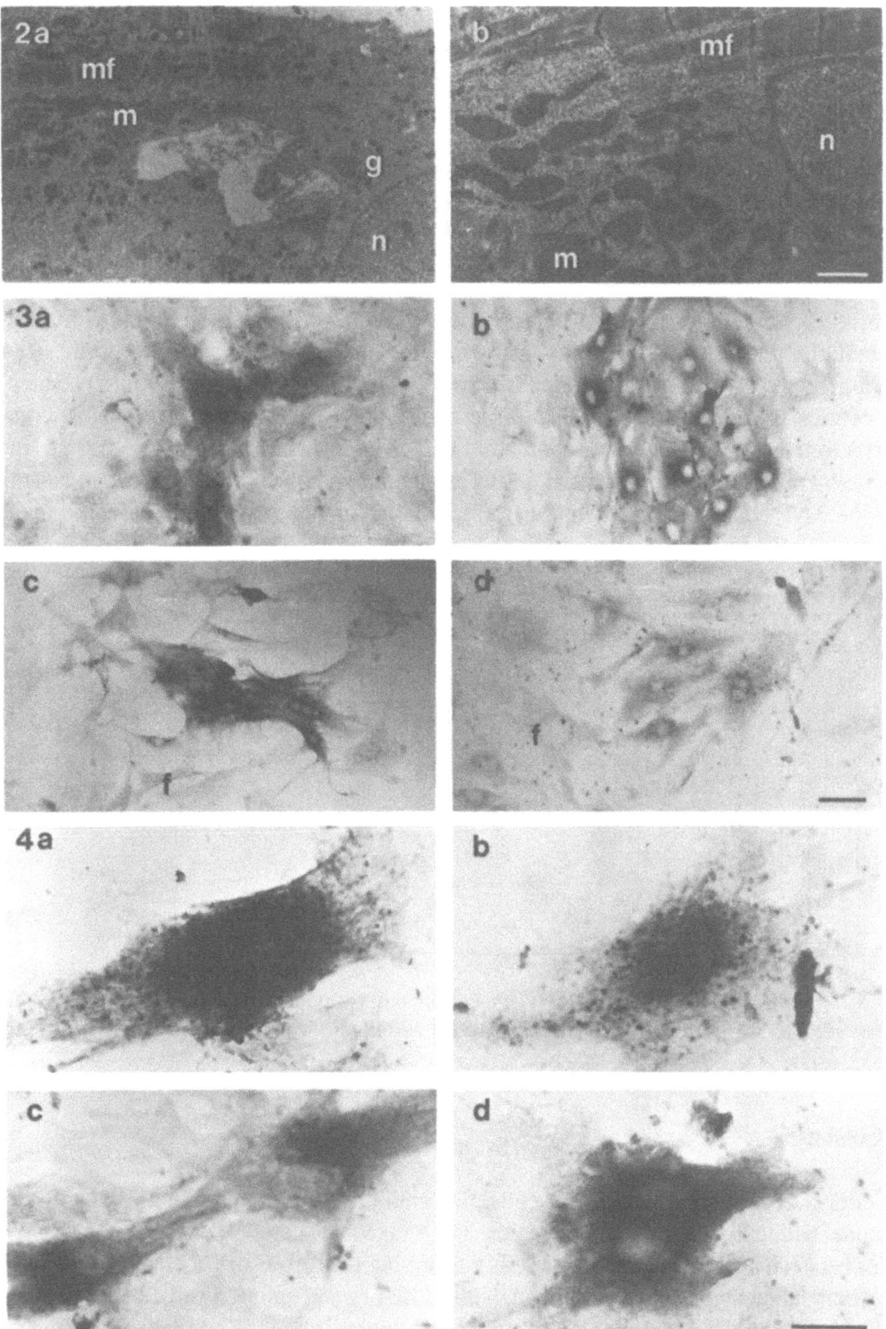

Fig. 2. Electron microscopy of cultured atrial and ventricular muscle cells (three days in culture). a) atrial cell; atrial granules (g), and myofibrils (mf); b) ventricular cell without granules (m) mitochondria; (n) nucleus; (7000x); scale bar: 0.5 μm

Fig. 3. Immunochemical staining for ANP of atrial and ventricular cells in culture (136 h). a,c) stained myocytes from right atria (AR) showing intense ANP immunoreactivity; fibroblasts (f) are unstained; b) stained cells of right ventricle (RV); d) weakly stained cells of left ventricle (VL). (700x); scale bar: 20 μm

Fig. 4. ANP staining patterns in atrial and ventricular cultures. a,b) cells from right atria with granular staining around the nucleus, c) two atrial muscle cells (AR) in contact; d) ventricular cell (VR) with diffuse immunoreactivity. (170x); scale bar: 50 μm

Thirty-six hours after plating atrial cells and cells from the right ventricle contained substantial amounts of ANP (20 - 40 ng ANP/mg protein) but, according to the RIA determination, ventricular cells stored little or no ANP. Ventricular cells with the exception of those from the right ventricle, secreted more ANP into the medium (in comparison to their cellular content) than atrial cells. However, with prolonged culture, the ANP content and secretion of atrial cells dropped drastically. Because during the same time the ANP content and secretion of ventricular cells increased moderately, atrial and ventricular cells became similar in this respect (Fig. 5).

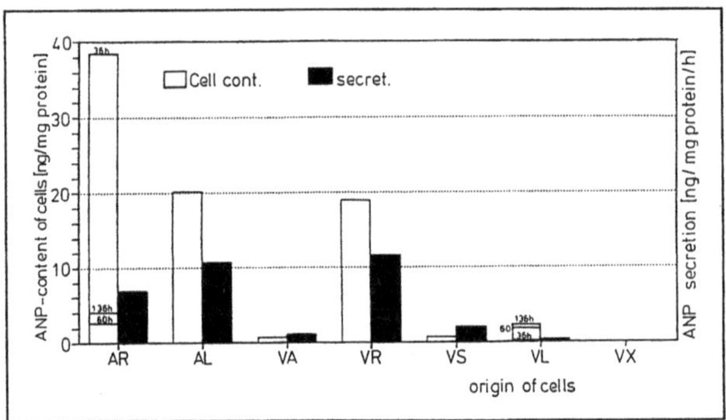

Fig. 5. Storage and secretion of ANP by cells from various cardiac regions (for nomenclature see Fig. 1). ANP contents of cell extracts and media were determined by RIA.

Discussion

In cultures of atrial and ventricular cells, myoendocrine ANP-containing cells could be immunochemically identified. The pattern of storage and secretion of ANP was found to differ between atrial and ventricular cells confirming the results of Bloch et al. (3). There is a discrepancy between the staining pattern of tissues and cultured cells. In the neonatal rat heart, only in the left ventricle could some immunoreactive cells be localized, whereas in our culture system cells from the right ventricle were immunoreactive. This unexpected result needs confirmation. The disappearance upon long-term culture of the difference in ANP content and secretion between atrial and ventricular cells could be a sign of dedifferentiation, possibly induced by fibroblast-overgrowth (9). The discrepancy between staining and RIA determinations suggests that the ANP immunoreactivity is more dependent on the form of storage than on the actual content of ANP. It is this property that distinguishes atrial from ventricular wall myocytes even after six days in culture.

This study corroborates our previous study on myosin light-and heavy chain expression which showed that atrial and ventricular cells retain their differential properties in short-term culture (4).

Acknowledgements: We thank U. Herbort for technical assistance at the EM facility of SFB 223. This work was supported by project "Herzmuskelzelle", University of Bielefeld.

References

1. Arjamaa, O., Vuolteenaho, O. (1985) Sodium ion stimulates the release of atrial natriuretic polypeptides (ANP) from rat atria. Biochem Biophys Res Commun 132: 375-381
2. Back, H., Stumpf, W.E., Ando, E., Nokihara, K., Forssmann, W.G. (1986) Immunocytochemical evidence for CDD/ANP-like peptides in strands of myoendocrine cells associated with the ventricular conduction system of the rat heart. Anat Embryol 175: 223-226
3. Bloch, K.D., Seidmann, J.G., Naftilan, J.D., Fallon, J.T., Seidman, C.E. (1986) Neonatal atria and ventricular secrete atrial natriuretic factor via tissue-specific secretory pathways. Cell 47: 695-702
4. Brand, Th., Jockusch, H. (1987) Differential properties of rat atrial and ventricular muscle cells in culture. Eur J Cell Biol 43: 8
5. Forssmann, W.G., Pickel, V., Reinecke, M., Hock, D., Metz, J. (1981) Immunohistochemistry and immunocytochemistry of nervous tissue. In: Heym, Ch., Forssmann, W.G. (eds) Techniques in neuroanatomical research. Springer, Berlin Heidelberg New York, pp 171-205
6. Jockusch, H., Füchtbauer, E.M., Füchtbauer, A., Léger, J.J., Léger, J., Maldonado, C.A., Forssmann, W.G. (1986) Long term expression of isomyosins and myoendocrine functions in ectopic grafts of atrial tissue. Proc Nat Acad Sci USA 83: 6697-6701
7. Jourdon, P. (1980) Ultrastructure and electrically activity of newborn rat heart reaggregates. Biol Cell 37: 149-154
8. Lang, R.E., Thölken, H., Ganten, D., Luft, F.C., Ruoskaoh, H., Unger, Th. (1985) Atrial natriuretic factor: a circulating hormone stimulated by volume loading. Nature 314: 264-266
9. Lompré, A.M., Poggioli, J., Vassort, G. (1979) Maintenance of fast Na-channels during primary culture of chick heart cells. J Mol Cell Cardiol 11: 813-825
10. Thompson, R.P., Simpson, J.A.V., Currie, M.G. (1986) Atriopeptin distribution in the developing rat heart. Anat Embryol 175: 227-233
11. Toshimori, H., Toshimori, K., Oura, C., Matsuo, H. (1987) Immunohistochemical study of atrial natriuretic polypeptides in the embryonic, foetal and neonatal rat heart. Cell Tissue Res 248-633

Combined quantitative morphological and biochemical study on atrial granules (AG)

W.M. Herbst, G. Mall, J. Weers, T. Mattfeldt, R.E. Lang, W.G. Forssmann

Clinic of Dermatology, University of Erlangen, and Departments of Pathology, Pharmacology, Anatomy and Cell Biology, University of Heidelberg, Heidelberg, FRG

Introduction

The first morphological study concerning the endocrine heart was published by Kisch (12) who had detected a peculiar cytoplasmic inclusion body in muscle cells of the guinea-pig atrium. Only a few years later, Jamieson and Palade (11) described that these structures were specific granules occuring in all mammalian species investigated, including man. Twelve years later, Marie et al. (13) analysed morphometrically the interference between the number of atrial granules and changes in salt and water balance. Recently, DeBold et al. (4) proved the natriuretic and diuretic properties of atrial extracts and postulated a new class of hormones, which are now known as the atrial natriuretic polypeptides (ANP). Further biological activities were detected, namely the vasorelaxant effect of atrial extracts (3,6). This vasorelaxant function is based on the polypeptide hormone, atriopeptin or cardiodilatin (CDD). It is now well known that the different polypeptides characterized in 1983 and 1984 are all expressed from the same gene and present as posttranslational or artificial cleavage products (2,7). In contrast to the large number of biochemical, physiological investigations, and morphological studies on cardiac hormones published so far, combined morphometric and biochemical studies are not available. Exact data based on both stereological-ultrastructural and radio-immunological examination of the endocrine heart, however, is of major importance for the understanding of the secretory activity of myoendocrine cells.

The aim of our study was to investigate the mass and number of atrial granules in order to obtain data on the biological variability of these structures. Furthermore, the obtained data were compared with radioimmunological parameters to determine the absolute amount of CDD/ANP in each average-sized atrial granule.

Materials and methods

Tissue preparation: 10 male Wistar rats (body weight: 250g) were subjected to retro-grade vascular perfusion via the abdominal aorta. The vascular system was flushed for 2 min with a dextran solution (Rheomacrodex^R) containing 0.5 g/1 procaine-HCl. Fixation was performed by exposure to 3% glutaraldehyde in 0.2 mol phosphate buffer for 12 min. Right ventricular auricles were excised carefully and the exact wet weight was determined with a precision balance. Subsequently, they were embedded in agar-agar and disected into slices of 0.5 mm. From these slices two random samples were obtained by systematic random sampling for electron microscopic morphometry. They were postfixed with 1% OsO_4, dehydrated with ethanol and embedded in Epon-Araldite. The remaining tissue was embedded in Paraplast for light microscopic morphometry.

Quantitative stereology: volume density (V_v) of atrial granules was estimated by point counting according to the standard equation

$$V_v = P_p \text{ (point density)} \quad (1)$$

Numerical density (N_v) of atrial granules was estimated according to DeHoff and Rhines (1961):

$$N_v = N_A / \bar{D} \quad (2)$$

N_A: numerical density of profiles on area

$\bar{D}$: mean caliper diameter

$\bar{D}$ cannot be obtained from planar sections without suppositions concerning the size and shape of particles. In the present investigation, atrial granules were considered spherical particles. The mean caliper diameter of spherical particles corresponds to the mean diameter $\bar{D}$ $(=2\bar{R})$ of spheres.

$\bar{R}$ was derived from the size distribution of the circular section profiles of atrial granules according to Bach (1967, cf. Weibel, 1979):

$$\bar{R} = \frac{1}{2} \frac{\sum\limits_{i=1}^{m} p(i)}{\frac{1}{t} \left(\frac{2}{\pi}\right)^{1/2} \sum\limits_{i=1}^{m} p(i)\, g(r_i/t)} - t/2 \quad (3)$$

r_i section profile radius in class i
$p(i)$ absolute frequency of measured profiles in class i

t: section thickness $(=70\ nm)$
$g(r_i/t)$: factors calculated by Bach (1967)

Alternatively, N_v of atrial granules was estimated according to the equation of Weibel and Gomez (1962)

$$N_v = K * \frac{1}{\beta} (N_A)^{3/2} * V_v^{-1/2} \quad (4)$$

β: shape coefficient $(=1.38$ for spherical particles$)$
K: coefficient of correction for size distribution $(=1.05)$

N_A and V_V were corrected for section thickness and truncation (Haug, 1967; Weibel and Elias, 1967).

The mean volume of atrial granules in the number of distribution (v_N) was estimated according to Bach (1967; see 15), mean volume weighted volume in the volume-weighted distribution was obtained from point sampled intercept lengths according to Gundersen and Jensen (1985): $v_V = \pi/3 * 1^3_0$.

Size distribution of the diameters of atrial granules was determined according to Bach (1967; cf Weibel 1979) and size distribution of particle volumes according to Gundersen and Jensen (1985) from v_N and v_V..

"Verified" biological variability of stereological parameters was derived from the observed variance between animals and the variance between sections (9). The stereological analysis was performed as a multi-stage sampling procedure.

Stage 1 (magnification 1.020 : 1; light microscopy): The volume density of myocardial cells was determined on paraffin sections (Masson-Goldner´s trichrome stain) according to standard methods. Reference volume was the total myocardial tissue of the right ventricular auricles.

Stage 2 (magnification 33.000 : 1; electron microscopy): The point density of atrial granules (P_p) and the numerical density of atrial granules on area (N_A) were determined on two ultrathin meshs. 30 meshs were systematically subsampled for evaluation. Counting was performed on-line with a television monitor. Reference volumes were the myocardial cells of the right auricles.

Stage 3 (magnification 80.000 : 1; electron microscopy) : The size distribution of the circular profiles ($n = 300$ per animal) and the lengths of point sampled intercepts (200 per animal) were determined at this stage. Measurements were performed on-line with semiautomatic image analyzing equipment (Videoplan, Kontron).

The total number of atrial granules in the right auricles was calculated by the following equation: Total auricle volume $*V_v$ myocardial cells $*N_v$ atrial granules.

Similary, the total volume of the atrial granules in the right auricles was determined. The total auricle volume was approximated by division of the weights with 1.02 g/cm^3.

Results

Stereological parameters: The results of our study are shown in Table 1. The most striking feature concerning atrial granules is the minimal deviation from the average values for all calculated parameters. Electron microscopy supports this analysis (see Fig. 1). Further information is obtained by immunohistochemistry on the ultrastructural level which shows constant distributions of immunogold labels in the respective areas (see Fig. 2).

Table 1.

means +/- SEM	
Stereological densities of atrial granules	
Volume density (Vv)	(mm^3/mm^3) $0.00387 \ +/- 0.00015$
Numerical density (Nv) [2]	$(1/mm^3)$ $303 * 10^9 \ +/- 13* 10^9$
Numerical density (Nv) [4]	$(1/mm^3)$ $301 * 10^9 \ +/- 16* 10^9$
Quantitative characteristics of atrial granules	
Mean diameter	(nm) $279 +/- 7$
Mean volume-weighted volume	(μm^3) $0.0162 \ +/- 0.001$
Mean volume	(mm^3) $0.0135 \ +/- 0.001$
(numerical distribution)	
Absolute quantities of atrial granules in right auricles	
Total volume	(mm^3) $0.0558 \ +/- 0.0079$
Total number	$4.46*10^9 \ +/- 0.67 * 10^9$
Size variability of atrial granules **mean +/- SD**	
Diameter	(nm) $279 +/- 74$
Mean volume	(mm^3) $0.0135 \ +/- 0.006$
"Verified" interindividual variance	
Parameter coeffiency of variation	
Diameter	4 %
Volume density	8 %
Numerical density	12 %

2 and 4: calculations according to equations (2) and (4), respectively.

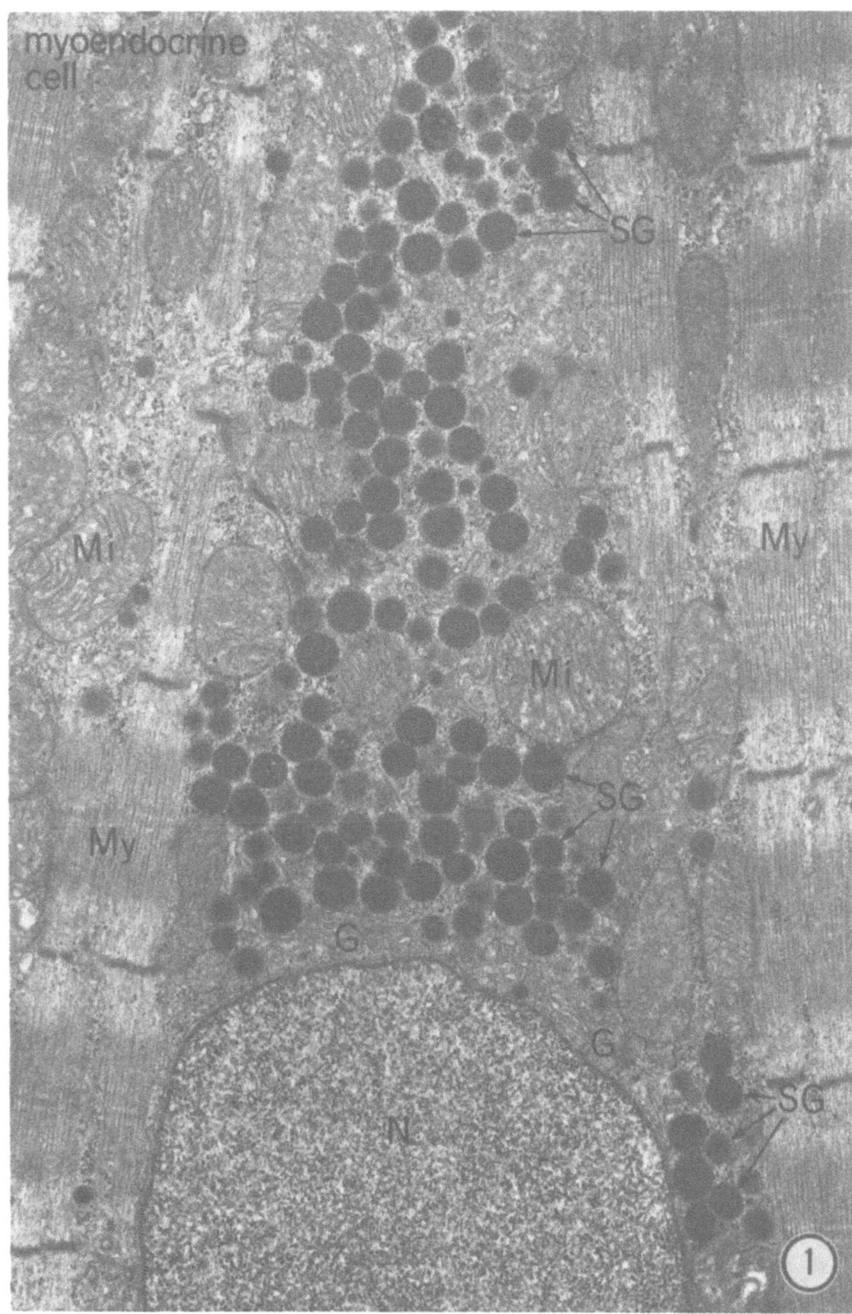

Fig.1. Electron micrograph of rat myoendocrine cell showing the perinuclear area with the secretory apparatus involved in the production and the secretory cycle of cardiac hormones. Note the Golgi-complex (G) adjacent to the nucleus (N), and agglomerations of secretory granules (SG); Mitochondria (Mi), myofibrils (My). (32000x)

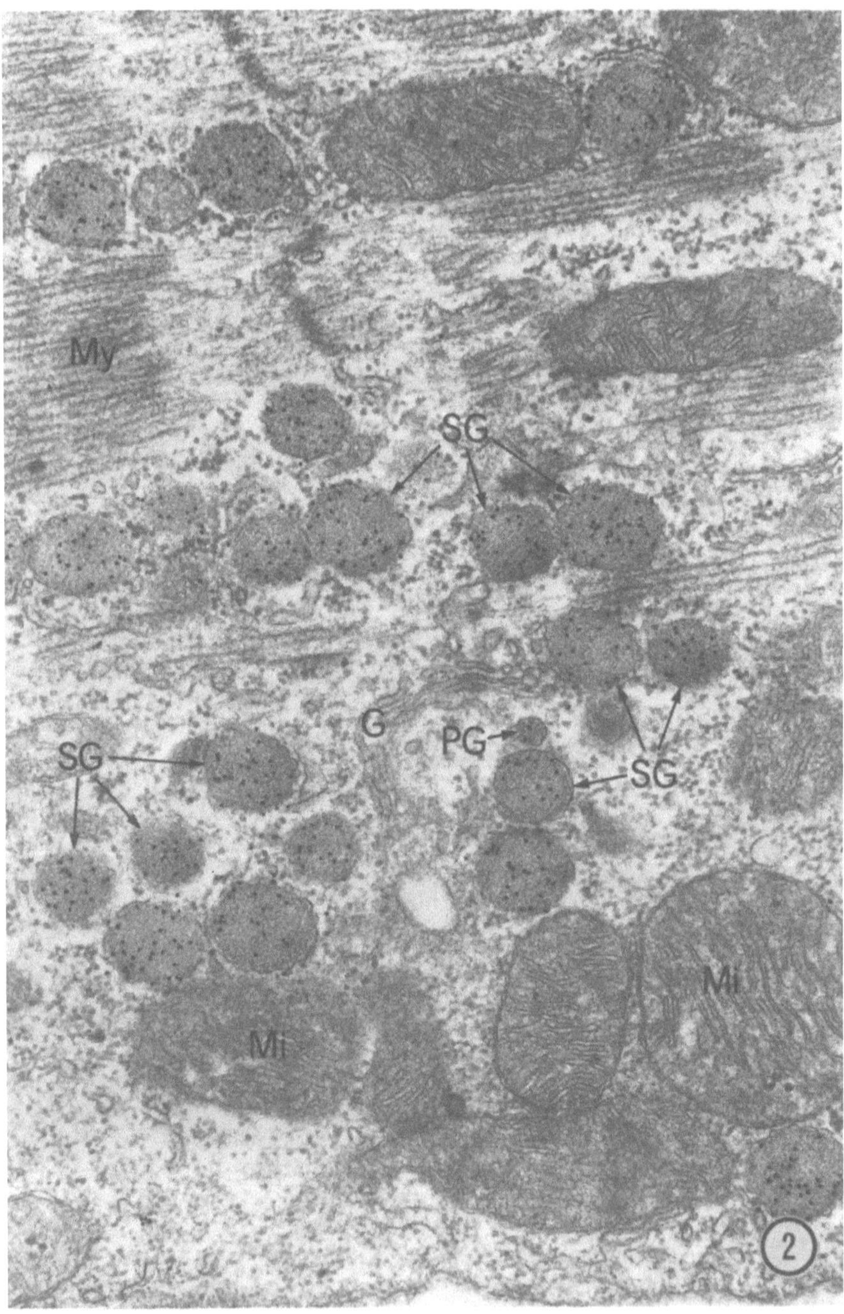

Fig. 2. Ultrastructural demonstration of CDD/ANP immunoreactivity related to the secretory granules of myoendocrine cells in the rat heart. Note that the number of colloidal gold particles is rather constant in relation to the sectioned area of the secretory granule. The Golgi-complex (G) is not stained, progranules (PG) exhibit little labels. Note further the low background in the sarcoplasma, the mitochondria (Mi), and the myofibrils (My). (41500x)

Comparison of biochemical and stereological parameters: Total mass of ANP in right auricles was determined by radioimmunology. In the right auricles of five male Wistar rats (body weight: 250 g) the mass amounted to 0.0085 mg, on average. Since the specific weight of proteins approximates 1.0 g/cm^3, the percentage of atrial granules which consist of CDD/ANP is approximated by the ratio: mass CDD/ANP/ volume AG = 0.0085/ 0.0558 * 100% = 15%. Furthermore, it can be deduced that 0.001 mg = 1 µg ANP is stored in 4.46 * 10^9/ 8.5 = 525 000 000 atrial granules.

Discussion

The highest amount of myoendocrine cells is found in the heart atrium. Specific granules (most of them located near the nucleus) are electron-dense, round to oval structures surrounded by a distinct membrane. The stereological variance between them is small and mean diameter weight and mean volume show remarkably little deviation from the mean value. This fact characterizes atrial granules as a very homogenous population. Immuno-electronmicroscopically, these structures show a constant immunoreactivity toward CDD/ANP-antibodies visualized by the uniform distribution of colloidal gold particles as seen in Fig. 2. It is suggested that this constant label in mature secretory granules may document the constant content of CDD/ANP-immunoreactive material as a minimum of inter individual variability of volume density, numeral density, and spherical radius of atrial granules that are characteristic for dynamic structures of biological significance, e.g., the basal lamina of the kidney (see Table 1). The comparison between RIA and stereology shows that 15 % of the total mass of atrial granules of the rat consists of immunoreactive CDD/ANP. In addition to CDD/ANP, atrial granules may contain matrix proteins and other, yet unknown, material such as more physiologically important polypeptides. Further studies are required to elucidate the entire composition of the atrial granules.

Acknowledgement: We are thankful to B. Brühl for carefully executing all the technical work.

References

1. Bach, G. (1967) Kugelgrößenverteilung der Schnittkreise; ihre wechselseitigen Beziehungen und Verfahren zur Bestimmung der einen aus der anderen. In: Weibel, E.R., Elias,H. (eds) Quantitative Methods in Morphology. Springer, Berlin Heidelberg New York, pp 23 - 45
2. Cantin, M., Genest. J. (1985) The heart and the atrial natriuretic factor. Endocr Rev 6: 107-127

3. Currie, M.G., Geller, D.M., Cole, B.R., Boylan, J.G., Yusheng, W., Holmberg, S.W., Needleman, P. (1983) Bioactive cardiac substances: potent vasorelaxant activity in mammalian atria. Science 221:71-73

4. DeBold, A.J., Borenstein, H.B., Veress, A.T., Sonnenberg, H. (1981) A rapid and potent natriuretic response to intravenous injection of atrial extracts in rats. Life Sci 28: 89-94

5. Dehoff, R.T., Rhines, F.N. (1961) Determination of the number of particles per unit volume from measurements made on random plane sections: the general cylinder and the ellipsoid. Trans AIME 221: 975-982

6. Forssmann, W.G., Hock, D., Lottspeich, F., Henschen A., Kreye, V., Christmann, M., Reinecke, M., Metz, J., Carlquist, M., Mutt, V. (1983) The right auricle of the heart is an endocrine organ. Cardiodilatin as a peptide hormone candidate. Anat Embryol 168: 307-313

7. Forssmann, W.G. (1986) Cardiac hormones. I Review on the morphology, biochemistry and molecular biology of the endocrine heart. Eur J Clin Invest 16: 439-451

8. Gundersen, H.J.G., Jensen, E.B. (1985) Stereological estimation of the volume-weighted mean volume of arbitrary particles observed on random sections. J Microsc 138: 127-142

9. Gundersen, H.J.G., Osterby, R. (1980) Sampling efficiency and biological variation in stereology. Mikroskopie 38: 143-148

10. Haug, H. (1967) Probleme und Methoden der Strukturzählung im Schnittpräparat In: Weibel, E.R., Elias, H. (eds) Quantitative Methods in Morphology. Springer, Berlin Heidelberg New York, S 58-78

11. Jamieson, J.D., Palade, G.E. (1964) Specific granules in atrial muscle cells. J Cell Biol 23: 151-172

12. Kisch, B. (1956) Electron microscopy of the atrium of the heart I. Guinea-pig. Exp Med Surg 14: 99-112

13. Marie, J.P., Guillemot, H., Hatt, P.Y. (1976) Le degré de granulation des cardiocytes auriculaires. Etudes planimetrique au cours de différents apports d´ eau et de sodium chez le rat. Path Biol 24: 549-554

14. Weibel, E.R. (1979) Stereological Methods, vol. I. Academic Press, London New York Toronto Sydney San Francisco

15. Weibel, E.R. and Gomez, D.M. (1962) A principle for counting tissue structures on random sections. J Appl Physiol 17: 343-348

Immunohistochemical evidence for a neurosecretory and neurotransmitter function of CDD/ANP-immunoreactive neurons in molluscs

M. Reinecke[1], H. Back[1], P. Jacobs[2], R. Schipp[2]

[1]Department of Anatomy and Cell Biology, University of Heidelberg, and [2]Department of Zoology, University of Giessen, FRG

Introduction

Recently, our group showed (5,6,7) that cardiac hormones of the atrial natriuretic poly-peptide (ANP)/cardiodilatin (CDD)-family (for review see 2) occur in myoendocrine cells of the hearts from representatives of all vertebrate classes. In the snail *Helix pomatia* (Gastropoda, Pulmonata), however, CDD/ANP-immunoreactivity was found exclusivly in neuroendocrine endings of the cardiac atrium while no CDD/ANP- immunoreactivity was present in cardiac myocytes (4,5). In correlation, no "specific" granules which characterize the cardiac myoendocrine system of all vertebrates (1,5,6,7) could be observed in heart muscle cells of the snail. Since furthermore, CDD/ANP-immunoreactivity was present in perikarya of the subesophageal ganglion and in nerve fibers of the intestinal nerve which supplies the heart, the cardiac hormone-like substances seem to constitute a neuroendocrine-cardiac axis in the snail *Helix pomatia* (4,5). In order to investigate this peculiar phylogenetic pattern in more detail we expanded our studies 1) by analyzing the exact distribution patterns of CDD/ANP-immunoreactive (-IR) neurons in the central ganglia of *Helix pomatia* and 2) by investigating the hearts and the central ganglia of other molluscan species, i.e., *Aplysia depilans* (Gastropoda, Opisthobranchia) and *Sepia officinalis* (Cephalopoda, Decabrachia) for the occurance of cardiac hormones. The study was carried out with Bouin-fixed paraffin-embedded tissue by means of immunohistochemistry applying several region-specific antisera against alpha-ANP (=CDD-28) and the peroxidase-antiperoxidase (PAP) technique.

Results

Helix pomatia: In the cardiac atrium, CDD/ANP-IR nerve endings were present in high densities (Fig. 1a), whereas no CDD/ANP-IR structures occurred in the ventricle. The CDD/ANP-IR was confined only to nerve endings which were characterized ultrastructurally by large heteromorphous neurosecretory granules (Fig. 1b). CDD/ANP-immunoreactivity was observed in perikarya of all portions of the central ganglia (meso- and metacerebrum of the cerebral ganglia as well as pleural, parietal, visceral and pedal ganglion) (Fig. 1c). On the average, the diameter of the CDD/ANP-IR neurons was 20 μm. Only in the visceral ganglion were additionally larger CDD/ANP-IR perikarya (40-200 μm dia-

This work was supported by the German Research Foundation (Re 520/3).

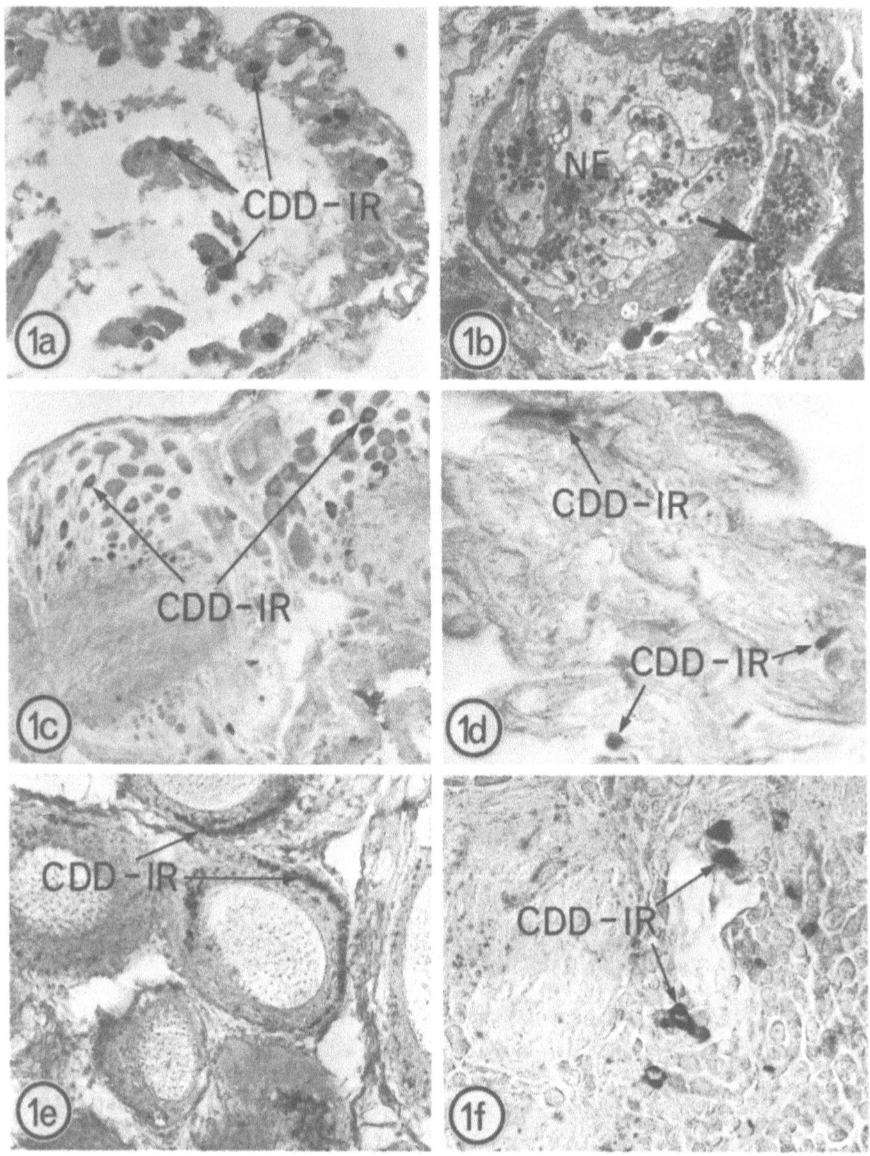

Fig. 1a. High density of CDD/ANP-IR nerve endings (arrows) in the cardiac atrium of *Helix pomatia*. (x450)

Fig. 1b. Electronmicrograph of a neurosecretory ending in the atrium of *Helix* which is characterized by large heteromorphous granules. This type of ending exhibits the same distribution pattern as the CDD/ANP-IR endings shown in Fig.1a. (x5 600)

Fig. 1c. CDD/ANP-IR perikarya (arrows) and nerve fibers in the subesophageal ganglion of *Helix*. (x250)

Fig. 1d. CDD/ANP-IR nerve endings (arrows) in the proximity of myocytes in the cardiac atrium of *Aplysia depilans*. (x560)

Fig. 1e. Large perikarya in the visceral ganglion of *Aplysia depilans* which exhibit CDD/ANP-immunoreactivity. (x530)

Fig. 1f. CDD/ANP-IR perikarya and fibers in the visceral lobe of the cephalopode *Sepia officinalis*. (x600)

meter) present. In the cerebral ganglia, numerous small perikarya (15 - 20 μm diameter) occurred which were most frequent in the pleural and the pedal lobe. In the mesocerebrum, CDD/ANP-IR perikarya were observed only infrequently while no CDD/ANP-IR perikarya were detected in the protocerebrum. CDD/ANP-IR nerve fibers occurred within the neuropile of all ganglia as well as around single non-CDD/ANP-IR perikarya. Furthermore, nerve fibers were present in the parietal-visceral and in the cerebro-pleural and cerebro-pedal connective tracts.

Aplysia and *Sepia:* Similar to the results obtained with *Helix*, in the cardiac atrium of *Aplysia* and in the atrium of the systemic heart of *Sepia* (8) CDD/ANP-IR nerve endings were observed but no CDD/ANP-IR myocytes were found. Most of the CDD/-ANP-IR nerve endings were found adjacent to cardiac myocytes (Fig. 1d). Compared to their density in *Helix* atrium CDD/ANP-IR nerve endings occurred only in- frequently in the atria of *Aplysia* and *Sepia*. CDD/ANP-IR perikarya and nerve fibers were present throughout the central ganglia of both species (Fig. 1e,f). Most of the CDD/ANP-IR nerve fibers were found in the neuropiles of the different ganglion lobes but in all parts of the central ganglia CDD/ANP/-IR fibers were also observed supplying non-CDD/-ANP-IR perikarya. While the majority of the CDD/ANP-IR perikarya were of small or medium size (up to 50 μm), large neurons (150-200 μm diameter) were present in the visceral lobes of both molluscan species. As was found in *Helix*, CDD/ANP-IR nerve fibres were observed running within the different connective tracts.

Conclusions and discussions

Our results indicate that the existence of an CDD/ANP-IR neurocardiac axis is not restricted to *Helix pomatia* but may be common in molluscs since representatives of different molluscan classes seem to possess a similar organized neurosecretory system. However, CDD/ANP-IR structures occur not only in the visceral lobes but in the intestinal nerves of molluscs as described before (4). Thus, they seem to also have other functional meanings than merely constituting one special neurosecretory pathway from the visceral lobe to the cardiac atrium. This hypothesis gains additional support by the observation that in another cephalopodan species, i.e.,*Octopus vulgaris*, CDD/ANP-IR was recently detected in nerve fibers of the neurosecretory system of the vena cava (NSV) where it seems to coexist with FMRFamide in identical neurohemal endings (3). Furthermore, not only CDD/ANP-IR perikarya occur throughout the central ganglia of the molluscan species studied but CDD/ANP-IR fibers are also found in an overall distribution, i.e., they occur within all ganglionic neuropiles, around perikarya, and in the connective tracts. This may suggest a neurotransmitter and/or neuromodulator function of the CDD/ANP-IR neurones within the central ganglia of molluscs.

References

1. Chapeau, C., Gutkowska, J., Schiller, P.W., Milne, R.W., Thibault, G., Garcia, R., Genest, J., Cantin, M. (1985) Localization of immunoreactive synthetic atrial natriuretic factor (ANF) in the heart of various animal species. J Histochem Cytochem 33: 541-550
2. Forssmann, W.G. (1986) Cardiac hormones. I. Review on the morphology, biochemistry and molecular biology of the endocrine heart. Eur J Clin Invest 16: 439-451
3. Martin, R., Voigt, K.H. (1987) The neurosecretory system of the octopus vena cava: A neurohaemal organ. Experientia 43: 537-543
4. Nehls, M., Reinecke, M., Lang, R.E., Forssmann, W.G. (1985) Biochemical and immunological evidence for a cardiodilatin-like substance in the snail neurocardiac axis. Proc Natl Acad Sci USA 82: 7762-7766
5. Reinecke, M., Nehls, M., Forssmann, W.G. (1985) Phylogenetic aspects of cardiac hormones as revealed by immunocytochemistry, electronmicroscopy and bioassay. Peptides 6 (suppl 3): 321-331
6. Reinecke, M., Betzler, D., Forssmann, W.G. (1987a) Immunocytochemistry of cardiac polypepide hormones (Cardiodilatin/atrial natriuretic polypeptide) in brain and hearts of Myxine glutinosa (Cyclostomata). Histochemistry 86: 233-239
7. Reinecke, M., Betzler, D., Forssmann, W.G., Thorndyke, M., Askensten, U., Falkmer, S. (1987b) Electromicroscopical, immunohistochemical, immunocytochemical and biological evidence for the occurence of cardiac hormones (ANP/CDD) in chondrichthyes. Histochemistry 87: 531-538
8. Schipp, R. (1987) General morphology and functional characteristics of the cephalopod circulatory system. An introduction. Experientia 43: 474-477

Dual distribution of cardiac hormones (CDD/ANP) in the heart and brain of vertebrates

M. Reinecke, D. Betzler, H. Segner, W.G. Forssmann

Department of Anatomy and Cell Biology, University of Heidelberg, Heidelberg, FRG

Introduction

About four years ago, several research groups independently isolated and characterized cardiac hormones from the atrial appendages of some mammalian species. As a consequence of the different isolation procedures and the accompaning bioassays (diuresis-natriuresis, vasodilation) used, a variety of cardiac hormone candidates were presented under different names (3,7,20). However, mainly based on cDNA studies (13,17,18,28) it recently became evident that the different peptides are members of the same family and are derived from a preprohormone, which is cleaved resulting in a peptide of 126 amino acids, i.e., gamma-ANP (15) or cardiodilatin- (CDD-) 126 (8,9). This prohormone contains in its C-terminal portion the final circulating form consisting of 28 amino acids, i.e. alpha-ANP (14) which is identical to CDD-28 (99-126) (7).

While there is increasing information on localization, pharmacology, and pathology of cardiac CDD/ANP in mammals, including man, only few studies have dealt so far with the occurrence of similar peptides in the hearts of submammalian vertebrates (5,6,21, 22,23). Similarly, the existence of CDD/ANP-immunoreactivity in the brain has so far been reported almost exclusively for mammals, i.e.*Tupaia* (10) and the rat (16,24,25, 27). In the rat brain, by the use of chromatographic methods, additional evidence has been obtained that the precursor molecule of alpha-ANP (=CDD-28) is produced in the perikarya and cleaved during axonal transport to its final small molecular form (11). To date, however, there is only one study which demonstrates the occurrence of CDD/ANP-immunoreactive (-IR) neurons in the central nervous system of a submammalian vertebrate: CDD/ANP-IR perikarya and fibers were detected in high densities throughout the brain of *Myxine glutinosa* (22), a representative of the most primitive vertebrates, the cyclostomes.

Thus, the present study deals with the occurrence of cardiac hormones from the CDD/ANP-family in the phylogeny of vertebrate heart and central nervous system, and summarizes earlier and new results. The investigation was carried out by means of electron microscopy and immunohistochemistry using a variety of antisera against different partial sequences of CDD/ANP (10,22) and the peroxidase-antiperoxidase (PAP) technique. The following species were investigated as representatives of the seven vertebrate classes: mammals (man, *Tupaia belangeri*, dog, rat, guinea pig), birds (*Gallus g. domesticus, Coturnix c. japonicus*), reptiles (*Anolis carolinensis*), amphibians (*Rana esculenta*), osteichthyes (fresh water: *Salmo gairdneri*, sea water: *Cottus scorpius*), chondrichthyes (*Squalus acanthias, Scyliorhinus canicula, Raja clavata, Chimaera monstrosa*), cyclostomes *(Myxine glutinosa)*.

This work was supported by the German Research Foundation (Re 502/3).

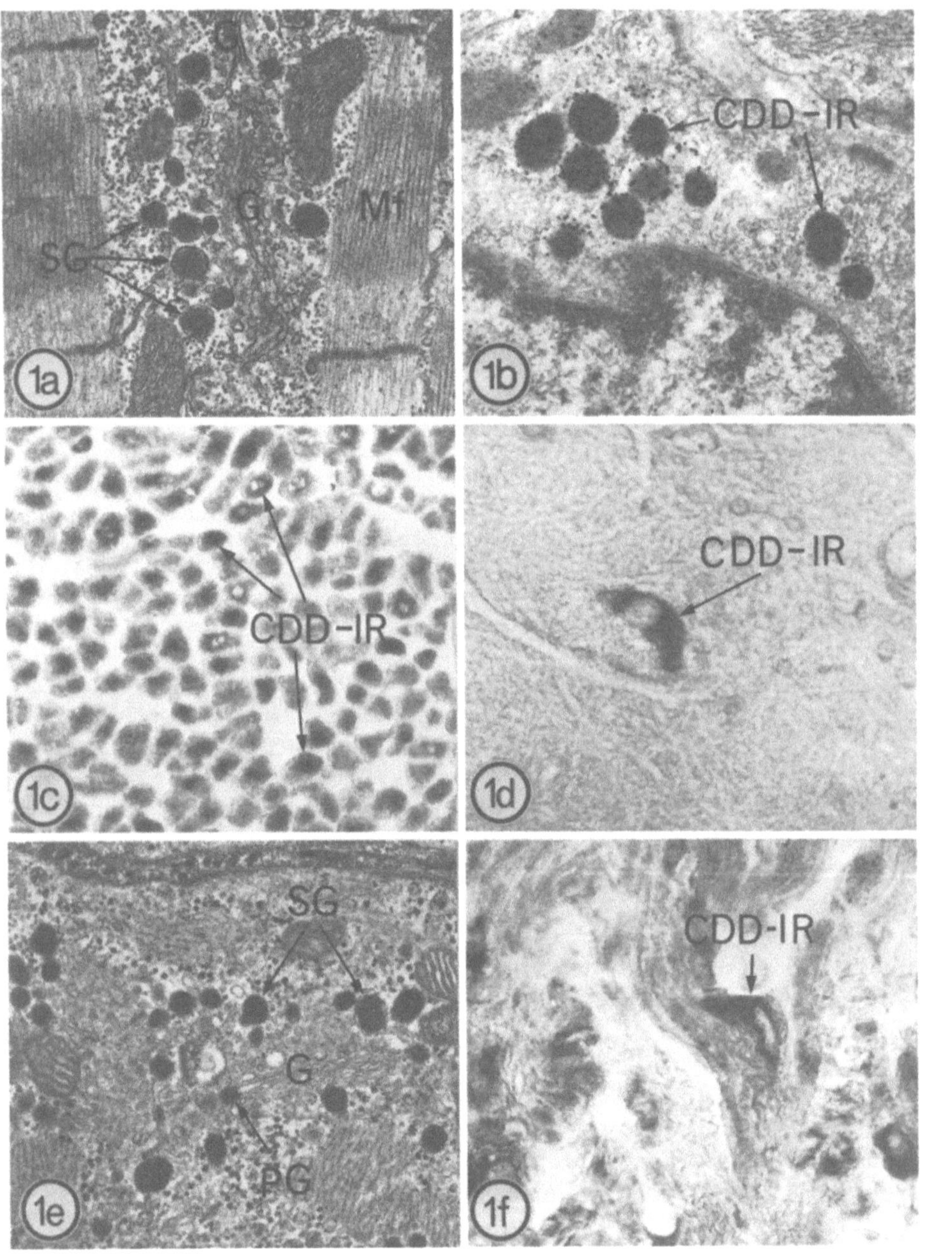

Fig. 1a. Myoendocrine cells from the pig right atrial appendage showing an accumulation of specific granules adjacent to the perinuclear Golgi apparatus. (x14 800)

Fig. 1b. CDD/ANP-IR staining of specific granules of a myoendocrine cell in the human atrium with the protein A-gold technique. The gold particles are almost exclusively located on the granules. (x28 400)

Fig. 1c. High density of CDD/ANP-IR cells in the right auricle of the pig heart. (x190)

88

Fig. 1d. CDD/ANP-IR perikarya and fibres in the hypothalamus of the primitive primate *Tupaia belangeri*. (x510)

Fig. 1e. Myoendocrine cell in the atrium of the frog *Rana esculenta*. Secretory granules are found in various locations. (x20 200)

Fig. 1f. Large numbers of CDD/ANP-IR cells are present in the atria of *Rana esculenta*. The CDD/ANP-immunoreactivity is found throughout the cytoplasm. (x420)

Results

Heart: By means of electron microscopy a large number of myoendocrine cells were identified in the atria of all investigated species (Figs. 1a, b, e, 2a). In the mammalian and avian species studied, the greatest density of myoendocrine cells occurred in the atrial appendages, whereas in the remaining vertebrate classes no obvious regional accumulations were found. In contrast to mammals, in the hearts of the avian species investigated, myoendocrine cells occurred only rarely. While in the ventricles of mammals, birds, and reptiles, myoendocrine cells were encountered only infrequently; in the ventricles of amphibians, ostheichthyes, chondrichthyes, and cyclostomes they were numerous. With increasing phylogenetic age of the species the number of ventricular myoendocrine cells was found to be also increased. Generally, the myoendocrine cells were similar to those found in mammals, i.e., they resembled ventricular myocardiocytes and were characterized by the occurrence of specific granules. Both, size and electron opacity of the specific granules varied slightly among the species studied (Figs. 1a, b, e; 2a). The granules were found in the perinuclear area, adjacent to the Golgi apparatus, in the interfibrillar spaces, and in subsarcolemmal location.

In all investigated species, in correspondence to the distribution patterns of the ultrastructurally identified myoendocrine cells in identical locations, CDD/ANP-IR cells were detected by means of immunohistochemistry (Figs. 1c, f; 2b, d, e). In submammalian species immunoreactivities were obtained only by the use of antisera directed against C-terminal sequences of gamma-ANP (=CDD-126), i.e., reacting with alpha-ANP (CDD-28); antisera dircincluded against N-terminal sequences of the prohormone reacted only with mammalian myoendocrine cells.

Brain: In the mammalian species studied, i.e.,*Tupaia belangeri*, CDD/ANP-IR perikarya were observed mainly in the periventricular nucleus (Fig. 1d), and in the supraoptic nucleus of the hypothalamus, and in the central amygdaloid nucleus, while high amounts of CDD/ANP-IR fibers occurred in the septal region, the periaqueducterial grey, in various regions of the medulla oblongata and the spinal cord.

While in the brain of the marine teleost bony fish *Cottus scorpius*, which was also studied for the occurrence of CDD/ANP-immunoreactivity, CDD/ANP-IR perikarya were scarce and observed unequivocally only in hypothalamic areas (Fig. 2c); CDD/ANP-IR fibers were present on all levels except for the telencephalon. The greatest density of CDD/ANP-IR fibres was found in the diencephalon, mainly in the hypothalamus, in the branchiomotoric nuclei, as well as in the fiber nerve tracts of the medulla oblongata and the spinal cord.

In the brain of the cyclostomian species *Myxine glutinosa* CDD/ANP-IR neurons were abundant at all levels. In the telencephalon, CDD/ANP-IR perikarya and fibers were present in the primordium hippocampi. In the diencephalon, CDD/ANP-IR perikarya occurred in the thalamic and hypothalamic (Fig. 2f) nuclei. Some small CDD/ANP-IR perikarya were observed in the medulla oblongata, which was mainly characterized by large CDD/ANP-IR fiber tracts proceeding into the spinal cord.

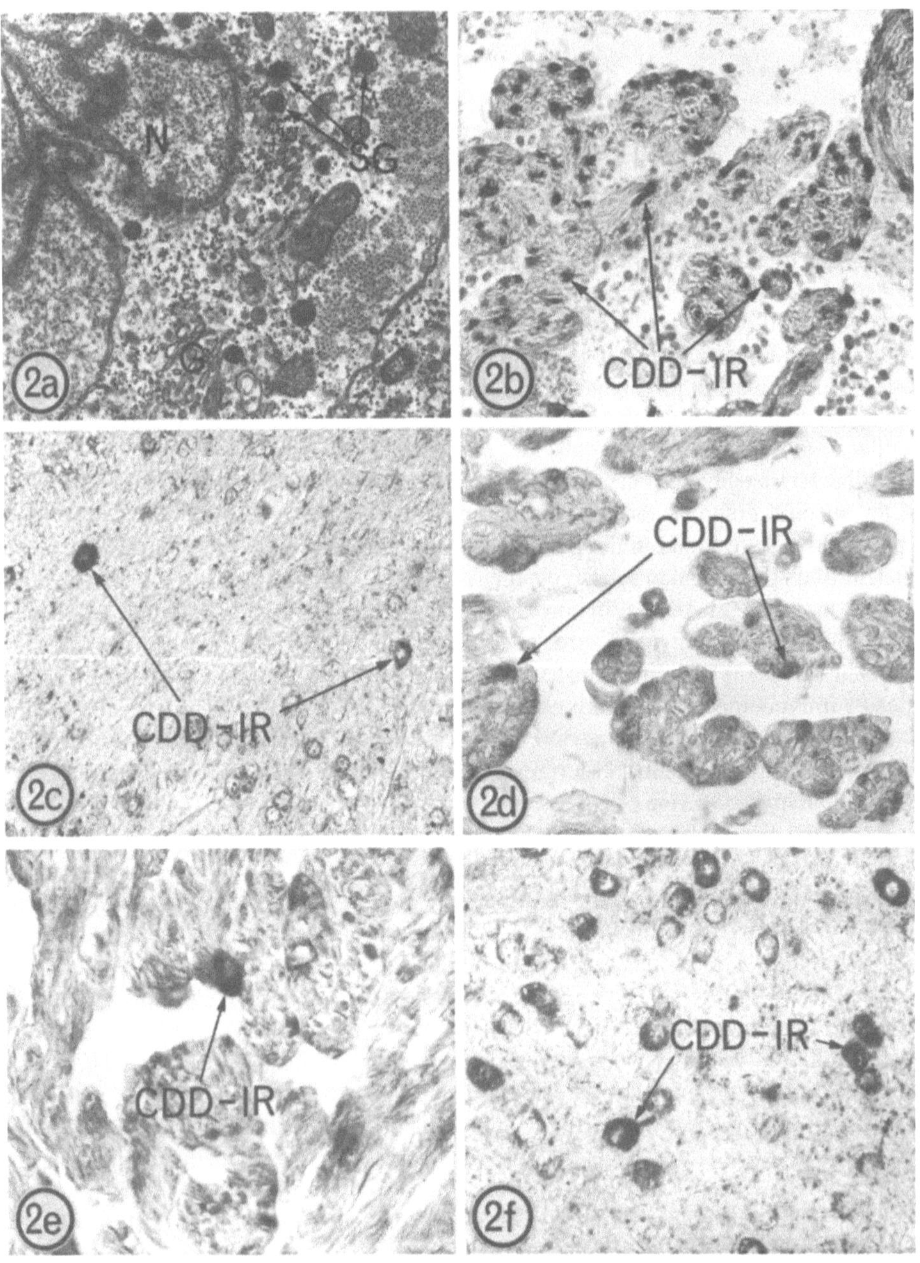

Fig. 2a. Myoendocrine cell from the atrium of the marine teleost bony fish *Cottus scorpius*. Dense-cored secretory granules occur in the interfibrillar spaces. (x11 200)
Fig. 2b. CDD/ANP-IR cells are found in large numbers throughout the atrium of *Cottus*. (x450)
Fig. 2c. CDD/ANP-IR perikarya in the diencephalon of *Cottus*. (x540)
Figs. 2d,e. CDD/ANP-IR cells in the atrium (a) and the ventricle (b) of the systemic heart of the cyclostomian species *Myxine glutinosa.*(a x400, b x640)
Fig. 2f. CDD/ANP-IR perikarya and fibres in the hypothalamus of *Myxine*. (x610)

Conclusions and discussion

Our results show that CDD/ANP-IR myoendocrine cells are present in the hearts from representatives of all vertebrate classes. Thus, at least the C-terminal bioactive portion of CDD/ANP seems to be highly preserved during phylogeny. Since for each vertebrate class only some species were investigated, we will not overemphasize our interpretations. However, some conclusions may be drawn. In birds, the cardiac hormones may have lost their physiological importance, since, in the avian hearts studied, CDD/ANP-IR myoendocrine cells were found only very infrequently compared to all other vertebrate classes. The small amount of CDD/ANP-like material in the avian heart has been demonstrated also by the use of bioassays (21,23) and may be indicative of a failure to detect cardiac hormone-like natriuretic activity in chicken heart (6) using a different extraction procedure. However, comparative pharmacological studies must be carried out to varify this hypothesis. Our result, that the presence of CDD/ANP-IR myoendocrine cells in the hearts of mammals, birds and reptiles is mainly restricted to the atria, while in the lower vertebrate classes they occur in considerable numbers also in the ventricles, is supported by other immunohistochemical (5) and bioassay (21,23) studies. Thus, there seems to be a phylogenetic trend to concentrate the cardiac-hormone producing cells with increasing complexitiy of the cardiovascular system in the atria. These observations may indicate that the discussed mammalian ventricular myoendocrine cells (1, 2,12,19,26, see also Göbel et al., this Symposium) may represent a phylogenetic remnant rather than having a particular physiological role in impulse conduction as suggested (26).

In the brain of all investigated species, i.e. representatives of the mammals, osteichthyes, and cyclostomes, CDD/ANP-IR perikarya and fibers were found. As reported for the rat (16,24,25,27) a particular high density of CDD/ANP-IR perikarya was present in hypothalamic regions. Thus, as suggested for mammals (10,11,16,24), in lower vertebrates CDD/ANP-like peptides may also be involved in the central regulation of fluid balance and cardiovascular function. However, the extremely high density of CDD/ANP-IR perikarya and fibers throughout all levels of the brain of *Myxine glutinosa* (22) may indicate additional specialized functions of these neurons in the phylogenetically unique species *Myxine*. Using chromatographic methods, for the rat brain, evidence has been obtained that a precursor form of alpha-CDD/ANP-28 is produced in the perikarya and cleaved during axonal transport to its final form (11). The low density of CDD/ANP-IR in the perikarya as compared to the large amount of CDD/ANP-IR fibers in the brain of

Cottus scorpius may indicate the existence of a so far unknown precursor molecule within the perikarya which does not react with the antisera applied.

References

1. Back, H., Stumpf, W.E., Ando, E., Nokihara, N., Forssmann, W.G. (1986) Immuno-cytochemical evidence for CDD/ANP-like peptides in strands of myoendocrine cells associated with the ventricular conduction system of the rat heart. Anat Embryol 175: 223-336
2. Bloch, K.D., Seidman, J.G., Naftilan, J.D., Fallon, J.T., Seidman, C.E. (1986) Neonatal atria and ventricles secrete atrial natriuretic factor via tissue-specific secretory pathways. Cell   47: 695-702
3. Cantin, M.,   Genest, J   (1985) The heart and the atrial natriuretic factor. Endocr Rev 6: 107-127
4. Cantin, M., Gutlkowska, J., Thibault, G., Milne, R.W., Ledoux, S., MinLi, S., Chapeau, C., Garcia, R., Hamet, P., Genest, J. (1984) Immunocytochemical localization of atrial natriuretic factor in the heart and salivary glands. Histochemistry 80: 113-127
5. Chapeau, C., Gutkowska, J., Schiller, P.W., Milne, R.W., Thibault, G., Garcia, R., Genest, J., Cantin, M. (1985) Localization of immunoreactive synthetic atrial natriuretic factor (ANF) in the heart of various animal species. J Histochem Cytochem 33: 541-550
6. DeBold, A.J., Salerno, T.A. (1983) Natriuretic activity of extracts obtained from hearts of different species and from various rat tissues. Can J Physiol Pharmacol 61: 127-130
7. Forssmann, W.G. (1986) Cardiac hormones. I. Review on the morphology, biochemistry and molecular biology of the endocrine heart.Eur J Clin Invest 16: 439-451
8. Forssmann, W.G., Birr, C., Carlquist, M., Christmann, M., Finke, R., Henschen, A., Hock, D., Kirchheim, H., Kreye, V., Lottspeich, F., Metz, J., Mutt, V., Reinecke, M. (1984) The auricular myocardiocytes of the heart constitute an endocrine organ. Characterization of a porcine cardiac peptide hormone, cardiodilatin-126. Cell Tissue Res 238: 425-430
9. Forssmann, W.G., Hock, D., Lottspeich, F., Henschen, A., Kreye, V., Christmann, M., Reinecke, M., Metz, J., Carlquist, M., Mutt, V. (1983) The right auricle of the heart is an endocrine organ. Cardiodilatin as a peptide hormone candidate. Anat Embryol 168: 307-313
10. Forssmann, W.G., Mutt, V. (1985) Cardiodilatin-immunoreactive neurons in the hypothalamus of Tupaia. Anat Embryol 172:1-5
11. Forssmann, W.G., Lang, R.E., Aoki, A., Reinecke, M., Rippegather, G., Hock, D. (1987) Cardiodilatin as a neuropeptide (cardiac hormones are also neuropeptides). In: Heym, C. (ed) Histochemistry and Cell Biology of Autonomic Neurons and Paraganglia. Springer, Berlin Heidelberg NewYork London Paris Tokyo, pp 43 - 50
12. Gardner, D.G., Descheppes, C.F., Ganong, W.F., Hane, S., Fiddes, J., Baxter, J.D., Lewicki, J. (1986) Extra-atrial expression of the gene for atrial natriuretic factor. Proc Natl Acad Sci USA 83: 6697-6701
13. Greenberg, B.D., Bencen, G.H., Seilhamer, J.J., Lewicki, J.A., Fiddes, J.C. (1984) Nucleotide sequence of the gene encoding human atrial natriuretic factor precursor. Nature 312: 656-658

14. Kangawa, K., Matsuo, H. (1984) Purification and complete amino acid sequence of alpha-human atrial natriuretic polypeptide (alpha-hANP). Biochem Biophys Res Comm 118: 131-139

15. Kangawa, K., Tawaragi, Y., Oikawa, S. Mizuno, A., Sakuragawa, Y., Nakazato, H., Fukuda, A., Minamino, N., Matsuo, H. (1984) Identification of rat gamma atrial natriuretic polypeptide and characterization of the cDNA encoding its precursor. Nature 312: 152-155

16. Kawata, M., Nakao, K., Morii, N., Kiso, Y., Yamashita, H., Imura, H., Sano, Y. (1985) Atrial natriuretic polypeptide: topographical distribution in the rat brain in radioimmunoassay and immunohistochemistry. Neuroscience 16: 521-546

17. Maki, M., Takayanagi, R., Misono, K.S., Pandey, K.N., Tibbetts, C., Inagami, T. (1984) Structure of rat atrial natriuretic factor precursor deduced from cDNA sequence. Nature 309: 722-724

18. Nakayama, K., Ohkubo, J., Hirose, T., Inayama, S., Nakanishi, S. (1984) mRNA sequences for human cardiodilatin-atrial-natriuretic factor precursor and regulation of precursor mRNA in rat atria. Nature 310: 699-701

19. Nemer, M., Lavigne, J.-P., Drouin, J., Thibault, G., Cannon, M., Antakly, T. (1986) Expression of atrial natriuretic factor gene in heart ventricular tissue. Peptides 7: 1147-1152

20. Palluk, R., Gaida, W., Hoefke, W. (1985) Minireview: Atrial natriuretic factor. Life Sci 36: 1415-1425

21. Reinecke, M., Nehls, M., Forssmann, W.G. (1985) Phylogenetic aspects of cardiac hormones as revealed by immunocytochemistry, electronmicroscopy and bioassay. Peptides 6 (suppl 3): 321-331

22. Reinecke, M., Betzler, D., Forssmann, W.G. (1987a) Immunocytochemistry of cardiac polypeptide hormones (Cardiodilatin/atrial natriuretic polypeptide) in brain and hearts of Myxine glutinosa (Cyclostomata). Histochemistry 86: 233-239

23. Reinecke, M., Betzler, D., Thorndyke, M., Forssmann, W.G., Askensten, U., Falkmer, S. (1987b) Electronmicroscopical, immunohistochemical, immunocyto-chemical and biological evidence for the occurrence of cardiac hormones (ANP/CDD) in chondrichthyes. Histochemistry 87: 531-538

24. Saper, C.B., Standaert, D.G., Currie, M.G., Schwartz, D., Geller, D.M., Needleman, P. (1985) Atriopeptin-immunoreactive neurons in the brain: presence in cardiovascular regulatory areas. Science 227: 1047-1049

25. Skofitsch, G., Jacobowitz, D.M., Eskay, R.L., Zamir, N. (1985) Distribution of atrial natriuretic factor-like immunoreactive neurons in the rat brain. Neuroscience 16: 917-948

26. Toshimori, H., Toshimori, K., Oura, C., Matsuo, H. (1987) Immunohistochemistry and immunocytochemistry of atrial natriuretic polypeptide in porcine heart. Histo-chemistry 86: 595-601

27. Zamir, N., Skofitsch, G., Eskay, R.L., Jacobowitz, D.M. (1985) Distribution of immunoreactive atrial natriuretic peptides in the central nervous system of the rat. Brain Res 365: 105-111

28. Zivin, R.A., Condra, J.H., Dixon., R.A.F., Seidah, N.G., Chrétien, M., Nemer, M., Chamberland, M., Drouin, J. (1984) Molecular cloning and characterization of DNA sequences encoding rat and human atrial natriuretic factors. Proc Natl Acad Sci USA 81: 6325-6329

Steroid hormones and the endocrine heart

W.E. Stumpf, H. Back, W.G. Forssmann

Department of Cell Biology and Anatomy, University of North Carolina, Chapel Hill, North Carolina, USA, and Department of Anatomy and Cell Biology, University of Heidelberg, Heidelberg, FRG

Summary

Evidence from autoradiographic studies with radiolabelled steroid hormones indicates the presence of nuclear receptors in cardiomyocytes for estrogen, androgen, gluco- and mineralcorticosteroids, as well as $1,25(OH)_2$ vitamin D_3. The results suggest that baseline manufacture and secretion of atrial peptide hormone and probably other functions of heart muscle are regulated by a direct genomic action of steroid hormones in a fashion similar to that known for brain and pituitary peptide and monoamine messengers. The specific effects of the individual hormones need to be studied and physiologic and possible therapeutic implications clarified.

The endocrine heart may include the atrium (5) and probably the ventricular conduction system (2). These structures contain myoendocrine cells that stain immunohistochemically with antibodies to a heart hormone, variably termed "atrial natriuretic polypeptide" (ANP) or "cardiodilatin" (CDD) (5). The involvement of the ventricular conduction system needs to be further established and was suggested from original observations in our laboratory in Chapel Hill, (2).

Indications for a link between steroid hormone(s) sites of action and cells that contain "atrial granules" were obtained before the discovery of atrial peptide hormone, when ^{3}H-estradiol was localized in atrial cardiomyocytes and an effect of the steroid hormone on the manufacture of a then putative atrial peptide hormone was considered (21). Since steroid hormones act as regulators of peptide hormone and monamine secretion in brain and pituitary (10,22) and in other tissues (3), this possibility must also be considered for the cardio-vascular system.

This is a brief review of histochemical evidence that suggests a direct action on heart muscle, endocrine or non-endocrine, of all of the steroid hormones, together with a consideration of the steroidal cardiac glycosides.

Estradiol

In autoradiograms prepared one or two hours after injection of ^{3}H-estradiol in rat and guinea pig, a characteristic nuclear concentration of the hormone is seen not only in classical reproductive tissues, such as uterine stroma, epithelium and muscle (20,23), but also in heart muscle (21). In the heart, as in the uterus, estradiol is bound to nuclei of muscle cells as well as connective tissue cells. In the uterus, estradiol binding to the inner and outer muscle layers is similar. In the heart, atrial myocytes show a much stronger or even exclusive nuclear uptake, when compared to the ventricular myocytes

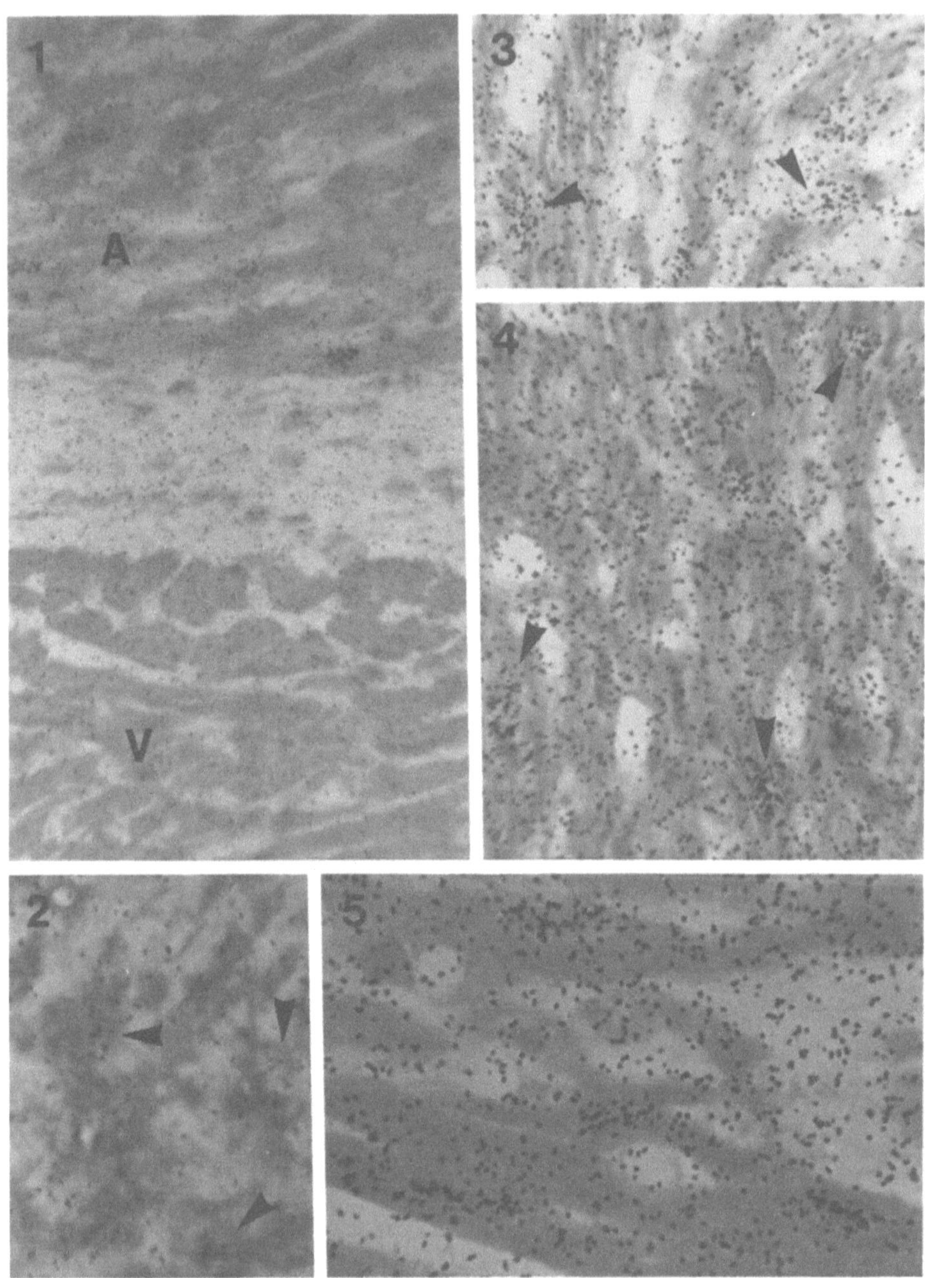

Figs. 1 - 5. Autoradiograms of rat heart after injection of ^{3}H-estradiol (Figs.1 and 2), ^{3}H-dexamethasone (Figs. 3 and 4), or ^{3}H-aldosterone. After ^{3}H-estradiol a distinct nuclear concentration of radioactivity is seen in atrial (A) myocytes, but not in ventricular (V) myocytes (Fig. 1); immunostaining with antibodies to atrial peptide hormone shows colocalization of ^{3}H-estradiol in nuclei (arrow heads) of cells that stain their cytoplasm

96

with anti-CDD/ANF (Fig. 2). After ^{3}H-dexamethasone nuclear concentration (arrow heads) or radioactivity can be seen in atrial (Fig. 3) and ventricular (Fig. 4) muscle cells, which is similar after 3H-aldosterone (Fig. 5, ventricle). Note the high cytoplasmic radioactivity after application of radiolabelled adrenal steroids. This may reflect the presence of occupied cytoplasmic receptors, or of metabolites, or of both. Four μm frozen sections, stained with methylgreen pyronin - except Fig. 2. Exposure times: 216 days (Figs. 1 and 2), 432 days (Figs. 3 and 4), 276 days (Fig. 5).

(Fig.1). Depending on the amount of radiolabelled estradiol administered and the length of autoradiographic exposure, there will be either no or little nuclear labelling detectable in ventricular muscle cells. The nuclear labelling in the atrium is not uniform and is most frequent and intense in auricular muscle. Nuclear labelling is also seen in certain connective tissue cells throughout the heart, including elements of the heart skeleton and the pericardium. No nuclear labelling could be detected in ganglion cells in the vicinity of the atrium. Whether or not cells of the ventricular conduction band are labelled remains to be investigated.

When autoradiograms are incubated with antibodies to atrial peptide by a combined technique of autoradiography-immunohistochemistry (18), atrial myocytes with nuclear radioactivity display cytoplasmic immunostaining (Fig.2), (2).

Progesterone

To date no evidence is available to support the existence of progesterone receptors in nuclei of heart muscle. Direct effects of progesterone on heart muscle can be expected if the general close relationship between estrogen and progestin actions is considered. In our autoradiographic studies with ovariectomized mice, no nuclear concentration of radiolabelled progestin was observed under conditions when nuclear labelling of uterine cells existed, although cytoplasmic radioactivity was high (Stumpf and Sar, unpublished).

Testosterone

Testosterone may not be acting as the original compound, which is probably a prohormone, but rather as the aromatized metabolite estradiol or the reduced metabolite 5-alpha dihydrotestosterone. Whether or not aromatization or 5-α–reduction or both occur in the heart is not yet clear. When radiolabelled dihydrotestosterone is injected into young female rhesus monkeys or baboons, evidence for the presence of androgen receptors is obtained in both ventricular and atrial cardiomyocytes (16).

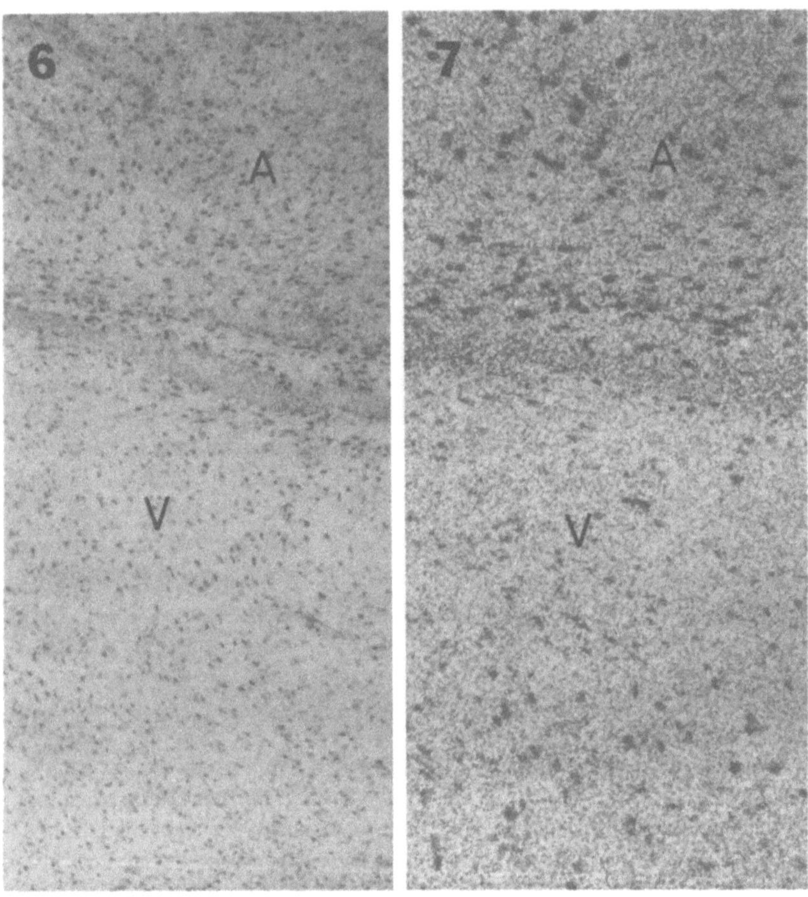

Fig. 6 and 7. Autoradiograms of rat heart after injection of ^{3}H-dexamethasone, showing nuclear and extranuclear radioactivity in both atrium and ventricle, which is however higher in the atrium (A), when compared to the ventricle (V). Left x 90, right x 220. Four µm, unstained. Exposure 402 days.

Adrenal steroid hormones

In preliminary autoradiographic studies with rats and mice, using ^{3}H-dexamethasone (Figs. 3 and 4), or ^{3}H-aldosterone (Fig. 5), nuclear concentration of radioactivity is seen in atrial and ventricular myocytes as well as in connective tissue (Stumpf and Sar, unpublished).

When autoradiograms from mice that were injected ^{3}H-dexamethasone were immuno-stained with antibodies to atrial peptide, a colocalization of nuclear radioactivity and cytoplasmic immuno precipitation in atrial muscle cells could be seen (1). This suggests a direct effect of dexamethasone on atrial peptide producing cells and is supported by observations published in the literature. In adrenalectomized animals, glucocorticoids increase the blood level and atrial content of atrial peptide (8). Also, it has been shown that certain nucleotide sequences in the CDD/ANP gene bind a glucocorticoid-receptor complex (11).

The results of biochemical studies with homogenized heart indicate specific binding of dexamethasone to cardiac cytosol in rat and dog (7), which is also suggestive of the presence of glucocorticoid receptor. In the same studies, aldosterone was shown to be taken up by the hearts of both species, however, the process was not found to be indicative of specific receptors.

Clinical effects for both gluco- and mineralcorticosteroids have been reported extensively in the literature (15), and the above mentioned experimental evidence suggests that direct effects exist on heart muscle by the adrenal steroid hormones. Results from recent unstained survey autoradiograms (Stumpf and Sar, unpublished), suggest a higher nuclear and extranuclear concentration of ^{3}H-dexamethasone in atrial tissue, when compared to ventricular tissue (Figs. 6 and 7).

Vitamin D

1,25(OH)$_2$vitamin D$_3$ is generally regarded as the steroid hormone that regulates calcium homeostasis and is therefore believed to exert a strong influence on muscle physiology through calcium blood levels. The first evidence to suggest the presence of nuclear receptors for this hormone in the heart for a direct effect on cardiomyocytes came from autoradiographic studies (24). Subsequent biochemical evidence supported the presence of 1,25-dihydroxyvitamin D$_3$ 4S receptors by sucrose gradient analysis of homogenized salt extracts of chromatin preparations from whole hearts (25). These biochemical studies did not distinguish between atrium and ventricle, nor between muscle cells and connective tissue cells. The function of 1,25(OH)$_2$vitamine D$_3$ in heart is unknown, but can be assumed to be related to the regulation of intracellular calcium levels. Although the available data are limited, direct effects of vitamine D on the endocrine heart are likely to exist.

Cardiac glycosides

Cardiac glycosides are not considered hormones or hormone analogs. Their molecular conformation differs from that of steroid hormones. Cardiac glycosides are, however, steroids that are not only known by their actions on heart muscle, but also by their estrogenic and uterotrophic effects, albeit mild compared to estradiol. Digitalis glycosides have been reported to induce vaginal cornification and to decrease urinary gonadotropin excretion in postmenopausal women, and to produce gynecomastia in elderly men (17). Digitoxin significantly increases uterine weight and inhibits the binding of estradiol to specific or saturable uterine binding sites (17). Digoxin, although eliciting weak uterotrophic effects (19), did not inhibit estradiol binding to receptor protein *in vitro* (17). While there is presently good evidence to support the concept of sodium-potassium ATPase-mediated effects of cardiac glycosides, an alternative mode may be considered in view of the above cited evidence and the information now available on heart endocrinology. Involvement of the heart endocrine system in cardiac glycoside action is possible and needs to be explored. Results from our preliminary autoradiographic studies with ^{3}H-digitoxin of low specific activity, did not provide evidence for a direct genomic action on atrial myocytes (Stumpf and Sar, unpublished), but neither did they exclude such a possibility. Further support for the concept that the action of digitalis may involve receptors for endogeneous peptides, such as the atrial peptide or other, may be supported by the presence of endogenous digitalis-like activity in the mammalian brain (4), the neural actions of digitalis (9) and the presence in the brain of CDD/ANF immunoreactivity (6,13,26).

Conclusions

Experimental and clinical evidence indicates that steroid hormones act on heart muscle. Since steroid hormones are known to exert in their target tissues regulatory functions on DNA transcription involving RNA, protein, and DNA synthesis, such a role can be assumed in the heart. In addition, non-genomic effects mediated through membrane or cytoplasmic receptors must also be considered, related to evidence for rapid *in vitro* effects and cytoplasmic accumulation of radiolabelled hormone, especially for progestin, adrenal steroids, and androgen. The specific effects of the different steroids remain to be clarified. Data from colocalization of ^{3}H-estradiol or ^{3}H-dexamethasone and antibodies to atrial peptide hormone suggests that manufacture and secretion of atrial peptide hormone, including extent and duration of response to reflex stimuli, are directly regulated by steroid hormones. Direct effects of aldosterone on atrial cardiomyocytes are also indicated from our autoradiographic data and suggest a direct feedback-loop between adrenal cortex and heart, since atrial natriuretic hormone affects aldosterone secretion (14) and aldosterone affects atrial peptide secretion (8). Probably the renin-angiotensin system, as well as vasopressin (12) are involved in this regulation.

The well-known protective effect of estradiol on the female heart and the adaptation to increased blood volume during pregnancy, may be mediated through direct actions of estradiol on atrial peptide producing-cardiomyocytes. In the male, the two main metabolites of testosterone, estradiol and dihydrotestosterone, appear both to act on heart muscle. It remains to be clarified, whether, and under which conditions, aromatization or 5-α-reduction predominates and what role these metabolites of testosterone play in health and disease. Since cardiac glycosides have weak estrogenic effects, some of the pharmacological actions of digitoxin may involve specific estradiol-like effects on atrial peptide hormone producing myocytes.

References

1. Back, H., Forssmann, W.G., Stumpf, W.E. (1988) Atrial myoendocrine cells (cardio-dilatin/ atrial natriuretic polypeptide containing myocardiocytes) are target cells for estradiol. Cell. Tiss. Res. (in press)
2. Back, H., Stumpf, W.E., Ando, E., Nokihara, K., Forssmann, W.G. (1986) Immuno-cytochemical evidence for CDD/ANP-like peptides in strands of myoendocrine cells associated with the ventricular conduction system of the rat heart. Anat Embryol 175: 223-226
3. Clark, S.A., Stumpf, W.E., Sar, M. (1986) Effect of 1,25 dihydroxyvitamin D_3 on insulin secretion. Diabetes 30: 382-386
4. Fishman, M.C. (1979) Endogenous digitalis-like activity in mammalian brain. Proc Nat Acad Sci USA 76: 4661-4663
5. Forssmann, W.G. (1986) Cardiac hormones: I. Review on the morphology, bio-chemistry and molecular biology of the endocrine heart. Eur J Clin Invest 16: 439-451
6. Forssmann, W.G., Mutt, V. (1985) Cardiodilatin -immunoreactive neurons in the hypothalamus of Tupaja. Anat Embryol 172: 1 - 5
7. Funder, J.W., Duval, D., Meyer, P. (1973) Cardiac glucocorticoid receptors: the binding of tritiated dexamethasone in rat and dog heart. Endocrinol 93: 1300-1308
8. Garcia, R., Debinski, W., Gutkowska, J., Kuchel, O., Thibault, G., Genest, J., Cantin, M. (1985) Gluco- and mineralcorticoids may regulate the natriuretic effect and the synthesis and release of atrial natriureticfactor by the rat atria *in vivo*. Biochem Biophys Res Comm 131: 806-814
9. Gillis, R.A., Quest, J.A. (1978) Neural actions of digitalis. Ann Rev Med 29: 73-79
10. Grant, L.D., Stumpf, W.E. (1975) Hormone uptake sites in relation to CNS biogenic amine systems. In: Stumpf, W.E., Grant, L.D. (eds) Anatomical neuroendocrinology. pp 445-464 Karger, Basel
11. Greenberg, B.D., Bencen, G.H., Seilhamer, J.J., Lewicki, J.A., Fiddes, J.C (1984) Nucleotide sequence of the gene encoding human atrial natriuretic factor precursor. Nature 312: 656-658
12. Gutkowska, J., Racz, K., Debinski, W., Thibault, G., Garcia, R., Kuchel, O., Cantin.M., Genest, J. (1987) An atrial natriuretic factor-like activity in rat posterior hypophysis. Peptides 8: 461-465
13. Jirikowski, G.F., Back, H., Forssmann, W.G., Stumpf, W.E. (1986) Coexistence of atrialnatriuretic factor (ANF) and oxytocin in neurons of the rat hypothalamus. Neuropeptides 8: 243-249
14. Kudo, T. Baird, A. (1984) Inhibition of aldosterone production in the adrenal glomerulosa by atrial natriuretic factor. Nature 312: 756-757
15. Lefer, A.M., Trachte, G.T. (1980) Effects of hormones on the heart. In: Bourne, G.H. (ed) Hearts and heart-like organs, Vol. 2. Academic Press, New York, pp 1-40
16. McGill, Jr. H.C., Anselmo, V.C., Buchanan, J.M., Sheridan, P.J. (1980) The heart is a target organ for androgen. Science 207: 775-777
17. Rifka, S.M., Pita, Jr. J.C., Loriaux, D.L. (1976) Mechanism of interaction of digitalis with estradiol binding sites in rat uteri. Endocrinol 99: 1091-1096
18. Sar, M., Stumpf, W.E. (1983) Simultaneous localization of steroid hormones and neuropeptides in the brain by combined autoradiography and immunocytochemistry. In: Conn, M. (ed) Methods in Enzymology. Vol. 103. Academic Press, New York, pp 631-638

19. Stoffer, S.S., Jian, N.-S. (1972) Digoxin-Estradiol-17beta interaction: a study of uterotropic effects in rats. Endocrinol 90: 843-846

20. Stumpf, W.E. (1968) Subcellular distribution of 3H-estradiol in rat uterus by quantitative autoradiography. Endocrinol. 83: 629-632

21. Stumpf, W.E. (1976) Techniques for the autoradiography of diffusible compounds. In: Prescott, D.M. (ed) Methods in cell biology, Vol. XIII, Academic Press, New York, pp 171-193

22. Stumpf, W.E., Jennes, L. (1984) The A-B-C (Allocortex-Brainstem-Core) circuitry of endocrine-autonomic integration and regulation. Relationships between estradiol sites of action and peptidergic-aminergic neuronal systems. Peptides 5 (suppl 1): 221-226

23. Stumpf, W.E., Sar, M. (1976) Autoradiographic localization of estrogen, androgen, progestin and glucocorticosteroid in "target tissues" and "non-target tissues." In: Pasqualini J. (ed) Receptors and mechanism of action of steroid hormones. Marcel Dekker, New York, pp 41 - 84

24. Stumpf, W.E., Sar, M., DeLuca, H.F. (1981) Sites of action of $1,25(OH)_2$ vitamin D_3 identified by thaw-mount autoradiography. In: Cohn, D.V., Talmage, R.V., Matthews, Jr. J.L., (eds) Hormonal control of calcium metabolism. Excerpta Medica, Amsterdam, pp 222-229

25. Walters, M.R., Wicker, D.C., Riggle, P.C. (1986) 1,25-dihydroxyvitamin D_3 receptors identified in the heart. J Molec Cell Cardiol 18: 67-72

26. Zamir, N., Skofitsch, G., Eskay, R.L., Jacobowitz, D.M. (1986) Distribution of immunoreactive natriuretic peptides in the central nervous system of the rat. Brain Res 365: 105-111

Biochemistry of cardiac hormones

D. Hock

Department of Anatomy and Cell Biology, University of Heidelberg, Heidelberg, FRG

Introduction

Since the discovery of an increased diuresis by left heart atria distension by Henry, et al. (15), and the first observation of Kisch (22) that the atria contain myocardial cells with particular inclusions, many attempts have been made to characterize the nature of hypothetized cardiac hormones. The first authors to emphasize an endocrine nature of atrial cells were Jamieson and Palade (19). However, for a long time it was suggested that the specific granules contain catecholamines. A direct involvement of atrial granularity in the regulatory process of electrolyte body fluid homeostasis was established by Marie et al. (26). Only since the early 1980s, have biological tests been available that allow the detection of specific functions of heart atrial extracts. DeBold and collaborators (5) introduced a diuretic test for rat atrial extracts. Deth et al. (6), Currie et al. (2), Grammer et al. (13), and Forssmann et al. (9), introduced tests to demonstrate the relaxation of smooth vascular muscle strips after the application of atrial extracts. However, it was not clear whether the atrial diuretic and vasorelaxant substances were all the same bioactive substances. Nevertheless, these tests incited the conception about the endocrine role of the heart in the regulation of extracellular fluid volume and blood pressure.

These assays served as a basis for several research groups to extract, isolate, and identify this substance. In this review, I will summarize the results from the preparation and isolation of porcine cardiac peptide; I will also present data on the preparation of bovine cardiac peptide 88, and in the first part, I will introduce you to a new method which allows large scale preparation on peptide hormones extracted from human body fluids.

Extraction and purification of Cardiodilatin/ANP from heart tissue

The nature of Cardiodilatin (CDD)/ANP can be characterized when sufficient amounts of atrial material are extracted and purified. We used porcine atria which were easily obtained in a local slaughterhouse; it was evident that thousands would have to be used to identify the substrate of cardiac specific granules. During the isolation, all steps of chromatography were tried out on two biotests:

1) A diuretic-natriuretic effect (5). During this bioassay partially purified extracts from various purification steps were used for physiological tests in trained conscious dogs. A bolus injection of 300-1000 µg of partially purified Cardiodilatin resulted in a dose-dependent increase in urine volume and sodium excretion.

2) A tissue-specific vasorelaxant effect of CDD/ANP (6): After a period of equilibration

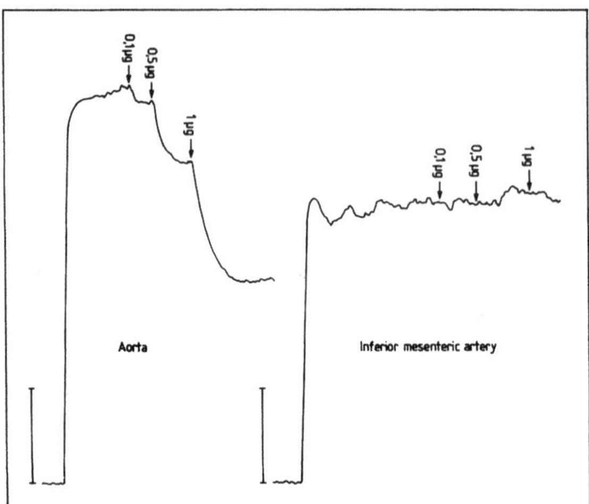

Fig.1. Tissue-specific vasorelaxant effect of cardiodilatin/ANP. After an equilibration period in a standard physiological salt solution, helical strips of rabbit aorta and inferior mesenteric artery are stimulated by 10^{-7} norepinephrine. Increasing amounts of prepurified CDD/ANP are added cumulatively. Bar: 5 mN of contractile force.

in a standard physiological salt solution, helical strips of vascular smooth muscle were stimulated by 10^{-7}M norepinephrine. The vasorelaxant effect of Cardiodilatin showed that the dilatory responses depended on the kind of vascular smooth muscle strips. The mesenteric arteries were weakly affected by cardiodilatin-126 while the smooth muscle of the aorta was more sensitive (Fig. 1).

For the first isolation of porcine Cardiodilatin-126, 40 kg of right atrial appendages were excised immediately after the animals thoracic cavities were opened in the slaughterhouse. The tissue was rapidly boiled in water and then frozen in liquid nitrogen. The tissue was collected and the extraction was carried out with 0. 2M acetic acid. The filtrate obtained was bound to alginic acid (27) and eluted by 0.2M hydrochloric acid. The eluate was salted out and submitted to a water-ethanol extraction. Three major fractions of peptides were obtained. The highest concentration of biological active material was detected in the water-ethanol-soluble fraction. Following precipitation at -20º C, this material was submitted to ion exchange chromatography (Fig. 2a). From this chromatography, several active fractions were obtained and desalted on a Sephadex G 25 column. The bioactive fractions were pooled and submitted to reversed-phase HPLC (Fig. 2b). In order to obtain purified material for the amino-acid analysis and sequence studies the bioactive fractions from the first HPLC purification step were rechromatographed twice on an analytical reversed-phase column (Fig. 2c).

For the biological activity, 1µl of each fraction from the final HPLC was tested. The aliquots were applied to the aortic muscle strips precontracted by norepinephrine. The fraction causing the strongest dilatation was used for the sequence-analysis. The amino acid sequence of CDD/ANP-126 was determined by sequencing the intact peptide and

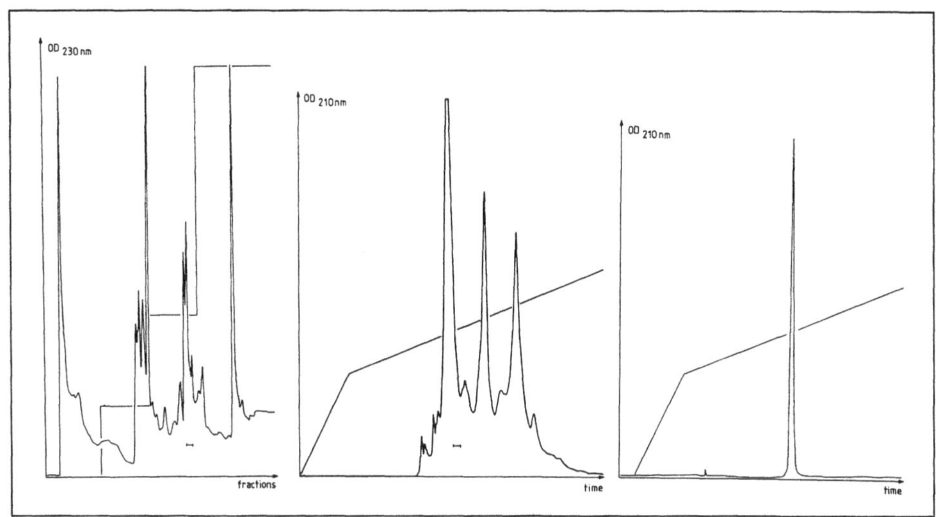

Fig.2.a. Separation of CDD/ANP on CM-cellulose. The lyophilized prepurified material is applied to a CM-cellulose column and eluted with a stepwise increase of the molarity from ammonium-bicarbonate concentration (0.01M, 0.08M, 0.2M and 0.5M). The CDD/ANP-containing fractions, indicated by a bar, are pooled and lyophilized

Fig.2.b. A typical HPLC elution profile of the material obtained by the procedure shown in Fig.2.a. The column (TSK-ODS-120T, 300 x 7.8mm) was eluted with a continuous gradient of acetonitrile with 0.1% TFA. The biologically active material is found in the first large peak indicated by a bar.

Fig.2.c. The final purification of CDD/ANP by reserve phase HPLC on an analytical colum (TSK-ODS-120T, 250 x 4.5mm); flow rate 0.5ml/min; solvent B 0.1% TFA in 80% acetonitrile.

```
ASN – PRO – VAL – TYR – GLY – SER – VAL – SER – ASN – ALA – ASP – LEU – MET – ASP – PHE –

LYS – ASN – LEU – LEU – ASP – HIS – LEU – GLU – ASP – LYS – MET – PRO – LEU – GLU – ASP –

GLU – ALA – MET – PRO – PRO – GLN – VAL – LEU – SER – GLU – GLN – ASN – GLU – GLU – VAL –

GLY – ALA – PRO – LEU – SER – PRO – LEU – LEU – GLU – VAL – PRO – PRO – TRP – THR – GLY –

GLU – VAL – ASN – PRO – ALA – GLN – ARG – ASP – GLY – GLY – ALA – LEU – GLY – ARG – GLY –

PRO – TRP – ASP – ALA – SER – ASP – ARG – SER – ALA – LEU – LEU – LYS – SER – LYS – LEU –

ARG – ALA – LEU – LEU – ALA – ALA – PRO – ARG – SER – LEU – ARG – ARG – SER – SER – CYS –

PHE – GLY – GLY – ARG – MET – ASP – ARG – ILE – GLY – ALA – GLN – SER – GLY – LEU – GLY –

CYS – ASN – SER – PHE – ARG – TYR – OH
```

Fig.3. Primary structure of cardiodilatin-126/ANP

fragments after cyanogen bromide cleavage on a solid-phase and a gas-phase sequenator. The PTH amino acids were identified by HPLC (Fig. 3)

We prepared antibodies against C- and N-terminal fragments of the peptide which react with the secretory granules in the myoendrocrine cells as demonstrated by ultrastructural immuno-histochemistry. The colloidal gold technique revealed the presence of a cardio-dilatin-immunoreactive substance within the specific secretory granules (9,10).

Amino acid sequence studies

There is a wide range of homology of cardiac hormones. The N-terminal parts of the cardiac hormone molecules in human (14,21,29,31), porcine (10), bovine (17,35), and rat (25,36) exhibit less homology in the amino acid sequence. A striking homology exists between the 28 amino acid residues at the C-terminal part of this peptide. The only difference is an isoleucine rat in the place of a methionine (human, porcine, bovine) at position 110.

Different amino acid sequences of low molecular weights of cardiac peptides have been reported from different laboratories (7,3,21,32). More or less, they all represented the C-terminal part of CDD/ANP-126 and have the same 17-residue disulphide-bonded core. However, the large variety of different atrial peptides extracted from atrial tissue in different laboratories suggested that at least some were artifacts due to extraction and purification procedures. Indeed, Kangawa et al., (21) showed that alpha-rANP- (5-27), a peptide identical to atriopeptin III (11), was isolated when a heat treatment was carried out after the extraction. The use of fresh tissue and heat treatment prior to homogenization was indispensable to prevent proteolysis during the extraction. As our procedure (9,16) is based on the techniques used by Mutt (28) for the isolation of gastrointestinal hormones combined with those from other laboratories, it is very difficult to say what particular step caused the artifactual proteolysis. The persistent application of extraction from quickly boiled atrial tissue led to high molecular weight forms of CDD/ANP only, suggesting that these peptides were the ones stored in the atrial tissue. However, it is still an open question what the ratio is of the different molecular forms of cardiac hormones in the normal atrial tissue and what it is under various physiological conditions.

Purification of bovine CDD/ANP-88

We continued our studies on cardiac hormones, including bovine Cardiodilatin, on large-scale preparations with the aim to simplify the purification method to obtain larger amounts of natural CDD for functional studies. The isolation procedure reported here is different from that originally used. Now we obtain cardiodilatin by a two-step purification. The crude extract is chromatographed on an automated cation exchange fast protein liquid chromatography (FPLC) and only one final step of reversed-phase HPLC is needed (17). The vasorelaxant peptide material from the FPLC is directly subjected to a reversed-phase column and all bioactive material is eluted in one peak. Using this method, we obtain 3mg highly purified CDD/ANP-88 from 10 kg bovine atria.

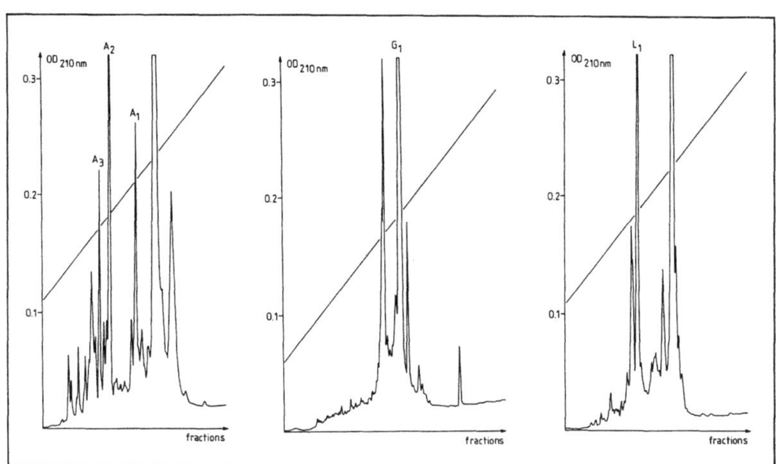

Fig.4. Separation of the enzymatically cleaved fragments of CDD-88/ANF-(39-126) by reversed-phase HPLC on a TSK-ODS-120 column (250 x 4.6mm). Solvent system, A: 0.01M hydrochloric acid, B: 0.01M hydrochloric acid in 80% acetonitrile. The enzymatic degradation was performed at 37° C for 3h. Enzymatic degradation with endoproteinase Arg C, Glu C and Lys C. The amino acid sequence of fragments A1, A2, A3, G1, and L1 was determined as illustrated in Fig. 5.

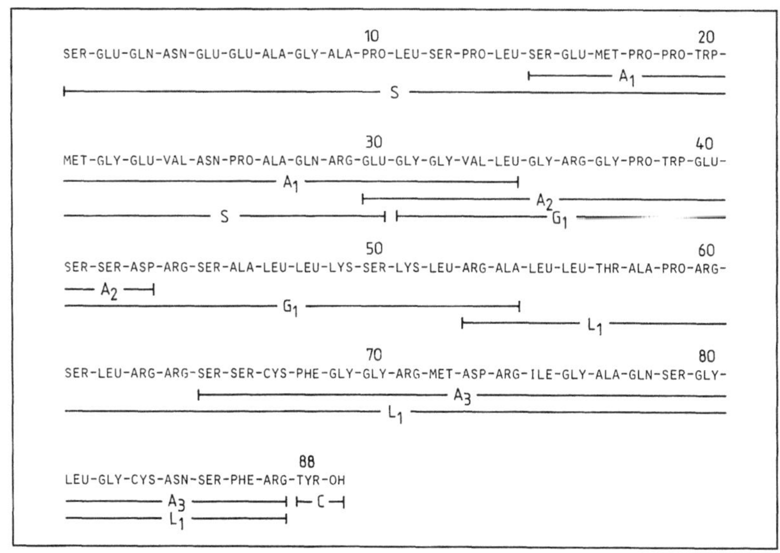

Fig.5. Complete amino acid sequence of bovine CDD-88/ANF-(39-126). Stepwise Edman degradation of the intact peptide (S) and the fragments (A1, A2, A3, G1, L1) from the enzymatic cleavage was performed on a 470A gas-phase sequencer from Applied Biosystems. The PTH-amino acids were identified by HPLC. The C-terminal tyrosine (C) was identified by enzymatic cleavage with carboxypeptidase A.

Amino acid sequence studies

The purified CDD/ANP-88 is subjected to enzyme digestion by endoproteinase Arg C, Glu C and Lys C. The fragments are separated by reversed-phase HPLC (Fig. 4). The structures of the distinct fragments are determined by a gas-phase protein sequencer. Furthermore, the primary structures of the fragments are confirmed by a sequence analysis of 5 nM of the undigested polypeptide. The primary structures of the entire CDD/ANP molecule are shown in Fig. 5. The C-terminal amino acid residue is identified following enzymatic cleavage with carboxypeptidase A. After 60 min, only tyrosine is found. A digestion of 180 min results in the cleavage of arginine, which is the second C-terminal residue.

Sequence studies showed that the last residue of all extracted CDD/ANP molecules was a tyrosine and that no atrial peptide extracted so far had a residue beyond this one. These findings, coupled with the fact that several peptides with varying N-terminal extensions had been sequenced, strongly suggested the existence of a high molecular form of a precursor peptide. Actually, it was shown that higher molecular weight forms could be processed into lower molecular weight forms by gentle proteolysis (4), or by mixing higher molecular forms of CDD/ANP with crude atrial extract (34).

Before the establishment of the CDD/ANP precursor by sequencing cDNA clones, several higher molecular weight forms of atrial peptides had been sequenced. The only complete sequence was the porcine CDD/ANP with 126 residues (104). Geller et al. (12) reported of a peptide with 111 residues - atriopeptigen; and another peptide, pronatrio-dilatin, with 106 amino acid residues had been isolated by Lazure et al. (23). Along with the detection of CDD/ANP-88 (18) it is now well established that according to genetic studies, a signal peptide of 25 amino acids is disconnected from a preprohormone of 151 amino acids, which results in the extractable form of CDD/ANP-126, and possibly CDD/ANP-88, the amino acid sequence from residue 39 to 126.

Isolation of the circulating human CDD/ANP

Several groups claimed that the circulating form of CDD/ANP is the 28 amino acid residue form, deduced from combined chromatography and RIA studies (33). Various molecular forms of this cardiac hormone family were reported to occur in human cardiovascular diseases (1).

We carried out a large scale extraction of this hormone from human blood using hemofiltrate, and for the first time we characterized the structure of the circulating form of this cardiac hormone in humans (8). Using this method, polypeptides from 1,000 l hemofiltrate with molecular weights of less than 20 k Daltons were absorbed by alginic acid, eluted, precipitated and desalted on a Sephadex G25 column. The bioactive polypep-tide material was further submitted to ion-exchange chromatography and then purified by reversed-phase HPLC. The final purified material, 20 nmol, was stepwise degraded on a gas-phase sequencer according to Edman. The PTH amino acids were identified by HPLC according to Lottspeich (24). The amino acid sequence of the circulating human CDD/ANP is illustrated in Fig. 6.

Our studies reveal that only the circulating and bioactive form of CDD/ANP in human

                          SER - LEU -
                              100
ARG - ARG - SER - SER - CYS· - PHE - GLY - GLY - ARG - MET -
                              110
ASP - ARG - ILE - GLY - ALA - GLN - SER - GLY - LEU - GLY -
                              120
CYS·· -ASN - SER - PHE - ARG - TYR·· - OH

· Confirmed by amino acid analysis
·· Confirmed by C-terminal end group analysis

Fig.6. Primary structure of circulating human CDD-(99-126)/α–hANF.

blood is identical with the molecule found by Kangawa and Matsuo (20) who performed the extraction from human atria. This molecule is included in the C-terminus of porcine and bovine CDD/ANP (8). This fact suggests that CDD/ANP-28 is processed during the exocytosis from the myoendocrine cells into the bloodstream. The characteristics of this circulating peptide are the two half-cystines indicative of the presence of a disulphide bond. A reduction of the disulphide bond abolishes all biological activity.

In summary, we may assume that the molecular biology and biochemistry of cardiac hormones are understood. Although different research groups may have different concepts of the processing of cardiac hormones, it is obvious from our large-scale extractions from porcine and bovine hearts, from the extraction of blood-born circulating CDD/ANP and from recent studies carried out by Nokihara et al. (30) on human atrial tissue that the heterogeneity of molecular forms reported is a result of inadequate tissue extractions. The molecules which do occur in atrial myoendocrine cells and are stored in the secretory granules according to ultrastructral immunocytochemistry are the preprohormone, the prohormone of 126 amino acids and possibly 88 amino acids; negligible amounts of CDD/ANP-28 are extractable from well-prepared tissues, while this form is the bioactive circulating polypeptide.

References

1. Arendt, R.M., Gerbes, A.L., Ritter, D., Stangl, E. (1986) Molecular weight heterogeneity of plasma ANF in cardiovascular disease. Klin Wochenschr 64 (Suppl VI) : 1

2. Currie, M.G., Geller, D.M., Cole, B.R., Boylan, J.G., YuSheng, W., Holmberg, S.W., Needleman, P. (1983). Bioactive cardiac substances: Potent vasorelaxant activity in mammalian atria. Science 221: 71 - 73

3. Currie, M.G., Geller, D.M., Cole, B.R., Siegel, N.R., Fok, K.F., Adams, S.P., Eubanks, S.R., Gallupi, G.R., Needleman, P. (1984a) Purification and sequence analysis of bioactive atrial peptides (atriopeptins). Science 223: 67 - 69

4. Currie, M.G., Sukin, B., Geller, D.M., Cole, B., Needleman, P. (1984b) Atriopeptin release from the isolated perfused rabbit heart. Biochem Biophys Res Commun 124: 711-715

5. DeBold, A.J., Borenstein, H.B., Vercss, A.T., Sonnenberg, H. (1981) A rapid and potent natriuretic response to intravenous injection of atrial myocardial extracts rats. Life Sci 28: 89-94

6. Deth, R.C., Wong, K., Fukagawa, S., Rocco, R., Smart, J.L., Lynch, C.J., Wawad, R (1982) Inhibition of rat aorta contractile response by natriuresis-inducing extract of rat atrium. Fed Proc Fed Am Soc Exp Biol 41: 983a

7. Flynn, T.G, DeBold, M.L., DeBold, A.J. (1983) The amino acid sequence of an atrial peptide with potent diuretic and natriuretic properties. Biochem Biophys Res Commun 117: 859-865

8. Forssmann, K., Hock, D., Herbst, F., Schulz-Knappe, P., Talartschik, J., Scheler, F., Forssmann, W.G. (1986) Isolation and structure of the circulating human cardiodilatin (alpha ANP). Klin Wochenschr 64: 1276-1280

9. Forssmann, W.G., Hock, D., Lottspeich, F., Henschen, A., Kreye, V., Christmann, M., Reinecke, M., Metz, J., Carlquist, M., Mutt, V. (1983) The right auricle of the heart is an endocrine organ. Anat Embryol 168: 307-313

10. Forssmann, W.G., Birr, C., Carlquist, M., Christmann, M., Finke, R., Henschen, A., Hock, D., Kirchheim, H., Kreye, V., Lottspeich, F., Metz, J., Mutt, V., Reinecke, M., (1984) The auricular myocardiocytes of the heart constitute anendocrine organ. Characterization of a porcine cardiacpeptide hormone, cardiodilatin-126. Cell Tissue Res 238: 425-430

11. Geller, D.M., Currie, M.G., Wakitani, K., Cole, B.R., Adams, S.P., Fok, K.F., Siegel, N.R., Eubanks, S.R., Galluppi, G.R., Needlemann, P. (1984a) Atriopeptins: A family of potent biologically active peptides derived from mammalian atria. Biochem Biophys Res Commun 120: 333-338

12. Geller, D.M., Currie, M.G., Siegel, M.R., Fok, K.F., Adams, S.P., Needleman, P., (1984b) The sequence of an atriopeptigen: A precursor of the bioactive atrial peptides. Biochem Biophys Res Commun 121: 802-807

13. Grammer, R.T., Fukumi, H., Inagami, T., Misono, K.S. (1983) Rat atrial natriuretic factor. Purification and vasorelaxant activity. Biochem Biophys Res Commun 116: 696-703

14. Greenberg, B.D., Bencen, G.H., Seilhamer, J.J., Lewicki, J.A., Fiddes, J.C. (1984) Nucleotide sequence of the gene encoding human atrial natriuretic factor precursor. Nature 312: 656-658

15. Henry, J.P., Gauer, O.H., Reeves, J.L. (1956) Evidence of the atrial location of receptors influencing urine flow. Circulating Res 4: 85

16. Hock, D., Herbst, F., Mutt, V., Forssmann, W.G. (1985) Endocrine peptides of the heart: Isolation and characterization of bovine cardiodilatin. Fifth International Washington Spring Symposium Abstract 31

17. Hock, D., Mayer, W., Back, H., Schriek, U., Herbst, F., Forssmann, W.G. (1986) Isolation and characterization of bovine cardiodilatin (CDD). Biol Chem Hoppe-Seyler 367: 267

18. Hock, D., Schriek, U., Fey, E., Forssmann, W.G., Mutt, V. (1986) Isolation of bovine cardiodilatin by fast protein liquid chromatography and reversed-phase high performance liquid chromatography. J Chromatogr 397: 347-353

19. Jamieson, J.D., Palade, G.E. (1964) Specific granules in atrial muscle. J Cell Biol 23: 151-172

20. Kangawa, K., Tawaragi, T., Oikawa, S., Mizuno, A., Sakuragawa, Y., Nakazato, H., Fukuda, A., Minamino, N., Matsuo, H. (1984a) Identification of rat atrial natriuretic polypeptide and characterization of the cDNA encoding its precursor. Nature 312: 152-155

21. Kangawa, K., Fukuda, A., Kubota, I., Hayash, Y., Matsuo, H. (1984b) Identification in rat atrial tissue of multiple forms of natriuretic polypeptides of about 3,000 daltons. Biochem Biophys Res Commun 121: 585-591

22. Kisch, B. (1956) Electron microscopy of the heart I. Guinea pig Exp Med Surg 14: 99-112

23. Lazure, C., Seidah, N.G., Chretien, M., Thibault, G., Garcia, R., Cantin, M., Genest, J. (1984) Atrial pronatriodilatin: A precursor for natriuretic factor and cardiodilatin. Amino acid sequence evidence. FEBS Lett 172: 80-86

24. Lottspeich, F. (1981) Identification of phenylthiohydantoin amino acids for micro sequencing Hoppe Seylers. Z Physiol Chem 361: 1829-1839

25. Maki, M., Takayanagi, R., Misono, K.S., Pandey, K.N., Tibbetts, C., Inagami, T. (1984) Structure of rat atrial natriuretic factor precursor deduced from cDNA sequence. Nature 309: 722-724

26. Marie, J.P., Guillemont, H., Hatt, P.Y. (1976) Le degre de granulation des cardiocytes auriculaires. Etude planimetric au cours de differents apports d › eau et de solium chez le rat. Path Biol 24: 549-554

27. Mutt, V. (1959) Preparation of highly purified secretin. Ark Kem 15: 69-74

28. Mutt, V. (1982) Chemistry of the gastrointestinal hormones and hormone-like peptides and a sketch of their physiology and pharmacology. In: Vitamins and Hormones, vol 39, Academic Press, New York - London. pp 231-42

29. Nakayama, K., Ohkubo, H., Hirose, T., Inayama, S., Nakanishi, S. (1984) mRNA sequence for human cardiodilatin-atrial natriuretic factor precursor and regulation of precursorb mRNA in rat atria. Nature 310: 699-701

30. Nokihara, K., Ando, E., Forssmann, W.G. (1987) Isolation of cardiodilatin from porcine atria using immunoaffinity chromatography and production of monoclonal antibody. Peptide Chemistry 1986: T. Miyazawa (ed) Protein Research Foundation, Osaka pp 7-11

31. Oikawa, S., Imai, M., Veno, A., Tanaka, S., Noguchi, T., Nakazato, H., Kanagawa, K., Fukuda, A., Matsua, H. (1984) Cloning and sequence analysis of cDNA encoding a precursorfor human atrial natriuretic polypeptide. Nature 309: 724-726

32. Seidah, N.G., Lazure, C., Chretien, M., Thibault, T., Garcia, R., Cantin, M., Genest, J., Nutt, R.F., Brady, S.F., Lyle, T.A., Palaveda, W.J., Colton, C.D., Ciccarone, T.M., Veber, D.F. (1984) Amino acid sequence of homologous rat atrial peptides: Natriuretic activity of native and synthetic forms. Proc Natl Acad Sci USA 81: 2640-2644

33. Thibault, G., Lazure, C., Schiffrin, E.L., Gutkowska, J., Chartier, L., Garcia, R., Seidah, N.G., Chretien, M., Genest, J., Cantin, M. (1985) Identification of a biologically active circulating form of rat atrial natriuretic factor. Biochem Biophys Res Commun 130: 981-986

34. Trippodo, N.C., Ghai, R.D., MacPhee, A.A., Cole, F.E. (1984) Atrial natriuretic factor: Atrial conversion of high to low molecular weight forms. Biochem Biophys Res Commun 119: 282-288

35. Vlasuk, G.P., Miller, J., Bencen, G.H., Lewicki, J.A. (1986) Structure and analysis of the bovine atrial natriuretic peptide precursor gene. Biochem Biophys Res Commun 136: 39

36. Yamanaka, M., Greenberg, G., Johnson, L., Seilhamer, J., Brewer, M., Freedemann, T., Miller, J., Atlas, S., Laragh, J., Lewicki, J., Fiddes, J. (1984) Cloning and sequence analysis of cDNA for the rat atrial natriuretic factor precursor. Nature 309: 719-72

The physiological importance of Atrial Natriuretic Peptide

J. Schnermann, J. P. Briggs

Departments of Physiology and Medicine, The University of Michigan Medical School, Ann Arbor, Michigan, USA.

Introduction

The enthusiasm with which the discovery of atrial natriuretic peptides (ANP) was met originated in part from the fact that their physiological role appeared to be immediately understandable. The impressive natriuretic potency of ANP together with its location in the cardiac atria suggested that ANP might be responsible for the increase in salt and water excretion known to result from an increase in atrial chamber pressure (25,26). Intense research performed over the past few years in many laboratories now suggests that this view is probably far too limited. Expression of ANP synthesis has been found in a number of tissues other than the cardiac atria and receptor binding studies have indicated target organs other than the kidneys. It is therefore safe to predict that our views of the physiological role of ANP are likely to require adaptations over the coming years. Nevertheless, at this time it seems reasonable to limit the discussion of the physiological importance of ANP to the original concept that it participates in body fluid homeostasis and blood pressure regulation. Figure 1 is a schematic representation of the functional role of ANP as far as it is discussed in this overview.

Regulation of total body sodium

A. ANP release. Since the early work of Lang et al. (40) it has been repeatedly confirmed that an acute expansion of the extracellular space by the infusion of saline or blood is associated with an increase in the plasma concentration of ANP (51,60). Most experimental evidence would indicate that ANP release is predominantly under the direct control of atrial pressure or some derivative thereof such as atrial wall distension. *In vitro* studies have shown that both the degree and the rate of atrial stretch determine the magnitude of ANP release (6). *In vivo*, an increase in atrial pressure by mitral valve stenosis or balloon inflation stimulates ANP release leading to an increase in ANP concentration in coronary sinus and peripheral blood (27,42). It is to be noted that, in general, studies of endogenous ANP release have used rather massive stimuli, large degrees of volume expansion or large shifts in fluid volumes. It will be critical to confirm that ANP release is also stimulated in response to small changes in extracellular volume like those produced, for example, by the ingestion of a salt-rich meal (60). Studies in dogs show an increase in plasma ANP by about 30 pg/ml for each mmHg increase in right atrial pressure and indicate that ANP release may in fact be extremely sensitive to changes in atrial pressure (27,50). While the precise sensitivity of ANP release to acute changes in fluid volumes still needs to be defined, the available evidence has corroborated the original hypothesis that an acute increase in extracellular volume

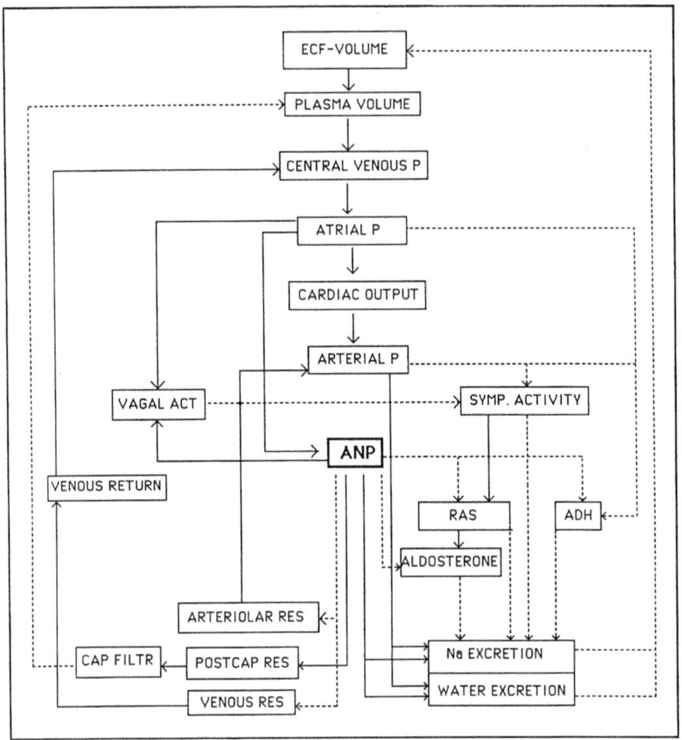

Fig. 1. Schematic representation of the role of atrial natriuretic peptides (ANP) in the regulation of Na and water excretion and of blood pressure. The relationship between two variables is direct when they are connected by solid lines and inverse when they are connected by broken lines. An uneven number of broken lines indicates a negative or homeostatic feedback relationship. The likely participation of a central nervous ANP system in some of the relationships is not indicated in this diagram.

leads to an increase in atrial pressure and an increased release of ANP into the circulation. In contrast to acute volume changes it appears that ANP release is not stimulated during long-term volume expansion induced by saline infusion or high salt diet administration over several days or weeks (44,51,54). On the other hand, in patients with congestive heart failure or in the cardiomyopathic hamster chronic elevations of plasma ANP have been found (3,5,19,21,54). In fact, ANP release per unit increase in atrial pressure may even be increased (5,19). These results indicate that during the administration of a high salt diet the atrial distension threshold for ANP release may adapt, but that pathological increments in atrial pressure are associated with increased ANP release and elevated ANP synthesis in both cardiac atria and ventricles (21).

B. Direct renal effects. Another critical step in validating the concept of a volume regulatory role of ANP is to establish that ANP is actually natriuretic at physiological concentrations. Several studies have examined the relationship between ANP infusion

114

rate and plasma ANP concentration, and it has become clear that the initial studies of the renal effects all have used pharmacological doses (27,60,63). Recent studies in our laboratory have examined the threshold infusion rate for the natriuretic action of ANP and measured the change in plasma ANP levels induced by threshold infusions (54a). We found that the threshold infusion rate in anesthetized rats lies between 10 and 20 ng/kg. min. This infusion is associated with a 1.6-fold increase in plasma ANP. Since a change in plasma ANP of this magnitude can be achieved by physiological stimuli such as an increase in atrial pressure, we conclude that variations in endogenous ANP are sufficiently large to induce a modest natriuresis. Additional results suggest that the physiological threshold may be even lower. In slightly volume-expanded animals, a significant natriuresis was produced by an infusion rate (10ng/kg·min) which was without effect in control rats. Thus, in volume expansion, the condition where endogenous ANP acts, smaller changes in plasma ANP may be required to induce natriuretic effects. In dogs, natriuresis in response to volume expansion was about twice as great than in response to exogenous ANP even though plasma ANP was the same in the two conditions (36). The opposite, a dramatic right shift in the threshold, has been observed during reductions in renal perfusion pressure (16,57). Shifts in the sensitivity of natriuresis to exogenous ANP are probably the consequence of simultaneous changes in the degree of stimulation or suppression of other factors affecting renal salt handling such as a renal perfusion pressure, the renin-angiotensin-aldosterone system, and the sympathetic tone. It is noteworthy that low dose infusion of ANP does not persistently alter renal potassium excretion, an appropriate property of a hormone involved in volume homeostasis (7,60,63).

Although the natriuretic effect of acute elevations of plasma ANP appears well established, chronic increases of ANP plasma levels do not seem to be associated with sustained natriuresis. For example, infusion of ANP over several days was accompanied by unchanged Na excretion (29). The mechanism for the failure of ANP to alter Na excretion during chronic elevations of plasma levels is unclear. The finding is consistent, however, with the observations in congestive heart failure, where resistance to ANP-induced natriuresis clearly develops (38).

The mechanism of the natriuretic action of ANP has stirred substantial debate. ANP induces well defined hemodynamic alterations in the kidney, and some investigators have suggested that these alterations are responsible for the natriuretic properties of ANP (33,45). The most commonly observed pattern is an increase in glomerular filtration rate with little or no change in renal plasma flow (7,14,22,45,55,61). The changes in vascular resistance underlying these effects have been clearly identified. ANP administration is associated with dilatation of preglomerular vessels and, particularly at higher doses, with constriction of postglomerular vessels (22,46). As a consequence of this rather unique dual action, total resistance and renal blood flow remain essentially unaltered while GFR and filtration fraction increase. The main cause for the increase in GFR is the marked rise in glomerular capillary pressure (22,53).

In a number of studies ANP has been shown to induce natriuresis without changes in GFR (9,55). In general with low "physiological" doses hemodynamic effects appear to be either less pronounced or absent (7,9). Natriuresis in the absence of changes in GFR indicates, of course, a reduction in the tubular absorption of water and salt. It is unclear if ANP affects fluid absorption in the proximal tubule. Micropuncture studies have not detected transport inhibition along the proximal convolution of superficial nephrons (9,33,55). Since fractional lithium clearance increases with ANP (11), an effect on proximal absorption of deep nephrons seems feasible (49). While basal transport in superficial proximal tubules may not be altered by ANP, recent experiments of Harris et al. (30) show that increasing peritubular concentrations of ANP block angiotensin-induced stimulation of proximal transport. Since the renin-angiotensin system is suppressed in states of volume expansion this finding may not have major physiological importance. The most likely site of a direct inhibitory effect of ANP on fluid absorption is the inner medullary collecting duct, a site where high affinity ANP receptors have been unequivocally demonstrated (13). Sonnenberg et al. (56) first demonstrated inhibition of NaCl transport in this part of the nephron by microcatheterization and Zeidel et al. (64) showed that oxygen consumption of collecting duct cells in culture was reduced by ANP. The mechanism of action is probably related to a decrease in Na permeability of the luminal membrane which would result in a reduction of intracellular Na available for active transport. An inhibition by ANP of Na uptake through the amiloride-sensitive channel has been observed in a cultured renal cell line (12).

In addition to direct transport effects in the inner medullary collecting duct it is possible that changes in medullary hemodynamics and tissue composition contribute to natriuresis. Takezawa et al. (57) recently showed that ANP increases papillary blood flow of rats without changing cortical flow. This finding may explain the results of Davis and Briggs that ANP infusion is followed by a very rapid dissipation of the medullary solute gradient; after only five minutes of exposure medullary hypertonicity was completely abolished (17). Loss of medullary solutes could impair absorption of salt and water in the loop of Henle and could conceivably contribute to natriuresis. Whether medullary hemodynamic changes and the concomitant changes in tissue composition occur in the physiological dose range remains to be established.

C. Effects on other salt and water regulating hormones. The direct renal effects of ANP on salt and water excretion appear to be supplemented by the inhibition of release or formation of salt-retaining hormones. Several groups of investigators demonstrated almost simultaneously that ANP inhibits aldosterone production by adrenal zona glomerulosa cells (4,18,28). The inhibitory effect is most clearly demonstrable during stimulation of aldosterone production by angiotensin, cyclic AMP, ACTH, or potassium (4,18,28), but basal production may be inhibited as well (4,28). Even though inhibition is demonstrable at relatively low doses, the precise contribution to salt-water balance is unclear. In view of the delayed onset of aldosterone, effects on Na transport inhibition of aldosterone production by ANP may play a role only during more prolonged increases in plasma ANP.

A fall in plasma renin activity or renin secretion rate following ANP administration is also sometimes observed (11,45,61). Thus, the natriuretic effect of ANP may be augmented by a suppression of the salt-retaining renin-angiotensin system. In view of the multiple controls of renin release it is difficult to clearly identify the mechanism utilized by ANP to inhibit renin secretion (61). Recent studies in *in vitro* systems indicate that ANP might interact directly with juxtaglomerular cells. Kurtz et al. (39) showed that ANP in rather low concentrations inhibited the spontaneous release of renin from primary cultures of juxtaglomerular cells. ANP also inhibited the basal and isoproterenol-stimulated release of renin from renal cortical slices (31,48).

Lastly, the inhibitory effect of ANP on the release of vasopressin is a further example of the interrelationship between ANP and other hormones involved in body fluid homeostasis. Samson first demonstrated that ANP reduces the elevated plasma levels of vasopressin during hemorrhage and dehydration (52). The potency of ANP to reduce vasopressin secretion is most pronounced when ANP is administered into the intra-cerebroventricular space (35,43). This finding, together with the demonstration of ANP receptors and ANP synthesis in the central nervous system and especially in the hypothalamus (47), suggests that ANP involved in the regulation of vasopressin secretion is of central nervous origin rather than of cardiac. Whatever the mechanism, it appears that suppression of vasopressin release and action is an important prerequisite for maximizing the water excretory effects of ANP. We have noted that in water diuretic rats coadministration of vasopressin and ANP can completely block the increase in water excretion seen with the administration of ANP alone (10). Furthermore, studies by Dillingham and Anderson in isolated collecting tubules indicate that ANP may also reduce the vasopressin-stimulated increase in water permeability (20).

Regulation of arterial blood pressure

In addition to its natriuretic effect, ANP has been noted from the first investigations to reduce arterial blood pressure promptly and regularly. Initially, it appeared likely that this fall in arterial pressure was caused by a decrease in peripheral vascular resistance since several groups of investigators had noted that ANP is able to relax the smooth muscle of isolated blood vessels (15,24). However, subsequent work in intact animals beginning with the studies of Ackermann et al. (1) has clearly shown that a decrease in total peripheral resistance is not normally seen in response to ANP administration (2, 14,22,37). Only in conditions of elevated baseline resistance has ANP been reported to consistently reduce the tone of resistance vessels. For example, in two kidney-one clip Goldblatt hypertensive rats total peripheral resistance fell with ANP infusion but it increased in DOCA-salt hypertensive animals (62).

The dominant, if not the sole reason, for the pressure reduction appears to be a fall in cardiac output (1,2,8,14,37). In conscious rats with chronically implanted catheters, Allen and Gellai have shown that ANP reduced arterial pressure, a change accompanied by a fall in cardiac output with unchanged or slightly increased peripheral resistance (2). The rise in the tone of resistance vessels is probably an expression of reflex activation of sympathetic efferent nerves. During sinoaortic denervation peripheral sympathetic nerve activity has been shown to decrease in response to ANP accompanied by a pronounced fall in blood pressure and heart rate (58).

The mechanism for the reduction in cardiac output is not completely clear. There is no convincing experimental evidence in support of the possibility that ANP has negative inotropic effects (14,32,41). Therefore, a reduced venous return is the most likely reason for the change in cardiac output. It has been suggested that ANP induces a preferential dilatation of large central veins, thereby increasing venous pooling of blood and reducing venous return (8,41). Alternatively, venous return may decrease because of a reduction in circulating plasma volume. This possibility is supported by the observation of several groups that ANP administration is followed by an increase in arterial hematocrit (1,22,

23,41,60,63). Since plasma volume fell also in nephrectomized animals it is not simply a consequence of the increased renal salt and water excretion (23,59). It is possible that ANP elevates postcapillary resistance and thereby promotes filtration in the capillary bed (59). This possibility is attractive since receptors coupled to vasoconstrictor events appear to be also present in renal post-glomerular vessels, another postcapillary vascular bed (22,46). In addition, studies in frog mesenteric capillaries have shown that ANP may also increase capillary hydraulic conductivity (34).

In view of the fall in arterial pressure it is remarkable, that at least in rats and dogs, heart rate often does not change during acute ANP infusion (2,22,23,37,45,58,60). The expected increase in heart rate and in peripheral sympathetic activity was noted when rats were vagotomized (1,58). Since atropine did not alter the acute ANP effect on heart rate and sympathetic activation it appears that ANP stimulates afferent vagal fibers which, by inhibiting sympathetic nervous output, counteract the sympathetic stimulation through baroreceptors (1,58). In studies on conscious rats prolonged infusion of ANP over 150 min. induced marked bradycardia later in the course of the experiment (2). Since atropine prevented the occurrence of late bradycardia the authors suggest that ANP administration may also be associated with stimulation of vagal efferent nerves (2). It is to be noted that in all studies related to the evaluation of circulatory effects of ANP, relatively high doses of exogenous ANP were applied. Thus, it is not possible at this time to determine with certainty to what extent circulatory adjustments are part of the response to acute changes in plasma ANP in the physiological concentration range. Further complicating this issue is the possibility that regulation of blood pressure may be a function of central nervous system actions of ANP. Thus, conclusions drawn from peripheral administration may not be fully justified.

Conclusion

Expansion of the extracellular space and subsequently, plasma volume, stimulates the secretion of ANP, an effect mediated predominantly by the increase in atrial pressure. ANP affects renal salt and water excretion directly by changes in glomerular filtration rate and by inhibition of salt absorbtion in the medullary collecting duct. ANP may also affect salt and water excretion indirectly by inhibiting the release and formation of salt-water retaining hormones such as aldosterone, angiotensin, and vasopressin.

Regulation of vascular filling pressures by direct vascular actions may be an additional homeostatic function of ANP. It may reduce peripheral resistance when baseline resistance is high. More importantly, it may produce venodilatation and reduce plasma volume, two events which, probably in conjunction with vagal activation and sympathetic inhibition, would decrease atrial pressure, cardiac output, and arterial pressure. The combination of renal and circulatory effects suggests that ANP plays an important physiological role as part of the mechanisms responsible for homeostasis of total body sodium and blood pressure.

References

1. Ackermann, U., rizawa, T.G., Milojevic, S., Sonnenberg, H. (1984) Cardiovascular effects of atrial extracts in anesthetized rats. Can J Physiol Pharm 62: 819-826
2. Allen, D.E., Gellai, M. (1987) Cardioinhibitory effects of atrial peptide in conscious rats. Am J Physiol 252: R610-R616
3. Arendt, R.M., Gerbes, A.L., Ritter, D., Stangl, E., Bach, P., Zaehringer, J., (1986) Atrial natriuretic factor in plasma of patients with arterial hypertension, heart failure or cirrhosis of the liver. J Hypertension 4 (suppl 2): S131-S135
4. Atarashi, K., Mulrow, P.J., Franco-Saez, R., Snajdar, R., Rapp, J. (1984) Inhibition of aldosterone production by an atrial extract. Science 224: 992-994
5. Bates, E.R., Shenker, Y., Grekin, R.J. (1986) The relationship between plasma levels of immunoreactive atrial natriuretic hormone and hemodynamic function in man. Circulation 73: 1155-1161
6. Bilder, G.E., Schofield, T.L., Blaine, E.H. (1986) Release of atrial natriuretic factor. Effects of repetitive stretch and temperature. Am J Physiol 251: F817-F821
7. Biollaz,J., Nussberger, J., Porchet, M., Brunner-Ferber, F., Otterbein, E.S., Gomez, H., Waeber, B., Brunner, H.R., (1986) Four-hour infusions of synthetic atrial natriuretic peptide in normal volunteers. Hypertension 8 (suppl II) : II-96-II-105
8. Breuhaus, B.A., Saneii, H.H., Brandt, M.A., Chimoskey, J.E. (1985) Atrio- peptin II lowers cardiac output in conscious sheep. Am J Physiol 249: R776-R780
9. Briggs, J.P., Steipe, B., Schubert, G., Schnermann, J. (1982) Micropuncture studies of the renal effects of atrial natriuretic substance. Pflügers Arch 395: 271-276
10. Briggs, J.P., Schnermann, J. (1987) Vasopressin dissociates the responses of Na and water excretion toatrialnatriuretic factor in water diuretic rats. In: Brenner B.M., Laragh J.E., Eds. Biologically Active Atrial Peptides. Raven Press, New York
11. Burnett, J.C., Granger, J.P., Opgenorth, T.J. (1984) Effect of synthetic atrial natriuretic factor on renal function and renin release. Am J Physiol 247: F863-F867
12. Cantiello, H.F., Ausiello, D.A. (1986) Atrial natriuretic factor and cGMP inhibit amiloride-sensitive Na transport in the cultured renal epithelial cell line, LLC-PK1. Biochem Biophys Res Comm 134: 852-860
13. Chai, S.Y., Sexton, P.M., Allen, A.M., Figdor, R., Mendelsohn, F.A.O. (1986) In vitro autoradiographic localization of ANP receptors in kidney and adrenal gland. Am J Physiol 250: F753-F757
14. Criscione, L., Burdet, R., Haenni, H., Kamber, B., Truog, A., Hofbauer, K.G. (1987) Systemic and regional hemodynamic effects of atriopeptin II in anesthetized rats. J Cardiovasc Pharm 9: 135-141
15. Currie, M.G., Geller, D.M., Cole, B.R., Boylan, J.G., Sheng, W.Y., Holmberg, S.W., Needleman, P. (1983) Bioactive cardiac substances: potent vasorelaxant activity in mammalian atria. Science 221: 71-73
16. Davis, C.L., Briggs, J.P. (1987) Effect of reduction in renal artery pressure on atrial natriuretic peptide-induced natriuresis. Am J Physiol 252: F146-F153
17. Davis, C.L., Briggs, J.P. (1987) Effect of the atrial natriuretic peptides on renal medullary solute gradients. Am J Physiol 253: F679-F684
18. DeLean, A., Racz, K., Gutkowska, J., Nguyen, T-T., Cantin, M., Genest, J. (1984) Specific receptor-mediated inhibition by synthetic atrial natriuretic factor of hormone-stimulated steroidogenesis in cultured bovin adrenal cells. Endocrinology 115: 1636-1638

19. Dietz, R., Purgaj, J., Lang, R.E., Schoemig, A. (1986) Pressure dependent release of atrial natriuretic peptide (ANP) in patients with chronic cardiac disease: does it reset? Klin Wochenschr (Suppl VI) : 42-46
20. Dillingham, M.A., Anderson, R.J. (1986) Inhibition of vasopressin action by atrial natriuretic factor. Science 231: 1572-1573
21. Ding, J., Thibault, G., Gutkowska, J., Garcia, R., Karabatsos, T., Jasmin, G., Genest, J., Cantin, M. (1987) Cardiac and plasma atrial natriuretic factor in experimental congestive heart failure. Endocrinology 121: 248-257
22. Dunn, B.R., Ichikawa, I., Pfeffer, J.M., Troy, J.L., Brenner, B.M. (1986) Renal and systemic hemodynamic effects of synthetic atrial natriuretic peptide in the anesthetized rat. Circ Res 59: 237-246
23. Flueckiger, J.P., Waeber, B., Matsueda, G., Delaloye, B., Nussberger, J., Brunner, H.R. (1986) Effect of atriopeptin III on hematocrit and volemia of nephrectomized rats. Am J Physiol 251: H880-H883
24. Forssmann, W.G., Hock, D., Lottspeich, F., Henschen, A., Kreye, V., Christmann, M., Reinecke, M., Metz, J., Carlquist, M., Mutt, V. (1983) The right auricle of the heart is an endocrine organ. Anat Embryol 168: 307-313
25. Gauer, O.H., Henry ,J.P., Reeves, J.L. (1956) Evidence of the atrial location of receptors influencing urine flow. Circ Res 4: 85-90
26. Gauer, O.H., Henry, J.P. (1976) Neurohormonal control of plasma volume. In: Guyton AC, Cowley AW (eds) International Review Physiology, Cardiovascular Physiology II, Volume 9: 145-190. Universtity Park Press, Baltimore
27. Goetz, K.L., Wang, B.C., Geer, P.G., Leadley, R.J., Reinhardt, H.W., (1986) Atrial stretch increases sodium excretion independently of release of atrial peptides. Am J Physiol 250: R946-R950
28. Goodfriend, T.L., Elliot, M., Atlas, S.A. (1984) Actions of synthetic atrial natriuretic factor on bovine adrenal glomerulosa. Life Sci 35: 1675-1682
29. Granger, J.P., Opgenorth, T.J., Salazar, J., Romero, J.C., Burnett, J.C. (1986) Long-term hypotensive and renal effects of atrial natriuretic peptide. Hypertension 8 (Suppl II) : 11-112-11-116
30. Harris, P.J., Thomas, D., Morgan, T.O. (1987) Atrial natriuretic peptide inhibits angiotensin-stimulated proximal tubular sodium and water reabsorption. Nature 326: 697-698
31. Henrich, W.L., Needleman, P., Campbell, W.B. (1986) Effect of atriopeptin III on renin release *in vitro*. Life Sci 39: 993-1001
32. Hiwatari, M., Satoh, K., Angus, J.A., Johnston, C.I., (1986) No effect of atrial natriuretic factor on cardiac rate, force, and transmitter release. Clin Exp Pharm Physiol 13: 163-168
33. Huang, C.L., Lewicki, J., Johnson, L.K., Cogan, M.G. (1985) Renal mechanism of action of rat atrial natriuretic factor. J. Clin Invest 75: 769-773
34. Huxley, V.H., Tucker, V.L., Verburg, K.M., Freeman, R.H. (1987) Increased capillary hydraulic conductivity induced by atrial natriuretic peptide. Circ Res 60: 304-307
35. Iitake, K., Share, L., Crofton, J.T., Brooks, D.P., Ouchi, Y., Blaine, E.H. (1986) Central atrial natriuretic factor reduces vasopressin secretion in the rat. Endocrinology 119: 438-440
36. Khraibi, A.A., Granger, J.P., Burnett, J.C., Walker, K.R., Knox, F.G. (1987) Role of atrial natriuretic factor in the natriuresis of acute volume expansion. Am J Physiol 252: R921-R924

37. Kleinert, H.D., Volpe, M., Odell, G., Marion, D., Atlas, S.A., Camergo, M.J., Laragh, J.E., Maack, T. (1986) Cardiovascular effects of atrial natriuretic factor in anesthetized and conscious dogs. Hypertension 8: 312-316

38. Koepke, J.R., DiBona, G.F. (1987) Blunted natriuresis to atrial natriuretic peptide in chronic sodium-retaining disorders. Am J Physiol 252: F865-F871

39. Kurtz, A., Della Bruna, R., Pfeilschifter, J., Bauer, C. (1986) Atrial natriuretic peptide inhibits renin release from juxtaglomerular cells by a c-GMP mediated process. Proc Natl Acad Sci USA 83: 4769-4773

40. Lang, R.E., Thoelken, H., Ganten, D., Luft F.C., Ruskoaho, H., Unger, T.H. (1985) Atrial natriuretic factor is a circulating hormone stimulated by volume loading. Nature 314: 264-266

41. Lappe, R.W., Smith, J.F.M., Todt, J.A., Debets J.J.M., Wendt, R.L. (1985) Failure of atriopeptin II to cause arterial vasodilation in conscious rats. Circ Res 56: 606-612

42. Ledsome, J.R., Wilson, N., Rankin, A.J., Courneya, C.A. (1986) Time course of release of atrial natriuretic peptide in the anaesthetized dog. Can J Physiol Pharm 64: 1017-1022

43. Lee, J., Feng, J.Q., Malvin, R.L., Huang, B-S., Grekin, R.J. (1987) Centrally administered atrial natriuretic factor increases renal water excretion. Am J Physiol 252: F1011-F1015

44. Luft, F.C., Sterzel, R.B., Lang, R.E., Trabold, E.M., Veelken, R., Ruskoaho, H., Gao, Y., Ganten, D., Unger, T. (1986) Atrial natriuretic factor determinations and chronic sodium homeostasis. Kidney Int 29: 1004-1010

45. Maack, T., Marion, D.N., Camargo, M.J.F., Kleinert, H.D., Laragh, J.H., Vaughan, D., Atlas, S.A. (1984) Effects of auriculin (atrial natriuretic factor) on blood pressure, renal function, and the renin-aldosterone system in dogs. Am J Med 77: 1069-1075

46. Marin-Grez, M., Fleming, J.T., Steinhausen, M. (1986) Atrial natriuretic peptide causes pre-glomerular vasodilatation and post-glomerular vasoconstriction in rat kidney. Nature 324: 473-476

47. Nakao, K., Morii, N., Yamada, T., Shiono, S., Sugawara, A., Saito, Y., Mukoyama, M., Arai, H., Sakamoto, M., Imura, H. (1987) Atrial natriuretic polypeptide in brain - implication of central cardiovascular control. Klin Wochenschr 65 (Suppl. VIII): 103-108

48. Obana, K., Naruse, M., Naruse, K., Sakurai, H., Demura, H., Inagami, T., Shizume, K. (1985) Synthetic atrial natriuretic factor inhibits *in vitro* and *in vivo* renin secretion in rats. Endocrinology 117: 1282-1284

49. Roy, D.R. (1986) Effect of synthetic ANP on renal and loop of Henle functions in the young rat. Am J Physiol 251: F220-F225

50. Salazar, F.J., Granger, J.P., Joyce, M.L.M., Burnett, J.C., Bove, A.A., Romero, J.C. (1986) Effects of hypertonic saline infusion and water drinking on atrial peptide. Am J Physiol 251: R1091-R1094

51. Salazar, F.J., Romero, J.C., Burnett, J.C., Schryver, S., Granger, J.P. (1986) Atrial natriuretic peptide levels during acute and chronic saline loading in conscious dogs. Am J Physiol 251: R499-R503

52. Samson, W.K. (1985) Atrial natriuretic factor inhibits dehydration and hemorrhage-induced vasopressin release. Neuroendocrinology 40: 277-279

53. Schnermann, J., Marin-Grez, M., Briggs, J.P. (1986) Filtration pressure response to infusion of atrial natriuretic peptides. Pflügers Arch 406: 237-239

54. Shenker, Y., Sider, R.S., Ostafin, E.A., Grekin, R.J. (1985) Plasma levels of immunoreactive atrial natriuretic factor in healthy subjects and in patients with edema. J Clin Invest 76: 1684-1687

54a. Soejima, H., Grekin, R.J., Briggs, J.P., Schnermann, J. (1988) Renal response of anasthetized rats to low-dose infusion of atrial natriuretic peptide. Am J Physiol (in press)

55. Sonnenberg, H., Cupples, W.A., DeBold, A.J., Veress, A.T. (1982) Intrarenal localization of the natriuretic effect of cardiac atrial extract. Can J Physiol Pharm 60: 1149-1152

56. Sonnenberg, H., Honrath, U., Chong, C.K., Wilson, D.R. (1986) Atrial natriuretic factor inhibits sodium transport in medullary collecting duct. Am J Physiol 250: F963-F966

57. Takezawa, K., Cowley, A.W., Skelton, M., Roman, R.J. (1987) Atriopeptin III alters renal medullary hemodynamics and the pressure-diuresis response in rats. Am J Physiol 252: F992-F1002

58. Thoren, P., Mark, A.L., Morgan, D.A., O'Neill, T.P., Needleman, P., Brody, M.J. (1986) Activation of vagal depressor reflexes by atriopeptins inhibits renal sympathetic nerve activity. Am J Physiol 251: H1252-H1259

59. Trippodo, N.C., Barbee, R.W. (1987) Atrial natriuretic factor decreases whole-body capillary absorption in rats. Am J Physiol 252: R915-R920

60. Verburg, K.M., Freeman, R.H., Davis, J.O., Villareal, D., Vari, R.C. (1986) Control of atrial natriuretic factor release in conscious dogs. Am J Physiol 251: R947-R956

61. Villareal, D., Freeman, R.H., Davis, J.O., Verburg, K.M., Vari, R.C. (1986) Renal mechanism for suppression of renin secretion by atrial natriuretic factor. Hypertension 8 (suppl II): II-28 - II-35

62. Volpe, M., Sosa, R.E., Mueller, F.B., Camargo, M.J.F., Glorioso, N., Laragh, J.H., Maack, T., Atlas, S.A. (1986) Differing hemodynamic responses to atrial natriuretic factor in two models of hypertension. Am J Physiol 250: H871-H878

63. Weidmann, P., Hasler, L., Gnaedinger, M.P., Lang, R.E., Uehlinger, D.E., Shaw, S., Rascher, W., Reubi, F.C. (1986) Blood levels and renal effects of atrial natriuretic peptide in normal man. J Clin Invest 77: 734-742

64. Zeidel, M.L., Seifter, J..L, Lear, S., Brenner, B.M., Silva, P. (1986) Atrial peptides inhibit oxygen consumption in kidney medullary collecting duct cells. Am J Physiol 251: F379-F383

Pharmacology of the cardiac endocrine system

M. G. Currie[1], C. E. Lofton[1], H. Schomer[1], W. H. Newman[1], R. W. Wrenn[2], W. F. Oehlenschlager[1]

[1]Department of Cell and Molecular Pharmacology and Experimental Therapeutics, Medical University of South Carolina, Charleston, South Carolina, USA, [2]Department of Anatomy, Medical College of Georgia, Augusta, Georgia, USA

Introduction

Over the past six years, data and information have accumulated which indicate that the atrial endocrine system and its secreted bioactive peptide(s) are important participants in the regulation of blood pressure and fluid and electrolyte homeostasis (21,12). The nomenclature for atrial peptides includes atrial natriuretic factor, cardionatrin, auriculin, cardiodilatin and the term that we developed, atriopeptin. For the sake of clarity, we will use the terms preproatriopeptin (preproAP), proatriopeptin (proAP), and atriopeptin (AP), when referring to the family of atrial peptides.

Atriopeptin is clearly a hypotensive hormone that exerts its effect on blood pressure through concerted actions on the kidney, pressor endocrine systems, the cardiovascular system, and the brain. To enhance our understanding of the pharmacology of atriopeptin, we have developed the following model (Fig. 1): Increased pressure results in a stretching of the atrial myocyte which in turn stimulates the release of atriopeptin; the circulating peptide then acts on the kidney to elicit a diuresis and natriuresis on pressor endocrine systems (e.g., renin-angiotensin-aldosterone and antidiuretic hormone) to inhibit synthesis and/or release pressor agents and promotors of fluid retention on the cardiovascular system to cause vasodilation and decreased cardiac output, and on the central nervous system to decrease thirst and salt appetite. The result of the pharmacological actions of atriopeptin on this diverse set of end organs is a reduction of plasma volume and blood pressure.

Many recent studies have been aimed at defining the role of atriopeptin in normal and various cardiovascular disease states. Our approach to understanding the cellular and molecular pharmacology of atriopeptin has been to focus on the regulation of the atrial endocrine system. On the basis of our work and that of others, we believe that the activity of this system is largely regulated by the following: 1) alterations in atriopeptin gene expression in the atria and ventricles; 2) post-translational modifications of proatriopeptin and atriopeptin (e.g., phosphorylation) that may affect bioactivity; 3) proteolytic processing of proatriopeptin to atriopeptin(s); 4) factors that influence the rate of secretion of the peptide; and 5) alterations in the expression of atriopeptin receptors. This concept of factors acting at several points or levels to regulate the activity of the atrial endocrine system is illustrated in Fig. 2. We believe that studying these regulatory sites may facilitate a better understanding of the physiology of this system, and also that this forthcoming information may be utilized to adapt strategies for pharmacological manipulation and experimental therapeutics.

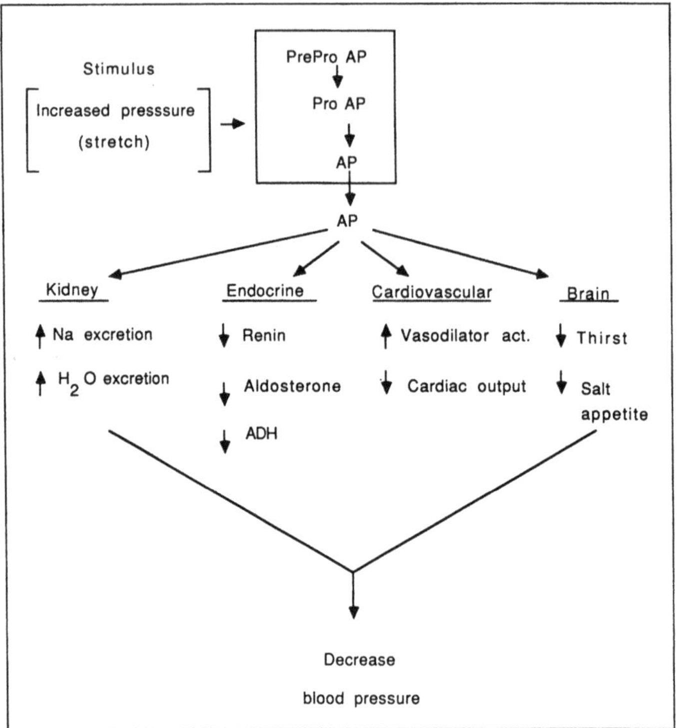

Fig. 1. Model of the role of the atrial endocrine system in the control of blood pressure

Regulation of the atriopeptin gene

The atriopeptin (AP) gene has been cloned and sequenced in several species including human and rat (18). The gene is believed to be present in a single copy which is comprised of three exons separated by two introns. Although the AP gene has been well-characterized structurally, little is known about its regulation. Analysis of the human gene sequence suggests the presence of a single, putative glucocorticoid receptor binding site within the second intron (18). Administration of pharmacological doses of dexamethasone to rats elicited a twofold increase in atriopeptin mRNA levels in cardiac tissue (14). This study supports a role for glucocorticoids in the regulation of the rat gene; however, evidence for a direct effect of glucocorticoids on the AP gene in any species is lacking.

Based on the postulated role of the atrial endocrine system in the regulation of fluid and electrolyte homeostasis, one might expect the states of salt and water balance to affect AP gene expression. Indeed, several groups have examined the effects of manipulations of salt and water balance on AP mRNA levels. Water deprivation (2-4 days) was shown to decrease atrial AP mRNA levels in rats to half the control levels (20). In contrast, elevated cardiac levels of AP mRNA were seen in rats subjected to DOCA-salt treatment,

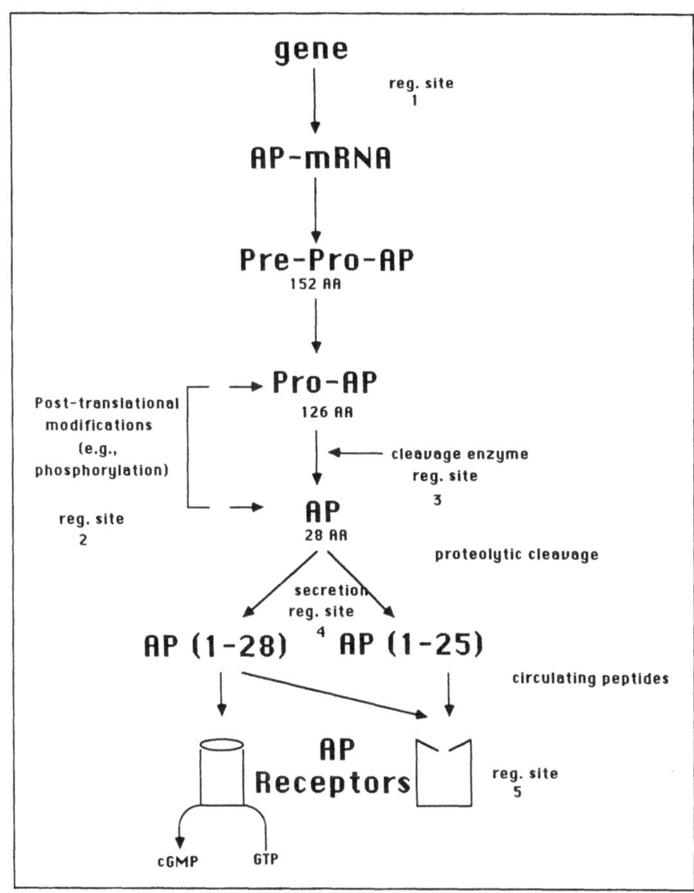

Fig. 2. Schematic depicting potential regulatory sites of the cardiac endocrine system

aorta-caval fistula, binephrectomy, and ureteral ligation. Thus, it appears that this system can respond to changes in salt and water balance by altering the level of expression of the AP gene.

Since the atrial endocrine system may play a role in the regulation of blood pressure, it is also plausible that changes in blood pressure influence AP gene expression. Elevated cardiac AP mRNA levels have been found in models of systemic and pulmonary hypertension. In the aortic-banded rat, a model of systematic hypertension, increased levels of AP mRNA were detected in the left ventricle (11). We have found increased levels of AP mRNA in both atrial and ventricular tissue from the monocrotaline-treated rat, a model of pulmonary hypertension (22). In both models the greatest enhancement of AP gene expression appears to be associated with the site of pressure overload and cardiac hypertrophy. Furthermore, we have observed elevated AP mRNA levels in the hypertrophied atria and ventricles of the BIO 14.6 hamster, a model of congestive heart failure and cardiac hypertrophy (10). Collectively, these studies suggest that the activity of the atrial endocrine system is markedly increased in various cardiovascular disease states by: 1) enhanced expression of the AP gene in the atria and, 2) recruitment of ventricular tissue for AP synthesis.

The signal for increased atrial and ventricular expression of the AP gene in response to sustained elevations of systemic and/or pulmonary arterial pressure is unknown. We propose that stretch of the cardiac myocyte may serve as a signal for enhanced expression of this gene. Acute stretch has been shown to stimulate the secretion of the peptide from the atria (12,21). It is plausible that chronic stretch may not only stimulate secretion of AP, but may also induce its gene. This mechanism would serve to maintain an adequate store of the hormone in the face of a chronic stimulus for secretion.

Post-translational modifications of the atrial peptides

Post-translational modifications of the atrial peptides constitute one of the least studied potential regulatory sites of the cardiac endocrine system. Initial *in vitro* studies demonstrated that atriopeptin is a high affinity substrate for the cAMP-dependent protein kinase (23). The resulting phosphorylation of atriopeptin enhanced its bioactivity as measured by Na-K-2C cotransport in cultured vascular smooth muscle cells (23). The relevance of the phosphorylation of synthetic atriopeptin by the cAMP-dependent protein kinase to the activity of the cardiac endocrine system remains to be determined. Efforts to detect phosphorylation of atrial peptides in cultured atrial myocytes have revealed the phosphorylation of proatriopeptin, but not of atriopeptin (2). Recently, we observed a similar profile of phosphorylation in *in vitro* studies utilizing protein kinase C (Wrenn and Currie, unpublished observation). The phosphorylation of proatriopeptin may represent another site at which protein kinase C influences the activity of the cardiac endocrine system, since activators of this enzyme have already been shown to enhance atriopeptin secretion (9).

Proteolytic processing of proatriopeptin

Initial purification efforts revealed the presence of two major forms of the atrial peptide - a higher molecular weight form and a lower molecular weight form (4). Treatment of the higher molecular weight form with serine proteases (trypsin, kallikrein) yields the lower molecular weight form (5,6). Concomitant with this conversion from higher to lower molecular weight is an enhancement of bioactivity. These studies suggest a precursor-product relationship between the two forms of the peptide. This relationship was confirmed by amino acid sequence analysis which revealed that the lower molecular weight form, a 28 amino acid peptide (atriopeptin) is derived from the C-terminus of the higher molecular weight form, a 126 amino acid peptide (proatriopeptin) (12,21). In contrast to other peptide endocrine systems in which the active hormone is the predominant storage form, the atrial endocrine system utilizes the prohormone (proatriopeptin) for this purpose. However, atriopeptin, not proatriopeptin, has been shown to be the major form present in normal plasma (25) and in the effluent of isolated perfused hearts (15)

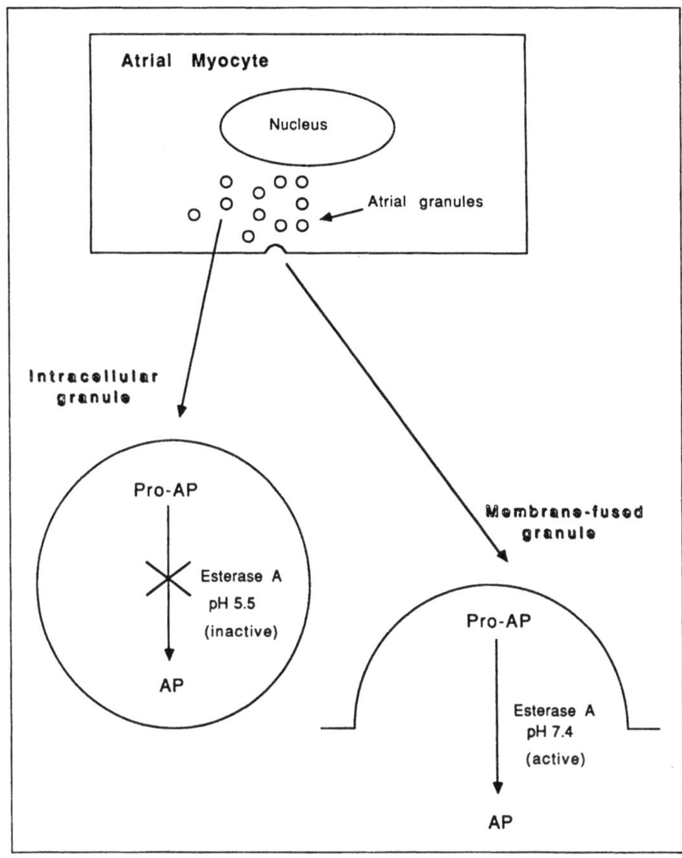

Fig. 3.   Model of the proteolytic processing of proatriopeptin by cardiac esterase A

of all species examined. This strongly suggests that the proteolytic processing of pro-atriopeptin to atriopeptin occurs within the heart during, or shortly after, the secretory event.

The cleavage of proatriopeptin to atriopeptin has been shown to be a site of activation of the hormone (5,6). The clinical importance of this regulatory site in the cardiac endocrine system is reflected by recent evidence indicating that a defect in this proteolytic processing may contribute to the pathophysiology of cardiovascular disease. In the plasma of control subjects, the predominant form of the peptide was found to be atriopeptin with insignificant levels of proatriopeptin. But hypertensive, cirrhotic, and congestive heart failure patients were shown to have considerable amounts of circulating proatriopeptin (1). The protease that performs this function has not been identified; however, preliminary evidence indicates that arginine esterase A may function as the cleavage enzyme (26). Esterase A, which has been purified from rat urine, is a serine protease and a member of the kallikrein gene family (3). Characterization of purified esterase A showed that the enzyme has an alkaline pH optimum and consists of two subunits with molecular weights of 21,000 and 10,000 (3). Specific monoclonal antibodies to esterase A have been developed that do not cross-react with other members of the kallikrein gene family despite the high degree of homology among these proteases. Immunocytochemical studies using a monoclonal antibody to esterase A reveal a co-localization of esterase A and proatriopeptin in rat atrial granules (26). Furthermore, we have preliminary evidence from HPLC analyses that esterase A cleaves proatriopeptin to produce the 2 amino acid, atriopeptin. Based upon these studies, we propose the following model for the cleavage of proatriopeptin to atriopeptin (Fig. 3): 1) Proatriopeptin and esterase A are both stored in the secretory granule, but cleavage is minimal because the protease is relatively inactive at the expected acidic pH of the granules; and, 2) esterase A catalyzed cleavage of proatriopeptin to atriopeptin occurs at time of secretion when the pH changes from that of the granule (pH = 5.5) to that of the extracellular space (pH = 7.4). This model should provide direction for pursuing future studies of the proteolytic conversion of proatriopeptin.

Studies examining the secretion of atriopeptin

We have utilized the isolated perfused rat heart as a model to study the secretion of atriopeptin. The problems that we have addressed with this model include: A) characterization of the molecular form(s) released by the heart; B) regulation of atriopeptin release; and, C) the effect of experimental diabetes on the release of atriopeptin. Each of these problems is addressed in the following sections.

A. Characterization of the molecular form(s) of atriopeptin released by the heart

Our initial studies addressed the question of what is the molecular form of the hormone released by the atria of the heart. Utilizing *in vitro* smooth muscle bioassays for quantitative measurement and HPLC for analysis of the form(s) of the atrial peptide released; we found that the isolated perfused heart releases predominantly a lower molecular weight form of the peptide (7). This is despite the fact that proatriopeptin is the predominant form stored in the atrial granules. A more detailed characterization of the peptide(s) released by the heart utilizing radioimmunassay for quantitation, HPLC for purification, and amino acid sequencing for structural analysis, indicates that perfused heart predominantly releases two lower molecular weight atriopeptins and less than 3% proatriopeptin. Determination of the amino acid sequences of the released atriopeptins indicates that the two peptides are a 28 amino acid peptide and a carboxyl-terminal truncated form of this peptide thought to consist of 25 amino acids (15). Definitive determination of the amino acid sequence of the truncated form will require carboxyl-terminal sequencing.

128

B. Regulation of atriopeptin release

Studies of the effect of adrenergic agents on atriopeptin release by the isolated perfused rat heart showed that norepinephrine, an alpha and beta adrenergic agonist, is an effective stimulus (8). Isoproterenol, a beta specific agonist, failed to elicit a response; however, the alpha-1 receptor agonist, phenylephrine, induced release in a dose-dependent manner. The stimulation of atriopeptin secretion by both norepinephrine and phenylephrine was inhibited by alpha-adrenergic blockade with phentolamine. Administration of BHT-920, a selective alpha-2 agonist, produced no effect. These data indicate that alpha-1 receptor activation enhances the secretion of atriopeptin by the atrial myocyte. Since the effects of alpha-1 receptor activation in many tissues appear to be mediated by stimulation of phosphatidylinositol turnover and subsequent activation of protein kinase C, one might expect activators of protein kinase C to stimulate release. Indeed, we have found that phorbol ester enhances the release of atriopeptin from the isolated perfused rat heart; whereas, 4-b phorbol, an inactive analog, had no effect (9). These findings suggest that the secretion of atriopeptin is modulated by the sympathetic nervous system.

C. Effect of experimental diabetes on atriopeptin release

We have examined the effect of streptozoticin-induced diabetes on norepinephrine-stimulated atriopeptin release in the isolated perfused rat heart model. Comparison of the responsiveness (atriopeptin release) of the hearts of control and diabetic animals to norepinephrine shows that diabetes causes a dramatic decrease in responsiveness. Treatment of diabetic animals with insulin returns the responsiveness to control levels. This finding indicates that a decrease in the activity of the atrial endocrine system may play a role in the cardiovascular pathophysiology associated with diabetes.

Potential relationship between atriopeptin and its receptor subtypes

The finding that the isolated heart releases two forms of atriopeptin, a 28 amino acid peptide and a putative 25 amino acid peptide, suggests the possibility that the heart can differentially regulate the production and release of these two peptides (15). We propose that the ratio of atriopeptin (1-28) to atriopeptin (1-25) could be altered under conditions spanning the spectrum of activity of the cardiac endocrine system from very high activity (e.g., hypertensive, salt loaded, volume expanded) to very low activity (e.g., volume depleted and salt depleted). An interesting extension of the concept of differential regulation involves the potential interaction between the forms of atriopeptin released by the heart and the distribution and expression of atriopeptin receptors. Based on structure-activity and receptor binding studies there appear to be two classes of atriopeptin receptors, one of which is associated with guanylate cyclase (13,17,19,24). These studies have shown that expression of both receptor subtypes may occur in the same organ and even in the same cell type. Furthermore, this work suggests that AP (1-28) is an effective agonist on both receptor subtypes; whereas, AP (1-25) would be expected to act only on the non-guanylate cyclase associated receptor (Fig. 4). Thus, the potential may exist for

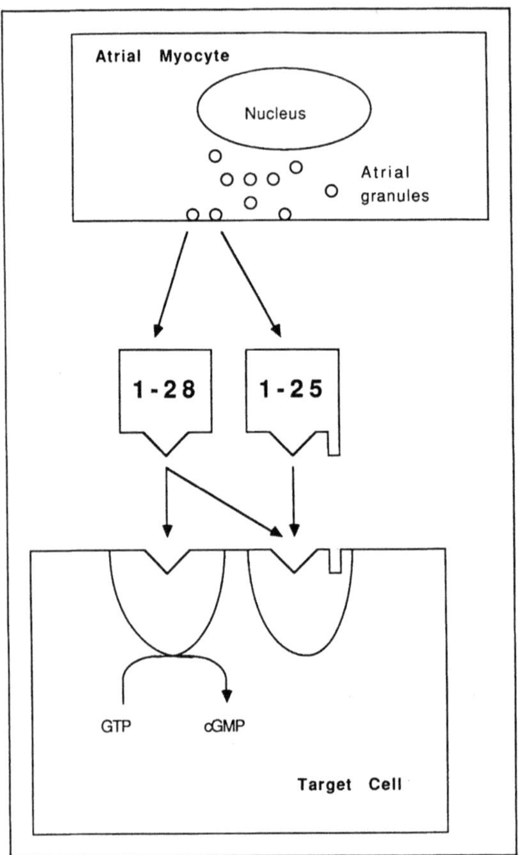

Fig. 4. Schematic of the potential relationship between the two secreted forms of atriopeptin (1-28, 1-25) and the atriopeptin receptor subtypes

the heart to target a specific population of receptors by selectively producing AP (1-25). This selective targeting of an atriopeptin receptor could obviously result in the production of selective actions on the various endorgans. In addition, we postulate that the target cells can alter their responsiveness to atriopeptin by altering the expression of the specific receptor subtypes. This expression of the receptor subtypes could be related to the potential regulation of the production of the two forms of atriopeptin, or these could be independent events that act separately to control the activity of the cardiac endocrine system.

Conclusion

An abundance of information exists concerning the pharmacological effects of atriopeptin on blood pressure and fluid and electrolyte homeostasis. The described actions of
atriopeptin suggest that this peptide may be a useful pharmacological agent for the
treatment of some cardiovascular diseases (e.g., hypertension and congestive heart
failure). In addition, an understanding of the cardiac endocrine system at the cellular and
molecular level may reveal regulatory sites that provide alternative opportunities for
pharmacological intervention. We feel that these sites are likely to include: 1) regulation
of AP gene expression; 2) post-translational modifications of the atrial peptides; 3)
proteolytic processing of the hormone; 4) secretion; and 5) distribution and expression
of AP receptors. It is our hope that this review will provide a framework for future
studies examining the role of the cardiac endocrine system in normal and disease states.

Acknowledgements: Supported in part by grants NIH HL29566 and American Heart
Association 851282. We thank Virginia Minchoff for her secretarial assistance.

References

1. Arendt, R. M., Gerbes, A. L., Ritter, D., Stangl, E. (1986) Molecular weight heterogeneity of plasma-ANF in cardiovascular disease. KIin Wochenschr 64: (suppl VI)
   97-102
2. Bloch, K. D., Jones, S. W., Seidman, C. E., Seidman, J. G. (1987) Proatrial natriuretic factor is phosphorylated by rat cardiocytes in culture. CIin Res 35: 264A
3. Chao, J. (1983) Purification and characterization of rat urinary esterase A, a plas-
   minogen activator. J Biol Chem 258: 4434-4439
4. Currie, M. G., Geller, D. M., Cole, B. R. et al. (1983) Bioactive cardiac substances:
   potent vasorelaxant activity in mammalian atria. Science 221: 72-73
5. Currie, M. G., Geller, D. M., Cole, B. R., Needleman, P. (1984a) Proteolytic activation of a bioactive cardiac peptide by *in vitro* trypsin cleavage. Proc Natl Acad
   Sci USA 81: 1230-1233
6. Currie, M. G., Geller, D. M., Chao, J., Margolius, H. S., Needleman, P. (1984b) Kallikrein activation of a high molecular weight atrial peptide. Biochem Biophys Res
   Commun 120: 451-455
7. Currie, M. G., Sukin, D., Geller, D.M., Cole, B. R., Needleman, P. (1984c) Atriopeptin release from the isolated perfused rabbit heart. Biochem Biophys Res Commun
   124: 711-717
8. Currie, M. G., Newman, W. H. (1986a) Evidence for a-1 adrenergic receptor regulation
   of atriopeptin release from the isolated rat heart. Biochem Biophys Res Commun
   137: 94-100
9. Currie, M. G., Newman, W. H. (1986b) The regulation of atriopeptin release from
   isolated rat heart. Fed Proc 45: 911
10. Currie, M. G., Oehlenschlager, W. F., Kurtz, D. T. (1987) Profound elevation of
    ventricular and pulmonary atriopeptin in a model of heart failure. (submitted)

11. Day, M. L., Schwartz, D., Wiegand, R. C., Currie, M. G., Standaert, D. G., Needleman, P. (1987) Ventricular atriopeptin: unmasking of mRNA and peptide synthesis by hypertrophy or dexamethasone. Hypertension 9: 485-489

12. DeBold, A. J. (1985) Atrial natriuretic factor: A hormone produced by the heart. Science 230: 767-770

13. Delean, A., Thibault, G., Seidah, N. G., Lazure, C., Gutkowska, J., Chreitien, M., Genest, J., Cantin, M. (1985) Structure activity relationships of atrial natriuretic factor (ANF). Biochem Biophys Res Commun 132: 360-367

14. Gardner, D. G., Hane, S., Tracheusky, D., Schenk, D., Baxter, J. D. (1986) Atrial natriuretic peptide mRNA is regulated by glucocorticoids *in vitro*. Biochem Biophys Res Commun 139: 1047-1054

15. Lanier, K. L., Baron, D. A., Currie, M. G. (1987) Biochemical analysis of the forms of atriopeptin released by neonatal and adult rat hearts. Fed Proc 46: 1294

16. Lattion, A.L., Michel, J. B., Arnauld, E., Corvol, P., Soubrier, F. (1986) Myocardial recruitment during ANF mRNA increase with volume overload in the rat. Am J Physiol 251: H890-H896

17. Leitman, D. C., Murad, F. (1986) Comparison of binding and cyclic GMP accumuation by atrial natriuretic peptides in endothelial cells. Biochem Biophys Acta 885: 74-79

18. Matsuo, H., Nakazato, H. (1987) Molecular biology of atrial natriuretic peptides. Endocrinology and Metabolism Clinics of North America 16: 43-61

19. Meloche, S., Ong, H., Cantin, M., Delean, A. (1986) Affinity cross-linking of atrial natriuretic factor to its receptor in bovine adrenal zona glomerulosa. J Biol Chem 261: 1525-1528

20. Nakayama, K., Ohkubo, H., Hirose, T., Inayama, S., Nakanishi, S. (1984) mRNA sequence for human cardiodilatin-atrial natriuretic factor precursor and regulation of precursor mRNA in rat atria. Nature 310: 699-701

21. Needleman, P., Adams, S. P., Cole, B. R., Currie, M. G., Geller, D. M., Michener, M. L., Saper, C. B., Schwartz, D., Standaert, D. G. (1985) Atriopeptins as cardiac hormones. Hypertension 7: 469-482

22. Oehlenschlager, W. F., Kurtz, D. T. Currie, M. G. (1988) Enhanced activity of the cardiac endocrine system during right ventricular hypertrophy. (submitted)

23. Rittenhouse, J., Moberly, L., O'Donnell, M. E., Owen, N. E., Marcus, F. (1986) Phosphorylation of atrial natriuretic peptides by cyclic AMP-dependent protein kinase. J Biol Chem 261: 7607-7610

24. Schenk, D. B., Phelps, M. N., Porter, J. G., Scarborough, R. M., McEnroe, G. A., Lewicki, J. A. (1985) Identification of the receptor for atrial natriuretic factor on cultured vascular cells. J Biol Chem 260: 14887-14890

25. Schwartz, D., Geller, D. M., Manning, P. T. et al. (1985) Ser-leu-arg-arg-atriopeptin III is the major circulating form of atrial peptide. Science 229: 397-400

26. Simson, J., Currie, M., Chao, L. (1988) Co-localization of esterase A and atriopeptin. Eur J Cell Biol (in press)

Sites of action of ANF as a circulating hormone

J. Gutkowska

Laboratory of the Biochemistry of Hypertension, Clinical Research Institute of Montreal, Montreal, Canada

Introduction

The physiological effect of ANF found its place in the pages of history about 2000 years ago. The ancient historian Joseph Flavius recorded in detail the contemporary construction of the Cesarea Maritima, the greatest harbor on the oriental shores of the Mediterranean. In doing so, he aptly referred to the trained divers who used to retrieve rocks and smooth the bottom of the sea as "urinatores", reflecting their physiological diuretic response to prolonged submersion in water.

More recent investigations have suggested that the heart may be a sensor of body fluid volume (61). Gauer, for example, demonstrated that mechanical distention of the left atrium was accompanied by rapid diuresis and natriuresis (39). These observations gave birth to the hypothesis of the existence of atrial stretch receptors, as suggested by Nonidez (58) many years earlier. It has also been postulated that the heart itself may be an organ which "translates the changes in volume and subsequently in the degree of stretching of the great vessels into changes in glomerular filtration rate and sodium balance" (25). Cardiac atria have been suggested to possess a blood volume regulatory mechanism.

For many years however, the regulation of the kidney by the heart remained unclear. In a landmark experiment DeBold illustrated that rat atrial homogenates contain a biologically active substance which produces diuresis and natriuresis (18). Enormous developments in peptide biochemistry, particularly purification techniques and molecular biology, facilitated rapid progress, which we are witnessing right now with ANF.

Early attempts to purify the substance responsible for this diuretic and natriuretic effect resulted in the isolation of several peptides composed of 21 to 35 amino acids (AA) (13). All these peptides of various lengths, appeared to be carboxy terminal fragments of prohormone which is stored in atrial granules.

Recently, cDNA cloning revealed that the ANF molecule is synthesized as a 151-AA precursor in humans and as a 152-AA precursor in rats. These precursors are composed of a prohormone which, in both humans and rats, contains 126-AA residues as well as a N-terminal signal peptide of 25-AA residues in humans and 24-AA in rats (4,45). A strong homology in the AA sequence of carboxy-terminal portions exists in various species; it decreases at the N-terminal portion of the molecule.

Radioimmunoassay

Various techniques have been developed for the measurement of ANF in biological fluids and tissue. The first attempts to quantify ANF were made via bioassays on the basis of its diuretic, natriuretic, and vasorelaxant properties (16,18,24). However, all these bioassays were found to be of insufficient sensitivity to accurately measure low concentrations of ANF in plasma. Then availability of synthetic, biologically-active ANF fragments permitted the development of very sensitive and specific RIA (34). Measurements of ANF in atria can now be performed directly from atrial homogenates (35).

Table 1. Effect of different conditions of blood withdrawal and various anesthetics on rat plasma ANF

	Dose per 100 g body weight		N		ANF pg/ml
Decapitation	-		88		46.1±4.2
Cannulation of jugular vein - 24h	-		30		93.9±7
Cannulation of jugular vein - 48h	-		29		81.9±11.5
Urethane	100	mg	9		42±8
Sodium pentobarbital	6	mg	10		59±7
Ketamine chlorhydrate	10	mg	51		195±20
Chloral hydrate	30	mg	20		224±80
Fentanyl	25	mg	6		693±114
Diethylether	-		64		806±75
Morphine	14	mg	10	mg	2443±282

However, nonspecific interference of plasma protein has made it necessary to extract ANF. Various methods of extraction have been established, viz., affinity chromatography, Sep-Pak cartridges, or Vycor glass beads (29).

The measurement of ANF in plasma generates a wide variety of results, from low or undetectable levels to 300 pg/ml by direct RIA, and from 6 pg/ml to 230 pg/ml by prior extraction in humans. In rats, even greater variations are observed, with values ranging from 50 to 3940 pg/ml (34). This variability of the results may be due to the procedure followed (direct assay or prior extraction). The values may even be influenced by the effectiveness of proteolytic enzyme inhibition during blood withdrawal and storage. The purity of standards, antibody specificity, and quality of radiolabelled tracer may alter the reliability of the results.

Other unrelated conditions (of the patient, for example) may affect the outcome of the assay. ANF values are highly dependent on water and salt intake (33), and even on the position of patients during blood withdrawal (42). Furthermore, in animals various anesthetics dramatically increase plasma ANF levels (Table 1).

Circulating forms

It is now well-accepted that atrial granules store ANF in the form of a 126-AA prohormone (77). The circulating form of ANF has also recently been identified. Because of its low concentration, rat plasma ANF has been purified from vasopressin- or morphine-stimulated animals (69,75). In both cases, the circulating form is identified as 28-AA carboxy-terminal (Ser 99-Tyr126), the same peptide which was primarily isolated by Flynn et al. (23) from rat atria. Ser 103-Tyr 126 ANF, which accounts for about 10%, and is the shorter form of ANF in rat plasma, has also been isolated (22,69). It is most probably a degradation product, not a secreted form. The incubation of synthetic (Ser 99-Tyr 126) ANF in rat plasma at 37°C generates only this shorter form, Ser 103-Tyr 126 (56).

In human blood, a more complex pattern of immunoreactive ANF has been found, suggesting that circulating human ANF is vulnerable to proteolytic enzyme degradation. The presence of at least three ANF peptides has been reported by several groups (32,79), and the structure of the main form has recently been elucidated it appears to be 28-AA carboxy-terminal Met 110 (Ser 99-Tyr 126) ANF (74).

Processing of pro-ANF

The biological activity of ANF resides in the C-terminal part of the prohormone. Therefore, *in vivo* processing of pro-ANF is required for the full expression of biological activity.

Several enzymes which process the pro-ANF have been identified in *in vitro* studies. Bovine atria contain membrane-bound metallo dipeptidyl carboxy hydroxylase, which can cleave the Ser-Phe-Arg bond to transform atriopeptin 11 to atriopeptin 1 (37). However, this cleavage does not seem to be involved in the production of circulating active ANF. The same group isolated another thiol-dependent protease from tissue that cleaves the Arg 98-Ser 99 bond, producing only 28-AA from the 31-AA residue peptide (Gly 96-Tyr 126) (6).

IRCM-SP1, considered to be a maturation enzyme for various prohormones, including proANF has been extracted from rat atria and ventricles (70). It yields several ANF peptides from prohormone ANF: (Ser 103-Tyr 126) ANF, (Ser 102-Tyr 126) ANF, and

(Ser 99 - Tyr 126) ANF. However, only small quantities of the circulating form are produced from pro-ANF by the action of this enzyme. Therefore, it is unlikely to be of physiological importance in the conversion of pro-ANF to its active form.

Blood contains enzymes capable of cleaving pro-ANF. Besides kallikrein and thrombin, platelet-associated enzymes from rat plasma produce short ANF peptides suggesting that circulating platelets might participate in the conversion of the ANF precursor (78). Fresh rat serum efficiently cleaves pro-ANF to its circulating form (10,26). However, further experiments are needed to determine which enzyme or enzymes are responsible for the physiological maturation of pro-ANF. It is not yet clear whether the prohormone is released into the bloodstream to give birth to the active form of ANF or whether the precursor is processed within atrial myocytes.

The confusion in this field was created by controversial results reported on the form of ANF secreted by rat cardiocytes. Pro-ANF is released from atrial or ventricular cardiocytes of newborn rats when the cells are cultivated in serum-free medium (9,27). In serum containing medium, only low molecular weight ANF is detected (31). Furthermore, serum protease generates low molecular weight ANF from pro-ANF. These findings indicate that the prohormone may be secreted from cardiocytes and that the maturation process occurs in the bloodstream. However, in contrast to these observations, only the 28-AA circulating form of ANF has been detected in perfusates from isolated, spontaneously-beating hearts of rats or rabbits in the absence of serum (17,76). Furthermore, HPLC of human coronary sinus plasma extracts does not indicate the presence of prohormone. These results favor the intracellular maturation process, indicating that plasma membrane-bound enzymes process pro-ANF.

ANF sites of action

Once ANF is released into the bloodstream it travels to various target organs, such as the kidney, adrenals, and pituitary to modify their regulatory activity on salt and water homeostasis, subsequently influencing blood pressure. Autoradiography has revealed ANF-binding sites in a variety of peripheral and brain tissue (7,8,62). Some of the binding sites have been characterized by kinetic studies and are considered as receptors, although the biological effect of ANF has not yet been established in these tissues (Table 2).

Kidney

The kidney appears to be a primary target organ for ANF. Injection of I^{125}ANF into rat aorta demonstrates binding sites in the renal cortical vasculature, mainly in glomeruli, outer medullary descending vasa recta, and inner medullary collecting duct cells. They are also present in the endothelium of smooth muscle cells of renal vessels (8).

ANF-binding studies on purified dog kidney membrane fractions show high- and low-affinity binding sites in glomeruli, high-affinity binding sites in the thick ascending limbs of Henle's loops, and in collecting ducts, no binding sites are found in proximal tubules (Table 2).

136

Table 2. ANF peripheral tissue receptor characteristics

Tissue	Kd pm	Bmax fmol/mgprotein	References
Dog kidney glomeruli	125	200	De Lean et al. (1985)
Thick ascending Henle´s loops	398	36	
Collecting ducts	398	150	
Rat glomeruli	27	390	Carrier et al. (1985)
Rat glomeruli	940 380	426 (low salt) 116 (high salt)	Ballerman et al. (1985)
Bovine adrenal cortex	1800	2500	Hirose et al. (1985)
Bovine adrenal cortex	1100	1000	Misono et al. (1985)
Rabbit aorta	130	96	Napier et al. (1984)
Rabbit mesenteric artery	102	110	Schiffrin et al. (1985)
Rat renal artery	44	18.5	
Bovine aortic endothelial cells in culture	100	16000 receptors/cell	Leitman et al. (1986)
Human platelets	30	10 receptors/cell	Schiffrin et al. (1986)
Human platelets	8-16	10-26 receptors/cell	Strom et al. (1987)
Rat thymocytes	170	218	Kurihara et. al.(1987)
Rat spleen	150	130	
Rabbit lung	320	166	Olins et al. (1986)
Rat lung fibroblast	660	216	Leitman et al. (1987)
Human placenta	2500	96	Sen (1986)

Two intracellular events parallel the effect of ANF on the kidney: 1) stimulation of particulate guanylate cyclase activity measured by cGMP formation (36), and 2) inhibition of adenylate cyclase activity (1). Physiologically, ANF has been found to increase the glomerular filtration rate (12) and reduce medullary hypertonicity (11). In the renal vasculature, it is a powerful antagonist of angiotensin II-induced vasoconstriction (53). ANF infusion suppresses renal renin secretion which, as a consequence, lowers the plasma renin level (52). *In vitro* studies on cultured juxtaglomerular cells have also revealed an inhibition of renin release by ANF (47).

Adrenals

Specific ANF receptors have been demonstrated in bovine adrenal cells high affinity (19) and in rat zona glomerulosa cells (67). The ANF which antagonizes the action of angiotensin II on the renal vasculature has a similar effect in the adrenal cortex. ANF has also been shown *in vitro* to inhibit angiotensin II, ACTH-, potassium-, and prostaglandin-stimulated aldosterone secretion (2,19). It even suppresses the response to intravenous infusion of angiotensin II (3) in rats while its action on basal aldosterone release is controversial; some studies have recorded little or no effect (15), or an inhibitory action (3,28).

Pituitary

A close relationship recently established between ANF and vasopressin. Both hormones are involved in body fluid regulation and they have the capacity to control renal and vascular function in opposite directions. Vasopressin has been found to stimulate ANF secretion. Using a bioassay, Sonnenberg and Veress (72) have reported the release of ANF from atrial slices by arginine vasopressin (AVP). Furthermore, Schwartz et al. (69) noted increased levels of circulating ANF after vasopressin injection, but in this *in vivo* experiment, only pressor doses of AVP induced ANF secretion; hemodynamic changes were most probably responsible for this effect. On the other hand, there is overwhelming evidence that ANF inhibits vasopressin secretion. It can suppress dehydration- or hemorrhage- stimulated vasopressin release in rats (63), indicating that the physiological control of fluid volume depends on a balance between ANF and AVP. In the superfusion system of posterior pituitaries (59), ANF (10^{-6}M - 10^{-9}M) has an inhibitory effect on basal-, KCl-, and angiotensin II- stimulated vasopressin release. This action parallels the increase of cGMP and the decrease of cAMP secretion. The intracerebroventricular injection of ANF has also been shown to suppress AVP release without hemodynamic changes (65). All effects of ANF on AVP are receptor-mediated.

In vitro binding studies have elaborated the presence of high-affinity receptors with pK of 9.9 and Bmax of 20 fmol/lobe (44). ANF-binding sites have been demonstrated in the rat anterior pituitary. There are relatively few reports on the effect of ANF on anterior pituitary hormone secretion. The involvement of adenylate cyclase/cAMP and guanylate cyclase/cGMP in the regulation of hormone release from the anterior pituitary has been established (48). Since ANF has been shown to have an effect on both second messengers, the modulation of anterior pituitary hormones by ANF should be expected. However, the reported data are not conclusive. Shibasaki et al. (71) documented an inhibitory effect of ANF on the basal and corticotropin releasing factor-stimulated secretion of peptide hormones derived from pro-opiomelanocortin, such as b-endorphin, aMSH and ACTH. Contrary to these results, however, Horvath et al. (43) have observed stimulated ACTH, LH, and FSH release in superfused rat pituitary cells. ANF has also been found ineffective in modifying anterior pituitary hormone secretion despite intracellular cGMP accumulation (38).

Vascular receptors

High-affinity ANF receptors have been demonstrated in the rabbit aorta (57), rat mesenteric and renal arteries (67), rat cultured smooth muscle cells (40), and bovine aortic endothelial and smooth muscle cells (66). All these tissues are sensitive to ANF′s vasorelaxant action, which is particularly effective in antagonizing contractions induced by angiotensin II. This vasorelaxant effect of ANF is associated with an increase in cGMP, suggesting that these receptors are coupled to guanylate cyclase (50).

ANF is now accepted as a new potent diuretic, natriuretic and vasorelaxant hormone. Its inhibition of the renin-aldosterone system, its opposite action to angiotensin II on the renal vasculature, and suppression of vasopressin release are mediated by specific receptors found in the kidney, adrenal, pituitary and vasculature, indicating the importance of this hormone in the regulation of water and salt homeostasis. However, its physiological role remains to be elucidated.

References

1. Anand-Strivastava, M.B., Vinay, P., Genest, J., Cantin, M. (1986) Effect of atrial natriuretic factor on adenylate cyclase in various nephron segments. Am J Physiol 251: F417-F423

2. Atarashi, K., Mulrow, P.J., Franco-Saenz, R., Snajdar, J. (1984) Inhibition of aldosterone production by an atrial extract. Science 224:992-994

3. Atarashi, K., Mulrow, P.J., Franco-Saenz, R. (1985) Effect of atrial peptide on aldosteron production. J Clin Invest 86: 1807-1811

4. Atlas, S. A., Kleinert, H.D., Camargo, M. J., Januszewicz, A., Sealey, J. E., Schilling, J. W., Lewicki, J. A., Johnson, L. K., Maack, T. (1984) Purification, sequencing and synthesis of natriuretic and vasoactive rat atrial peptide. Nature 309: 717-719

5. Ballerman, B.J., Hoover, R.L., Karnovsky, M.J., Brenner, B.M. (1985) Physiologic regulation of atrial natriuretic peptide receptors in rat renal glomeruli. J Clin Invest 76: 2049-2056

6. Baxter, J.H., Wilson, I. B., Harris, R.B. (1986) Identification of an endogenous protease that processes atrial natriuretic peptide at its amino terminus. Peptides 7: 407-411

7. Bianchi, C., Gutkowska, J., Ballak, M., Thibault, G., Garcia, R., Cantin, M (1986) Radioautographic localization of ^{125}I-atrial natriuretic factor binding sites in the brain. Neuroendocrinology 44: 365-372

8. Bianchi, C., Gutkowska, J., Garcia, R., Thibault, G., Genest, J., Cantin, M. (in press 1987) Localization of ^{125}I-atrial natriuretic factor (ANF) binding sites in ratrenal medulla. A light and electron microscope radioautographic study. J Histochem Cytochem 35: 149-153

9. Bloch, K.D., Scott, J.A., Zisfein, J.B., Fallon, J.T., Margolies, M.N., Seidman, C.E., Matsueda, G.R., Homey, C.J., Graham, R.M., Seidman, J.G. (1985) Biosynthesis and secretion of proatrial natriuretic factor by cultured rat cardiocytes. Science 230: 1168-1171

10. Bloch, K.D., Zisfein, J.B., Margolies, M.N., Homcy, C.J., Seidman, J.G., Graham, R.M. (1987) A serum protease cleaves pro-ANF into a 14-kilodalton peptide and ANF. Am J Physiol 252: E147-E151, 1987

11. Borenstein, H.B., Cupples, W.A., Sonnenberg, H., Veress, A.T. (1983) The effect of a natriuretic atrial extract on renal hemodynamics and urinary excretion in anaesthetized rats. J Physiol 334: 133-140

12. Camargo, M.J., Kleinert, H.D., Atlas, S.A., Sealey, J.E., Laragh, J.H., Maack, T. (1984) Ca-dependent hemodynamic and natriuretic effects of atrial extraction in isolated rat kidney. Am J Physiol 246: F447-F456

13. Cantin, M., Genest, J. (1985) The heart and the atrial natriuretic factor. Endocrine Rev 6: 107-127

14. Carrier, F., Thibault, G., Schiffrin, E.L., Garcia, R., Gutkowska, J., Cantin, M., Genest, J. (1985) Partial characterization and solubilization of receptors for atrial natriuretic factor in rat glomeruli. Biochem Biophys Res Comm 132: 666-673

15. Chartier, L., Schiffrin, E.L., Thibault, G. (1984) Effect of atrial natriuretic factor (ANF)-related peptides on aldosterone secretion by adrenal glomerulosa cells: Critical role of the intramolecular disulphide bond. Biochem Biophys Res Comm 122: 171-174

16. Currie, M.G., Geller, D.M., Cole, B.R., Boylan, J.G., Yu Sheng, W., Holmberg, S.W., Needleman, P. (1983) Bioactive cardiac substances: potent vasorelaxant activity in mammalian atria. Science 221: 71-73

17. Currie, M.G., Geller, D.M., Cole, B.R., Siegel, N.R., Fok, K.K., Adams, S.P., Eubanks, S.R., Galluppi, G.R., Needleman, P. (1984) Purification and sequence analysis of bioactive atrial peptides (atriopeptins). Science 223: 67-69

18. DeBold, A.J., Borenstein, H.B., Veress, A.T., Sonnenberg, H. (1981) A rapid and potent natriuretic response to intravenous injection of atrial myocardial extract in rats. Life Sci 28: 89-94

19. De Léan, A., Gutkowska, J., McNicoll, N., Schiller, P.W., Cantin, M., Genest, J. (1984a) Characterization of specific receptors for atrial natriuretic factor in bovine adrenal zona glomerulosa. Life Sci 35: 2311-2318

20. De Léan, A., Racz, K., Gutkowska, J., Nguyen, T.T., Cantin, M., Genest, J. (1984b) Specific receptor-mediated inhibition by synthetic atrial natriuretic factor of hormone-stimulated steroidogenesis in cultured bovine adrenal cells. Endocrinology 115: 1636-1638

21. De Léan, A., Vinay, P., Cantin, M. (1985) Distribution of atrial natriuretic factor receptors in dog kidney fractions. FEBS 193: 239-242

22. Eskay, R., Zukowska-Grojec, Z., Haass, M., Dave, JR., Zamir, N. (1986) Circulating atrial natriuretic peptide in conscious rats: Regulation of release by multiple factors. Science 232: 636-639

23. Flynn, T.G., DeBold, M.L., DeBold, A.J. (1983) The amino acid sequence of an atrial peptide with potent diuretic and natriuretic properties. Biochem Biophys Res Comm 117: 859-865

24. Garcia, R., Thibault, G., Cantin, M., Genest, J. (1984) Effect of purified atrial natriuretic factor on rat and rabbit vascular strips and vascular beds. Am Physiol 247: R34-R39

25. Gauer, O.H., Henry, J.P., Sieker, H.O. (1961) Cardiac receptors and fluid volume control. Progress in Cardiovascular Diseases 4: 1-2

26. Gibson, T.R., Shields, P.P., Glembotski, C.C. (1987) The conversion of atrial natriuretic peptide (ANP)-(1-126) to ANP-(99-126) by rat serum: Contribution to ANP cleavage in isolated perfused rat hearts. Endocrinology 120: 764-772

27. Glembotski, C.C., Gibson, T.R. (1985) Molecular forms of immunoreactive atrial natriuretic peptide released from cultured rat atrial myocytes. Biochem Biophys Res Comm 132: 1008-1017

28. Goodfriend T.L., Elliott, M.E., Atlas, S.A. (1984) Actions of synthetic atrial natriuretic factor on bovine adrenal glomerulosa. Life Sci 35: 1675-1682

29. Gutkowska, J., Horky, K., Thibault, G., Januszewicz, P., Cantin, M., Genest, J. (1984a) Atrial natriuretic factor is a circulating hormone. Biochem Biophys Res Comm 125: 315-323

30. Gutkowska, J., Thibault, G., Januszewicz, P., Cantin, M., Genest, J. (1984b) Direct radioimmunoassay of atrial natriuretic factor. Biochem Biophys Res Comm 122: 593-601

31. Gutkowska, J., Thibault, G., Racz, K., Lazure, C, Garcia, R., Seidah, N.G., Chrétien, M., Genest, J., Cantin, M. (1985) ANF (Arg 101 - Tyr 126) is the peptide secreted by rat atrial cardiocytes in culture. Biochem Biophys Res Comm 130: 1217-1225

32. Gutkowska, J., Bonan, R., Roy, D., Bourassa, M., Garcia, R., Thibault, G., Genest, J., Cantin, M. (1986a) Atrial natriuretic factor in human plasma. Biochem Biophys Res Comm 139: 287-295

33. Gutkowska, J., Schiffrin, E.L., Cantin, M., Genest, J. (1986b) Effect of dietary sodium on plasma concentration of immunoreactive atrial natriuretic factor in normal humans. Clin Invest Med 9: 222-224

34. Gutkowska, J. (1987a) Radioimmunoassay for atrial natriuretic factor. Nucl Med 14: 323-331

35. Gutkowska, J., Genest, J., Thibault, G., Garcia, R., Larochelle, P., Cusson, J.R., Kuchel, O., Hamet, P., De Léan, A., Cantin, M. (1987b) Circulating forms and radioimmunoassay of atrial natriuretic factor. Endocrinol Metab Clin N Am 16: 184-198

36. Hamet, P., Tremblay, J., Pang, S.C., Garcia, R., Thibault, G., Gutkowska, J., Cantin, M., Genest, J. (1984) Effect of native and synthetic atrial natriuretic factor on cyclic GMP. Biochem Biophys Res Comm 123: 515-527

37. Harris, R.B., Wilson, I.B. (1985) Conversion of atriopeptin II to atriopeptin I by atrial dipeptidyl carboxy hydrolase. Peptides 6: 393-396

38. Heisler, S., Simard, J., Assayag, E., Mehri, Y., Labrie, F. (1986) Atrial natriuretic factor does not affect basal, forskolin- and CRF-stimulated adenylate cyclase activity, cAMP formation or ACTH secretion, but does stimulate cGMP synthesis in anterior pituitary. Molecul Cell Endocr 44: 125-131

39. Henry, J. P., Gauer, O.H., Reeves, J.L. (1956) Evidence of the atrial location of receptors influencing urine flow. Circ Res :85-90

40. Hirata, Y., Ganguli, M., Tobian, L., Iwai, J. (1984) Dahl S rats have increased natriuretic factor in atria but are markedly hyporesponsive to it. Hypertension 6 (suppl 1): 1-148-1-155

41. Hirose, S., Akiyama, F., Shinjo, M., Ohno, H., Murakami, K. (1985) Solubilization and molecular weight estimation of atrial natriuretic factor receptor from bovine adrenal cortex. Biochem Biophys Res Comm 130: 574-579

42. Hollister, A.S., Tanaka, I., Imada, T., Onrot, J., Biaggioni, I., Robertson, D., Inagami, T. (1986) Sodium loading and posture modulate human atrial natriuretic factor plasma levels. Hypertension (suppl II) 8: II-106-II-111

43. Horvath, J., Ertl, T., Schally, A.V. (1986) Effect of atrial natriuretic peptide on gonadotropin release in superfused rat pituitary cells. Proc Natl Acad Sci USA 83: 3444-3446

44. Januszewicz, P., Thibault, G., Garcia, R., Gutkowska, J., Genest, J., Cantin, M. (1986) Effect of synthetic atrial natriuretic factor on arginine vasopressin release by the rat hypothalamo-neurohypophysial complex in organ culture. Biochem Biophys Res Commun 134: 652-658

45. Kangawa, K., Tawaragi, Y., Oikawa, S., Miziano, A., Sakuragawa, Y., Nakazato, H., Fukuda, A., Minamino, N., Matsuo, H. (1984) Identification of rat gamma atrial natriuretic polypeptide and characterization of the cDNA encoding its precursor. Nature 312: 152-155

46. Kurihara, M., Katamine, S., Saavedra, J.M. (1987) Atrial natriuretic peptide, ANP (99-126), receptors in rat thymocytes and spleen cells. Biochem Biophys Res Commun 145: 789-796

47. Kurtz, A., Bruna, R.D., Pfeilschifter, J., Taugner, R., Bauer, C. (1986) Atrial natriuretic peptide inhibits renin release from juxtaglomerular cells by a cGMP-mediated process. Proc Natl Acad Sci USA 83: 4769-4773

48. Labrie, F., Borgeat, P., Lemay, A., Lemaire, S., Barden, N., Drouin, J., Lemaire, I., Jolicoeur, P., Bélanger, A. (1975) Role of cyclic AMP in the action of hypothalamic regulatory hormones. Adv Cyclic Nucleotide Res 5: 787-801, 1975

49. Lemaire, I., Jolicoeur, P., Belanger, A. (1975) Role of cyclic AMP in the action of hypothalamic regulatory hormones. Adv Cyclic Nucleotide Res 5: 787-801, 1975

50. Leitman, D.C., Murad, F. (1986) Comparison of binding and cyclic GMP accumulation by atrial natriuretic peptides in endothelial cells. Biochem Biophys Acta 885: 74-79

51. Leitman, C., Agnost, V.L., Tuan, J.J., Andresen, J.W., Murad, F. (1987) Atrial natriuretic factor and sodium nitroprusside increase cyclic GMP in cultured rat lung fibroblasts by activating different forms of guanylate cyclase. Biochem J 244: 69-74

52. Maack, T., Marion, D.N., Camargo, M.J.F., Kleinert, H.D., Laragh, J.H., Vaughan, E.D. Jr., Atlas S.A. (1984) Effects of auriculin (atrial natriuretic factor) on blood pressure, renal function and the renin-aldosterone system in dogs. Am J Med 77: 1069-1075

53. Maack, T., Kleinert, H.D. (1986) Renal and cardiovascular effect of atrial natriuretic factor. Biochem Pharmacol 35: 2057-2064

54. Meloche, S., Ong, H., Cantin, M., De Léan, A. (1986) Molecular characterization of the solubilized atrial natriuretic factor receptor from bovine adrenal zona glomerulosa. Mol Pharmacol 30: 537-543

55. Misono, K.S., Grammer, R.T., Rigby, J.W., Inagami, T. (1985) Photoaffinity labeling of atrial natriuretic factor receptor in bovine and rat adrenal cortical membranes. Biochem Biophys Res Commun 130: 994-1001

56. Murthy, K.K., Thibault, G., Garcia, R., Gutkowska, J., Genest, J., Cantin, M. (1987) Degradation of atrial natriuretic factor in the rat. Biochem J 240: 461-469

57. Napier, M.A., Vandlen, R.L., Albers-Schonberg, G., Nutt, R.F., Brady, S., Lyle, T., Winquist, R., Faison, E.P., Heinel, L.A., Blaine, E.H. (1984) Specific membrane receptors for atrial natriuretic factor in renal and vascular tissue. Proc Natl Acad Sci USA 81: 5946-5950

58. Nonidez, J.F. (1937) Identification of the receptor areas in the venae cavae and pulmonary veins which initiate reflex cardiac acceleration (Bainbridge's reflex). Am J Anat 61: 203-231

59. Obana, K., Naruse, M., Inagami, T., Brown, A.B., Naruse, K., Kurimoto, F., Sakurai, H., Demura, H., Shizume, K. (1985) Atrial natriuretic factor inhibits vasopressin secretion from rat posterior pituitary. Biochem Biophys Res Commun 132: 1088-1094

60. Olins, G.M., Patton, D.R., Tjoeng, F.S., Blehm, D.J. (1986) Specific receptors for atriopeptin III in rabbit lung. Biochem Biophys Res Comm 140: 302-307

61. Peters, J.P. (1962) The problem of cardiac edema. Am J Med 12: 66-76

62. Quirion, R., De Léan, A., Gutkowska, J., Cantin, M., Genest, J. (1984) Receptor/acceptors for the atrial natriuretic factor (ANF) in brain and related structures. Peptides 5: 1167-1172

63. Samson, W.K. (1985) Atrial natriuretic factor inhibits dehydration and hemorrhage-induced vasopressin release. Neuroendocrinology 40: 277-279

64. Sen, I. (1986) Identification and solubilization of atrial natriuretic factor receptors in human placenta. Biochem Biophys Res Comm 135: 480-486

65. Share, L., Iitake, K., Crofton, J.T., Brooks, D.P., Ouchi, Y., Blaine, E. (1986) Central atrial natriuretic factor induces vasopressin release in conscious rats. Fed Proc 45: 166

66. Schenk, D.B., Johnson, L.K., Schwartz, K., Sista, H., Scarborough, R.M., Lewicki, J.A. (1985) Distinct atrial natriuretic factor receptor sites on cultured bovine aortic smooth muscle and endothelial cells. Biochem Biophys Res Comm 127: 433-442

67. Schiffrin, E.L., Chartier, L., Thibault, G., St-Louis, J., Cantin, M., Genest, J. (1985) Vascular and adrenal receptors for atrial natriuretic factor in the rat. Circ Res 56: 801-807

68. Schiffrin, E.L., Deslongchamps, M., Thibault, G. (1986) Platelet binding sites for atrial natriuretic factor in humans. Characterization and effects of sodium intake. Hypertension 8 (suppl II):II-6-II-10

69. Schwartz, D., Geller, D.M., Manning, P.T., Siegel, N.R., Fok, K.F., Smith, C.F., Needleman, P. (1985) Ser-Leu-Arg-Arg-atriopeptin III: The major circulating form of atrial peptide. Science 229: 397-400

70. Seidah, N.G., Cromlish, J.A., Hamelin, J., Thibault, G., Chrétien, M. (1986) Homologous IRCM-serine protease 1 from pituitary, heart atrium and ventricle: A common pro-hormone maturation enzyme? Biosci Rep 6: 835-844

71. Shibasaki, T., Naruse, M., Yamauchi, N., Masuda, A., Imaki, T., Naruse, K., Demura, H., Ling, N., Inagami, T., Shizume, K. (1986) Rat atrial natriuretic factor supresses proopiomelanocortin-derived peptides secretion from both anterior and intermediate lobe cells pituitary *in vitro*. Biochem Biophys Res Commun 135: 1035-1041

72. Sonnenberg, H., Veress, A.T. (1984) Cellular mechanism of release of atrial natriuretic factor. Biochem Biophys Res Commun 124: 443-449

73. Strom, T.M., Weil, J., Bidlingmaier, F. (1987) Platelet receptors for atrial natriuretic peptide in man. Life Sci 40: 769-773

74. Theiss, G., Morich, A.J.F., Neuser, D., Schroder, W., Stasch, J-P., Wohlfeil, S. (1987) α-h-ANP is the only form of circulating ANP in humans. FEBS 218: 159-162

75. Thibault, G., Lazure, C., Schiffrin, E.L., Gutkowska, J., Chartier, L., Garcia, R., Seidah, N.G., Chrétien, M., Genest, J.,Cantin, M. (1985) Identification of a biologically active circulating form of rat atrial natriuretic factor. Biochem Biophys Res Comm 130: 981-986

76. Thibault, G., Garcia, R., Gutkowska, J, Lazure, C., Seidah, N.G., Chrétien, M., Genest, J., Cantin, M. (1986) Identification of the released form of atrial natriuretic factor by the perfused rat heart. Proc Soc Exp Biol Med 182: 137-141

77. Thibault, G., Garcia, R., Gutkowska, J., Bilodeau, J., Lazure, C., Seidah, N.G., Chrétien, M., Genest, J., Cantin, M. (1987) The propeptide Asn 1-Tyr 126 is the storage form of rat atrial natriuretic factor. Biochem J 241: 265-272

78. Trippodo, N.C., Januszewicz, A., Pegram, B.L., Cole, F.E., Kohashi, N., Kardon, M.B., MacPhee, A.A., Frohlich, E.D. (1985) Rat platelets activate high molecular weight atrial natriuretic peptide *in vitro*. Hypertension 7: 905-912

79. Yandle, T.G., Richards, A.M., Nicholls, M.G., Cuneo, R., Espiner, E.A., Livisey, J.H. (1986) Metabolic clearance rate and plasma half-life of alpha-human atrial natriuretic peptide in man. Life Sci 38: 1827-1833

Pathomorphology of the endocrine heart

G. Rippegather

Department of Anatomy and Cell Biology, University of Heidelberg, Heidelberg, FRG

Summary

Our present investigations show that the heart with its myoendocrine cells will probably become an important tool in the evaluation of the status of cardiovascular and renal diseases. Tissue of the endocrine heart, blood plasma, and urine from patients undergoing open heart surgery for aorto-coronary bypass and correction of valve and septum diseases was examined. The right auricular myoendocrine cells were studied with immunohistochemistry and electron microscopy for cardiac hormones (Cardiodilatin = CDD 99-126/alpha human ANP). Depending on the status of the patient, differing quantities of CDD-IR (Cardiodilatin immunoreactivity) in the myoendocrine cells were detected by light microscopy, which is confirmed by the granularity or, by autophagolytic processes as seen in the electron microscope. Furthermore, pre- and postoperative blood samples as well as 24 h urine samples were analyzed by radioimmunoassay to measure the concentrations of CDD 99-126/alpha hANP. The concentrations of CDD/ANP measured by radioimmunoassay in plasma and urine samples vary according to different heart diseases. The highest concentrations were obtained in patients with mitral valve diseases (plasma: 67.3 +/- 31.9 fmol/ml, urine: 138.1 +/- 42.6 fmol/ml and aorto-coronary bypass patients with hypertension (plasma: 50.1 +/- 18.5, urine: 93 +/- 25.9 fmol/ml). Similar conditions of plasma and urine samples were detected in normotensive ACB-patients (plasma: 27.7 +/- 9.3, urine: 48.9 +/- 13.7 fmol/ml) and patients with aortic stenosis (plasma: 15.7 +/- 5.9, urine: 43.7 +/- 21.7). Blood and urine samples were also examined by high pressure liquid chromatography (HPLC) to determine the molecular forms of their circulating CDD/hANP, which were then compared. The present work shows that pathomorphological analysis plays an important role in the diagnosis of cardiovascular diseases.

Since the discovery of the heart as an endocrine organ, its role in the regulation of blood pressure and electrolyte-water threshold (3,9) has become evident. In spite of numerous publications which show the importance of the endocrine heart (1,2,8) there is little information about ultrastructural changes of the atrial endocrine organ during various cardiovascular diseases. Thus, we studied the parameters available for diagnostic and prognostic tools in patients undergoing cardiac surgery, i.e., aorta-coronary bypass, valve replacement or repair, and interventions in cases of cardiac malformations.

Methods

The auricular tissue, which was removed when the right atrium was opened to introduce the tube for extracorporal circulation, was collected. This tissue was divided into several parts for examination by electron and light microscopic immunohistochemistry (6) in order to analyze the cardiac hormone content. Pre- and postoperative venous blood samples as well as 24 h urine samples were taken from the same patients. These samples were extracted by sep-pak C 18 cartridges (Waters Associates, Milford, Massachusetts, USA) in order to determine the concentrations and forms of heart hormones in plasma and urine by radioimmunoassay and high pressure liquid chromatography (HPLC). Using the radioimmunoassay with an antibody against the C terminus of the Cardiodilatin 99-126/alpha human ANP peptide as previously described (9), we were able to measure the CDD/ANP levels. HPLC studies were carried out on a reversed-phase column (Orpegen HD-Sil. 18-5s.80). The column was prepared by the use of a linear gradient of acetonitrile (4-80%) containing 80 mM phosphoric acid. Absorption was measured at 210 nm. Fractions of 1 ml were eluted and used directly for radioimmunoassay. In order to localize the CDD 99-126/ alpha human ANP peak in the HPLC diagram, we used synthetic alpha human ANP (Bissendorf) as a control.

Results

All myoendocrine cells of the heart exhibit several common features: a) they can be stained selectively by antibodies directed specifically against cardiac hormones of the CDD/ANP family; b) their ultrastructure can be distinguished from other myoendocrine cells of the heart and conductive cells by its endocrine secretory apparatus containing all stages of pro-granule formation and granule maturation; and c) the Golgi apparatus is found to be at the poles of the elongated nuclei, (the peri-nuclei region), lateral to the elongated nuclei, (the paranuclear region), and at the periphery of the cells under the sarcolemma, (the telenuclear region); as well as d) the presence of specific granules is primarily in the perinuclear and telenuclear regions of the myoendocrine cells.

The CDD/ANP containing granules were stained either by immunohistochemistry (peroxidase-anti-peroxidase method, (15)) or by immunohistochemistry combined with electron microscopy using the immunogold method (10) with antibodies against the C and N terminus of cardiodilatin/atrial natriuretic peptide.

Pathomorphological features by light microscopical studies and immunohistochemistry showed a concomitantly high stainability for CDD/ANP on light microscopic sections when the tissue content is high; a light granularity is seen with in the electron microscope, as in the case of aorta-coronary bypass patients with hypertension. An extremely high tissue content of CDD/ANP was observed in the atria with mitral valve stenosis in light microscopic sections. Immunoreactivity could be found mainly in the periphery of the tissue sections. Compared to this, CDD immunoreactivity is much lower in patients with mitral valve insufficiency. In the case of patients with multiple valve diseases, i.e., mitral valve stenosis, aortic stenosis, and tricusspidal insufficiency, we were also able to detect a high amount of CDD/ANP immunoreactivity in the endocrine auricular tissue. ACB patients with normotension exhibited moderate CDD/ANP immunoreactivity and there was an equal distribution of immunoreactivity in the whole tissue.

Light microscopic immunohistochemical investigations of auricular tissue of patients with aortic stenosis display similar characteristics as ACB patients with normotension, with the exception of those aortic stenosis patients with fibrosis where the tissue is damaged and a high CDD/ANP immunoreactivity is found.

Pathomorphological features of electron microscopic studies

Electron microscopy shows a heterogenous image of the ultrastructure of myoendocrine cells (Figs. 1-5). Signs of inhibition or acceleration of the secretory cycle as well as variable pathomorphological changes (Figs. 1,2) in the myoendocrine cells depend on the age and status of the patients. Reasons for the different morphological and ultrastructural appearances of the endocrine heart seem to be the absence (as in the case of ACB patients with normotension) or the presence (as in the case of ACB patients with hypertension) of a pathological overload of the atria. Usually normotensive ABC patients show a little altered and age dependent structure of the myoendocrine cells (Fig. 3).

In hypertensive ACB patients, an increased development of secretory granules can be observed. One can find high amounts of progranula and granula as well as a hypertrophy of the Golgi apparatus which is confirmed by an increased rate of secretion in the cells. Various types of ultrastructure are detected in the myoendocrine cells of patients with mitral valve diseases and aortic stenosis. A marked destruction of the myoendocrine cells by an autophagolytic process (Fig. 1,2), which indicates a necrosis of the heart tissue, is seen as well as some normal or hypergranulated myoendocrine cells (Figs. 4,5). Furthermore, a hypertrophy of mitochondria can be observed, which indicates the increased use of energy by increased atrial pressure, and high amounts of empty vesicles are found (Fig. 1).

Results of radioimmunoassay

The CDD/ANP concentrations of pre- and postoperative plasma samples and 24 h urine samples were measured by radioimmunoassay. The cardiac peptide levels of normotensive patients undergoing open heart surgery for aorto-coronary bypass served as a control and varied from 27.7 +/- 9.3 fmol/ml in plasma and from 48.9 +/- 13.7 fmol/ml in urine samples. The levels of atrial natriuretic peptide in plasma as well as in the urine of ACB patients with hypertension however rose above normal by a factor of 2 or 6 and were higher in patients with more severe disease (50.1 +/- 18.5 fmol/ml in plasma and 93 +/- 25.9 fmol/ml in urine). The highest levels of CDD/ANP were found in patients suffering from mitral valve stenosis. The concentrations detected by radioimmunoassay were 67.3 +/- 31.9 fmol/ml in plasma and 138.1 +/- 42.6 fmol/ml in urine samples. The concentration of CDD/ANP in plasma and urine of patients with aortic stenosis are almost in the same range as those of normotensive ACB-control patients (15.7 +/- 5.9 fmol/ml in plasma and 43.7 +/- 21.7 fmol/ml in urine samples).

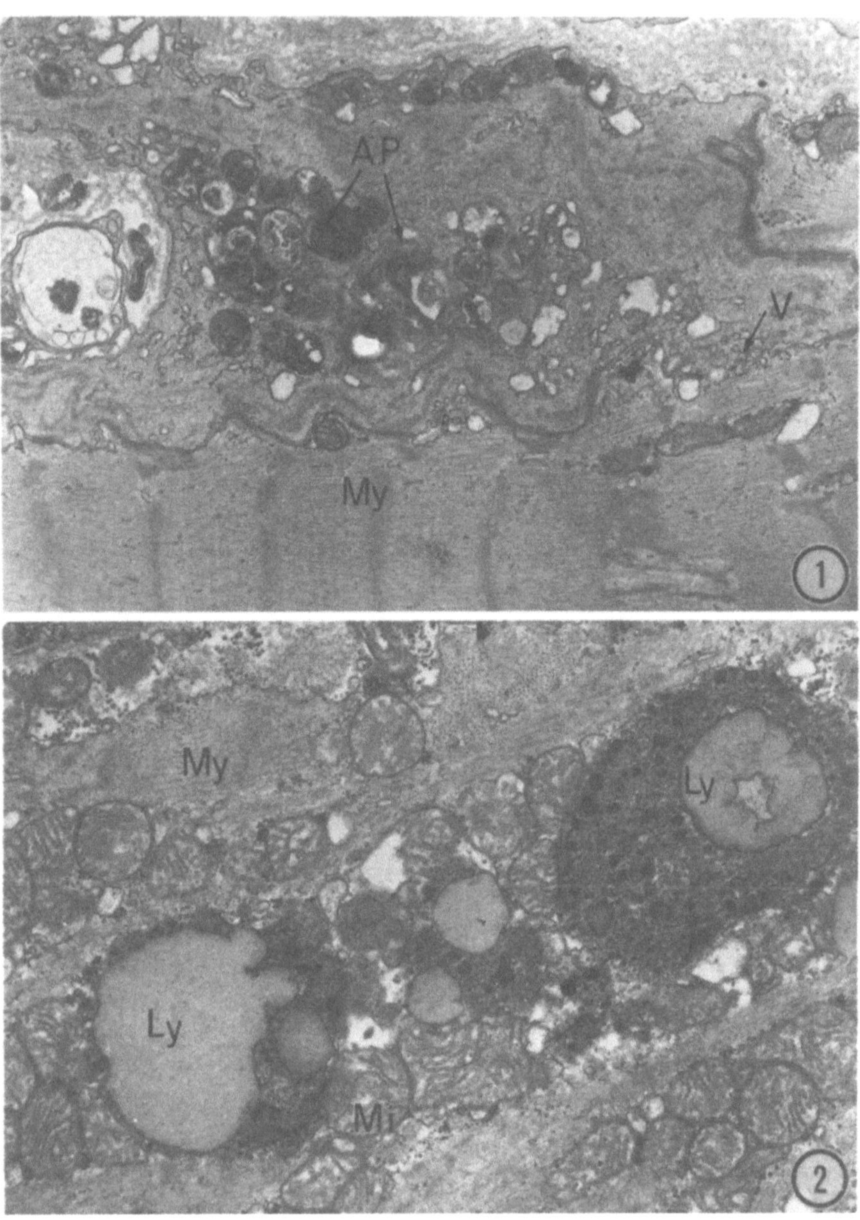

Figs. 1 and 2. Ultrastructure of myoendocrine cells of patients suffering from aortic stenosis. 1. autophagolytic processes (AP) showing a strong degeneration of atrial myocardiocyte were localized. Empty vesicles (V) were found. (x15 000) 2. Degenerating myoendocrine cell filled with lamallar and heteromorphic lysosomes (Ly). The cell contains intact myofibrilles (My) and mitochondria (Mi). (x24 200)

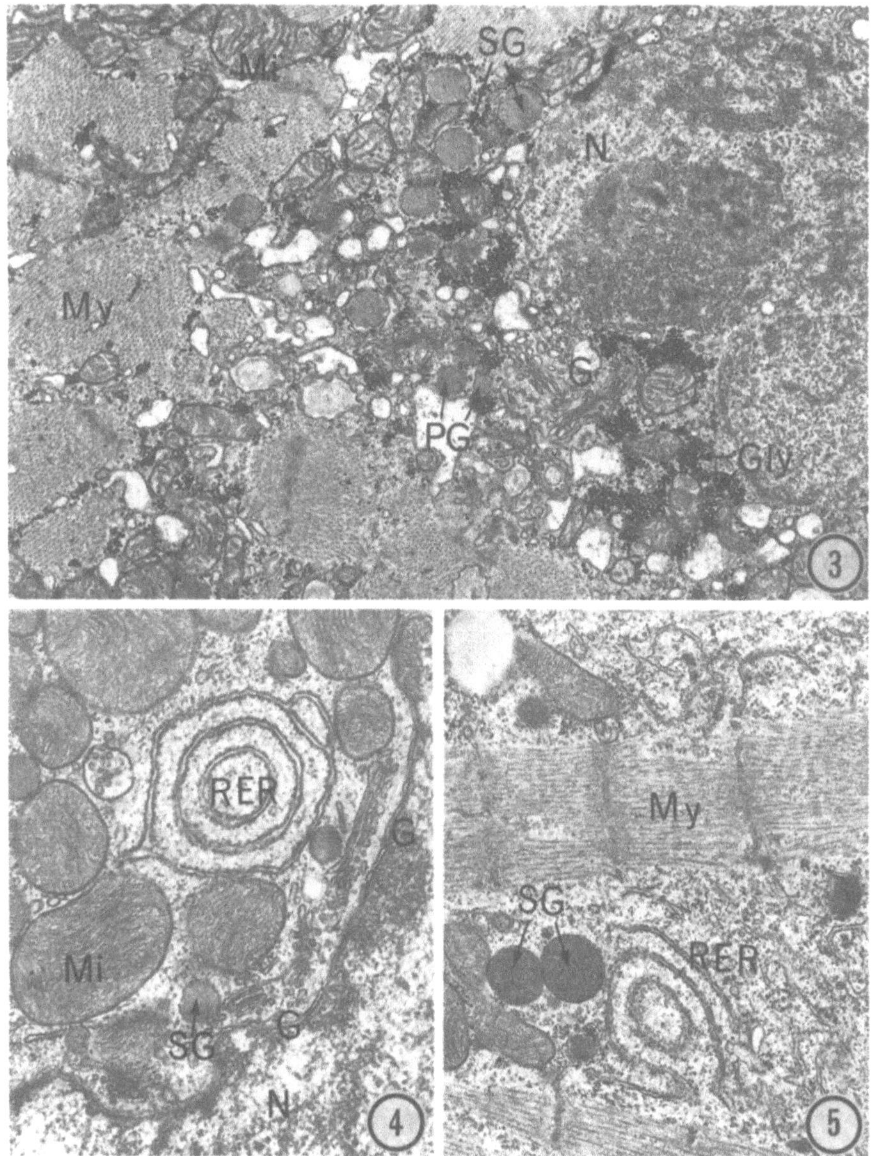

Fig. 3. Ultrastructure of the atrial myocardium of a normotensive ACB-patient (aorto-coronary bypass). Localization of the perinuclear region containing the Golgi apparatus (G) with different stages of granule naturation, progranules (PG) and secretory granules (SG) as well as mitochondria (Mi) and myofibrilles (My). Cytoplasmatic glycogen (Gly) is concentrated in the perinuclear region. N = nucleus. (x95 200)
Figs. 4 and 5. Ultrastructure of myoendocrine cells of patients suffering from mitral valve stenosis. Localization of rough endoplasmatic reticulum (RER), high amounts of secretory granules (SG) and a developed Golgi apparatus (G) indicating an increased protein synthesis. 4. (x94 500); 5. (x15 500)

A further important factor seems to be that the urine concentrations of CDD/ANP are higher than those measured in plasma. This was true in all cases of heart failure except for those patients suffering from mitral valve insuffiency, where we obtained nearly equivalent amounts of CDD/ANP in both plasma and urine samples.

If one compares the CDD/ANP concentrations in pre- and postoperative plasma samples a tendency towards a slight decrease in CDD/ANP plasma levels becomes evident. If the preoperative amount of CDD/ANP is very high, as in the case of ACB patients with hypertension or patients with mitral valve stenosis, the observed postoperative decrease is more significant (14).

Discussion

In this study, the results of the pathomorphology and related biochemistry on the status of the endocrine heart demonstrate that in various cardiovascular diseases (aorto-coronary bypass with normotension, aorto-coronary bypass with hypertension, aortic stenosis, mitral valve insufficiency, mitral valve stenosism, tricusspidal insufficiency and Fallot tetralogy) the analysis of CDD/ANP concentrations contributes to a better understanding of the outcome of operative interventions in these patients.

We obtained characteristic results from nearly 800 patients operated on since 1983. Thus, we were able to show a strong homogeneity in morphological appearances of the endocrine apparatus of the heart and the detectable amounts of cardiac hormones in blood and urine samples. The differing concentrations of CDD/ANP in even one case of heart failure may be substantiated by our morphological studies.

In the interesting case of patients suffering from mitral valve diseases, different and scattering levels of CDD/ANP were measured in plasma and urine by radioimmunoassay. Correspondingly, we found a heterogeneous status of the myoendocrine cells by using an electron microscope. Depending on the degree of granularity in the cells or the status of the secretory apparatus with its Golgi apparatus, progranules and granula, we found equivalently high or low concentrations of CDD/ANP in plasma and urine samples of the same patient. Another typical example is the case of aorto-coronary bypass patients with hypertension or normotension. As Dietz et al. (4) published, we observed a dependency of blood pressure and cardiac hormone content in tissue, plasma, and urine. A pressure overload of the atrium and its distention are accepted to be among the responsible factors for cardiac hormone release (9,14). Considering this, we were not surprised to observe a hypertrophy of the Golgi apparatus and energy-producing system of mitochondria as well as a high amount of CDD/ANP immunoreactivity and, an increase in CDD/ANP concentrations in blood and urine samples in ACB patients with hypertension and mitral valve stenosis. This result does not speak in favour of a down-regulation of peptide release during a chronic pressure overload. The tendency towards a slight decrease in CDD/ANP plasma concentrations after operative interventions in nearly every case of heart failure demonstrates a relief of chronic pressure overload.

Another fact worth considering is the increase in CDD/ANP concentrations in urine samples in comparison to the values of CDD/ANP measured in plasma samples of the same patients is not correlated. Thus, CDD/ANP in the urine may represent a compartment independent from blood plasma.

Conclusion

The combined morphological and biochemical investigations show that the plasma and urine contents of radioimmunoassayable CDD/ANP and the morphological status of the endocrine heart may be a tool for the clinical evaluation of the individual patient. Our study emphasizes that cardiac hormones are released in order to protect the cardiovascular system against overload and high blood pressure and that ANP could be a hormone marker of the hemodynamic overload in clinical heart diseases. It has been shown that the circulating CDD/ANP detected in the combined HPLC/RIA-analysis shows one molecular form (5) which was isolated and sequenced in our laboratory and was mentioned before by Tanaka et al. (16), Yamanaka et al. (17) and Neeldeman et al. (12). Some studies with plasma and urine of patients suffering from aortic stenosis point out the existence of another molecular form of the cardiac peptide which could result from pathoogical processes in the heart. We are just beginning to understand the significance of the endocrine heart.

References

1. Cantin, M., Genest, J. (1985) The heart and the atrail natriuretic factor. Endocr Rev 6: 107-127
2. Currie, M.G., Geller, D.M., Cole, B.R., Boylan, J.G., YuSheng, W., Holmberg, S.W., Needleman, P. (1983) Bioactive cardiac substances: potent vasorelaxant activity in mammalian atria. Science 221: 71-73
3. DeBold, A.J., Borenstein, H.B., Veress, A.T., Sonnenberg, H. (1981) A rapid and potent natriuretic response to intravenous injection of atrial extracts in rats. Life Sci 28: 89-94
4. Dietz, R., Purgaj, J., Lang, R.E., Schoemig, A. (1986) Pressure dependent release of atrial natriuretic peptide (ANP) in patients with chronic cardiac disease: does it reset? Klin Wochenschr 64 (Suppl VI): 42-46
5. Forssmann, K., Hock, D., Herbst, F., Schulz-Kappe, P., Talartschik, J., Scheler, F., Forssmann, W.G. (1986) Isolation and structure of the circulating human cardiodilatin (alpha ANP). Klin Wochenschr 64: 1276-1280
6. Forssmann, W.G., Pickel, V., Reinecke, M., Hock, D., Metz, J. (1981) Immunohistochemistry and immunocytochemistry of nervous tissue In: Heym, Ch., and Forssmann, W.G. (eds.), Techniques in neuroanatomical research. Springer, Berlin Heidelberg New York, pp 171-205
7. Forssmann, W.G., Birr, C., Carlquist, M., Christmann, M., Finke, R., Henschen, A., Hock, D., Kirchheim, H., Kreye, V., Lottspeich, F., Metz, J., Mutt, V., Reinecke, M. (1984) The auricular myocardiocytes of the heart constitute an endocrine organ. Characterization of a porcine peptide hormone, cardiodilatin-126. Cell Tissue Res 238: 425-430
8. Forssmann, W.G. (1986) Cardiac hormones. I. Review on the morphology, biochemistry, and molecular biology ofthe endocrine heart. Eur J Clin Invest 16: 439-451
9. Lang, R.E., Thoelken, H., Ganten, D., Luft, P.C., Ruskoaho, H., Unger, Th. (1985) Atrial natriuretic factor - a circulating hormone stimulated by volume loading. Nature 314: 264-266

10. Maldonado, C.A., Saggau, W., Forssmann, W.G. (1986) Cardiodilatin-immunoreactivity in specific atrial granules of human heart revealed by the immunogold stain. Anat Embryol 173: 295-298

11. Marie, J.P., Guillemot, H. Hatt, P.Y. (1976) Le degrá de granulation des cardiocytes auriculaires. Études planimétrique au cours de différents apports d'eau et de sodium chez le rat. Path Biol 24: 549-554

12. Needlemann, P. (1986) Atriopeptin biochemical pharmacology. Fed Proc 45: 46-67

13. Rippegather, G., Maldonado, C.A., Schulz-Knappe, P., Hock, D., Lang, R.E., Forssmann, W.G. (1986) Morphologie der myoendokrinen Zellen des Rattenherzens bei akuten Veränderungen des Blutvolumens. Verh Anat Ges 81: 109-113

14. Rippegather, G., Saggau, W., Forssmann, W.G. (1987) Pathomorphologie des endokrinen Herzens In: Kreye, V.A.W., Bussmann, W.D. (Hrsg) ANP - Atriales natriuretisches Peptid und das kardiovaskuläre System. Steinkopff, Darmstadt, S 169-178

15. Sternberger, L.A. (1979) Cytochemistry. 2nd edn John Wiley and Sons, New York

16. Tanaka, H., Shindo, M., Gutkowska, J., Kinoshita, A., Urata, H., Ikeda, M., Arakawa, K. (1986) Effect of acute excercise on plasma immunoreactive atrial natriuretic factor. Life Sci 39: 1685-1693

17. Yamanaka, M., Greenberg, B., Johnson, L., Seilhammer, J., Brewer, M., Friedman, T., Miller, J., Atlas, S., Laragh, J., Lewicki, J., Fiddes, J. (1984) Cloning and sequence analysis of the cDNA for the rat atrial natriureticfactor precursor. Nature 309: 719-722

Endocrine heart in experimental cardiomyopathy

A.J.G. Riegger, D. Elsner, F. Muders, E. Pascher, E.P. Kromer

Medical University Clinic Würzburg, Würzburg, FRG

Summary

The pathophysiological importance of atrial natriuretic peptide (ANP) in the development of congestive heart failure is unknown.

We studied hemodynamic and renal changes in an animal model of congestive heart failure in the conscious dog following incremental intravenous doses of ANP (0.01, 0.03, 0.1, 0.3, 0.6 mg/kg/min) before and after chronic heart failure was established. Heart failure was induced by rapid right ventricular pacing (260 beats/min) for 10 days. After 10 days cardiac output and aortic pressure were significantly reduced, mean pulmonary and right atrial pressure significantly increased. ANP increased from 7.9 + 3.0 to 49.1 + 11.4 pg/ml (p<0.01). In contrast to healthy dogs, incremental intravenous infusions of ANP did not change cardiac output, pulmonary, or right atrial pressure. After heart failure was induced ANP reduced only aortic pressure. In chronic heart failure we found a striking attenuation of renal effects following ANP infusions. Plasma renin concentration was significantly suppressed by ANP at baseline, in heart failure it was not influenced.

Our results show a marked attenuation of hemodynamic and renal effects of ANP in heart failure in comparison to a normal cardiocirculatory status. This suggests a less important role of ANP in long-term regulation of chronic heart failure. This may be due to a down-regulation of ANP-receptors, a defect of post-receptor intracellular mechanisms, or may be due to counter-regulating neuro-humoral mechanisms that override it the effects of ANP.

Introduction

In 1981 DeBold et al. demonstrated that intravenous administration of atrial extracts to intact rats induces profound natriuresis and diuresis (1). Atrial natriuretic peptide (ANP) is elevated in heart failure in proportion to the severity of the disease (2). Pharmacological doses of ANP-infusions have beneficial effects on cardiac function in patients with severe chronic heart failure by reducing pre- and afterload (3,4,5). In contrast to healthy volunteers a considerable attenuation of the renal effects of ANP was found in patients with heart failure.

Until now no investigations have tried to precisely define hemodynamic, renal and hormonal responses to ANP using a whole dose response in an experimental preparation of chronic heart failure. Therefore, we performed dose-response studies in healthy animals and dogs with congestive heart failure by infusing ANP in incremental doses.

This study was supported by the "Deutsche Forschungsgemeinschaft"

Methods

In six female mongrel dogs (mean weight 17.5 kg) we induced chronic low output heart failure by rapid right ventricular pacing. The technique is described in detail in two previous publications (6,7). The animal preparation allows the measurement of cardiac output, mean aortic pressure, right atrial pressure, and mean pulmonary arterial pressure. All measurements were performed in conscious dogs trained to stand quietly on a table in a cotton mesh support sling. Urine was collected using a Foley catheter which was inserted into the bladder under short anaesthesia (thiopental sodium 12.5 mg/kg). Synthetic sterile 28-amino acid alpha-ANP (Bissendorf, FRG) dissolved in 0.9% saline, was administered into the pulmonary artery by constant infusions (each dose lasting for 30 min.). Hemodynamic, hormonal, and renal measurements were performed at the end of each infusion period. ANP infusions covered a whole dose response curve using 0.01, 0.03, 0.1, 0.3 and 0.6mg ANP/kg/min. All measurements were performed in the conscious healthy dog and after 10 days of rapid right ventricular pacing when congestive heart failure was established.

Plasma ANP was measured by radioimmunoassay (Peninsula Lab., Inc.) previously extracted on C_{18} Sep Pak cartridge (Waters Ass., Massachusetts). Plasma renin concentration was measured as described previously (7). Electrolytes were measured by flame photometry (automatic analyzer SMAC Technicon).

Statistical analysis

Data are presented as means ± standard error of the mean. Analysis of variance for repeated measurements was performed to determine the significance of changes of hemodynamic, renal, and humoral parameters. To test the significance of single comparisons a non-parameteric procedure (Wilcoxon matched pairs signed-ranks test) was applied. Correlation coefficients were obtained by linear regression analysis (method of least squares). Statistical significance was accepted at the p<0.05 levels.

Results

The hemodynamic changes and changes in plasma ANP-levels during the development of congestive heart failure due to rapid right ventricular pacing are summarized in Fig. 1. Rapid right ventricular pacing resulted in a significant fall of cardiac output and mean arterial blood pressure, a significant increase of mean pulmonary arterial pressure, right atrial pressure and plasma levels of ANP. Plasma ANP values increased during the development of congestive heart failure about six to sevenfold. A significant positive linear correlation was found between right atrial pressure and plasma ANP (r = 0.75; p<0.001).

Figure 2 shows hemodynamic data during the dose-response study in relation to ANP plasma levels at base-line and at the end of each infusion period. ANP infusions decreased mean arterial blood pressure in healthy dogs and in dogs with congestive heart failure. Stroke volume was reduced by ANP infusions in healthy animals due to a significant

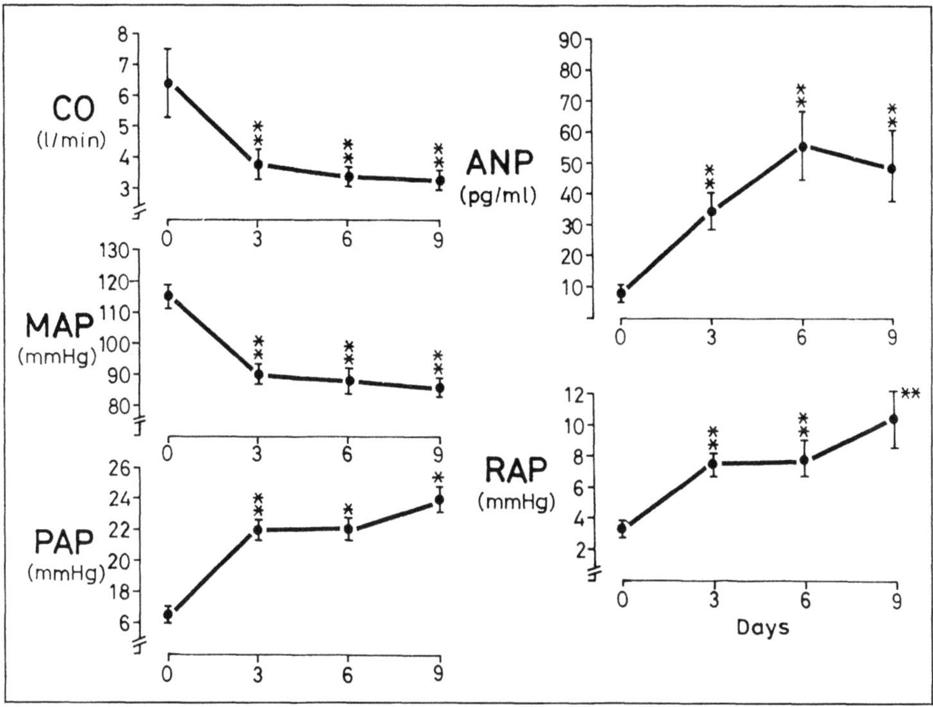

Fig.1. Cardiac output (C0), mean arterial pressure (MAP), mean pulmonary arterial pressure (PAP), mean right atrial pressure (RAP), and plasma levels of atrial natriuretic peptide (ANP) at baseline and during the development of congestive heart failure. (*p< 0.05; **p<0.03)

reduction of preload, documented by a significant fall of mean right atrial pressure. In heart failure, when right atrial pressure was significantly (p<0.03) elevated, incremental infusions of ANP had no effect on stroke volume and preload. Heart rate and calculated total peripheral vascular resistance did not change under base-line conditions and in heart failure.

Figure 3 shows the changes of urine flow and urinary sodium excretion in healthy dogs, and dogs with congestive heart failure in relation to the plasma levels of ANP. As expected we found a striking increase of urinary flow and sodium excretion in healthy animals. In dogs with congestive heart failure the renal effects of ANP were markedly attenuated. Nearly identical responses were found for urinary chloride, magnesium and calcium excretion and to a lesser extent for urinary potassium excretion.

Figure 4 shows plasma renin concentration before and after cardiac pacing. ANP infusions reduced plasma renin concentration in healthy animals significantly. In heart failure, when plasma renin concentration was significantly elevated (p<0.03), incremental doses of ANP did not alter plasma renin concentration significantly.

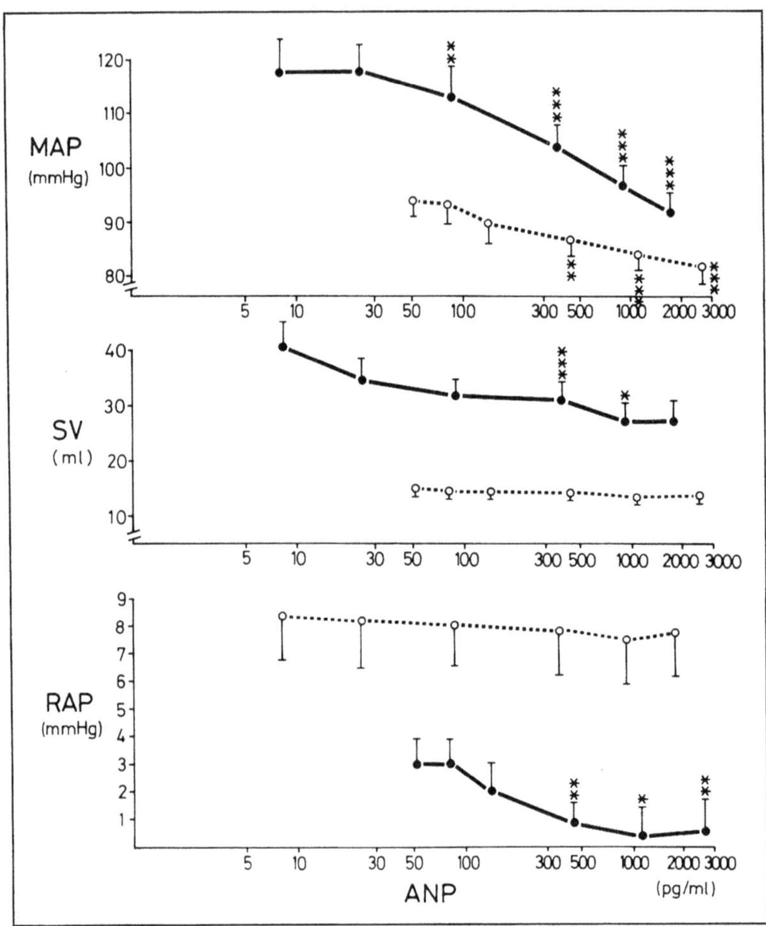

Fig.2. Mean arterial pressure (MAP), stroke volume (SV) and mean right atrial pressure (RAP) in healthy dogs (•---•), and in dogs with heart failure (0---0) in relation to plasma levels of atrial natriuretic peptide (ANP) at baseline and during incremental infusions of ANP. (*p<0.05; **p<0.04; ***p<0.03)

Discussion

In our study we demonstrated an increase of ANP in relation to right atrial pressure in an animal model of congestive heart failure during the development of the disease. We investigated for the first time in an animal preparation of chronic heart failure effects of ANP in healthy and diseased dogs by infusing incremental doses of ANP covering a whole-dose response. In healthy control dogs we found a significant decrease of mean arterial blood pressure, cardiac output, stroke volume and right atrial pressure without

156

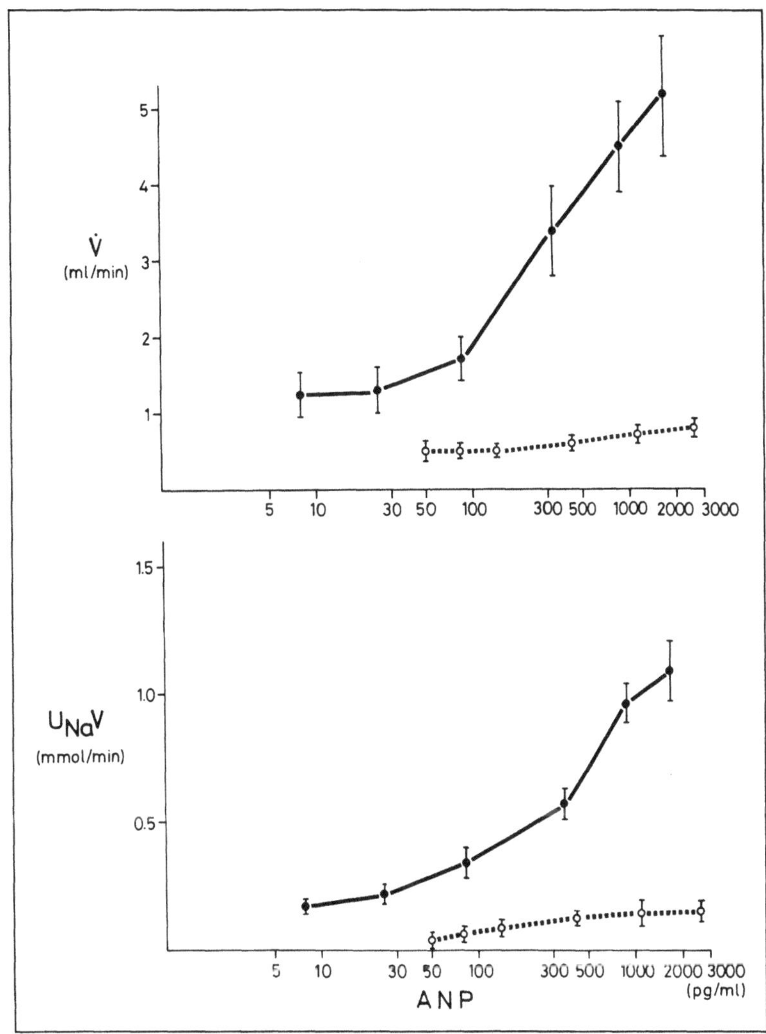

Fig.3. Urine flow (V) and urinary sodium excretion (UNaV) in dogs before (•—--•), and after established heart failure (0---0) in relation to plasma atrial natriuretic hormone (ANP) at baseline and during the dose response study

any changes on total peripheral vascular resistance and heart rate. These data confirm previous findings (8,9), that ANP decreases cardiac output by a reduction of stroke volume due to a reduced venous return. The explanation for this phenomenon may be a venodilatation and/or a reduced circulating blood volume. As total peripheral resistance did not change, our data indicate that mean arterial pressure is lowered by ANP by a reduction of stroke volume and cardiac output. In established congestive heart failure all

157

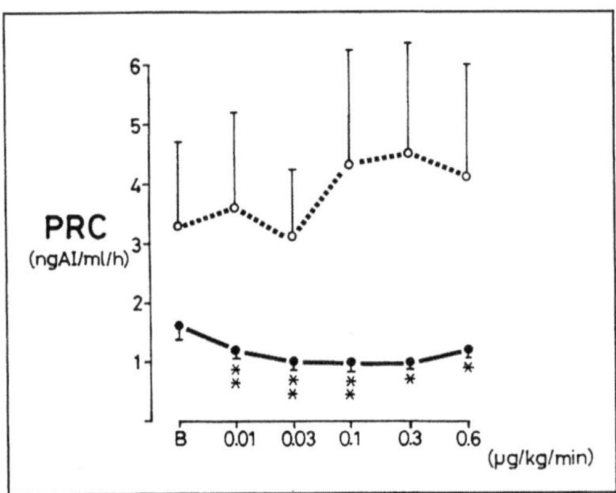

Fig.4. Plasma renin concentraion (PRC) before (•---•) and in heart failure (0---0) at baseline (B) and during intravenous infusion of atrial natriuretic peptide (ANP). (*p<0.05; **p<0.03)

hemodynamic changes induced by ANP infusions found in healthy control dogs were completely abolished with the exception of a small reduction of mean arterial blood pressure.

The increase of ANP during the development of heart failure can be explained by an increase in atrial wall tension (10,11,12). Similar to findings in patients with severe congestive heart failure (3,4,5) we found in our experiment a marked attenuation of the renal response to exogenous ANP in dogs with heart failure. We found significant suppression of renin secretion in the control animals, as has been reported in man (5). The suppression of renin production may be mediated through renal vascular receptors and receptors in the macula densa for ANP. Renin may also be suppressed by a direct inhibitory action on the juxtaglomerular cells (13). Similar to results in patients with severe heart failure (3) we found no significant effects of ANP on plasma renin secretion when heart failure was induced. The marked diminution of hemodynamic, renal and hormonal responses to exogenous ANP in heart failure may be due to a down-regulation of binding sites for ANP which has been demonstrated in platelets of patients with congestive heart failure (14). A reduction of ANP binding sites has also been shown in cultured vascular smooth muscle cells after exposure with ANP (15) and in rats after mineralocorticoid administration (16). The decreased responsiveness of the vascular system and the nearly complete absence of renal excretory effects of ANP in heart failure may also be due to a possible defect of post-receptor intracellular mechanisms or may be due to counter regulating neurohumoral mechanisms which may be able to override the effects of ANP.

The attenuated effects of ANP in chronic heart failure suggest a less important role of ANP in the long-term regulation of the disease as a counter-regulating system to vasoconstrictor, and sodium and water retaining mechanisms like the renin-angiotensin-aldosterone system, the sympathetic nerve activity, and vasopressin.

References

1. DeBold, A.J., Borenstein, H.B., Veress, A.T. (1981) A rapid and potent natriuretic response to intravenous injection of atrial myocardial extract in rats.Life Sci 28: 89-94
2. Tikkanen, I., Fyhrquist, I.R., Metsärinne, K., Leidenius, R. (1985) Plasma atrial natriuretic peptide in cardiac disease and during infusion in healthy volunteers. Lancet II: 66-69
3. Riegger, A.J.G., Kromer, E.P., Kochsiek, K., (1986) Human atrial natriuretic peptide: plasma levels, haemodynamic, hormonal and renal effects in patients with severe congestive heart failure. J Cardiovasc Pharmacol 8: 1107-1112
4. Crozier, I.G., Nicholls, M.G., Ikram, H., Espiner, E.A., Gometz, H.J., Warner, N.J. (1986) Haemodynamic effects of atrial peptide infusion in heart failure. Lancet II: 1242-1245
5. Cody, R.J., Atlas, S.A., Laragh, J.H., Kubo, S.H., Covit, A.B., Ryman,K.S., Shaknovich, A., Pondolfino, K., Clark, M., Camargo, M.J.F., Scarborough, R.M. (1986) Atrial natriuretic factor in normal subjects and heart failure patients. J Clin Invest 78: 1362-1374
6. Riegger, A.J.G., Liebau,G. (1982) The renin-angiotensin-aldosterone system, antidiuretic hormone and sympathetic nerve activity in an experimental model of congestive heart failure in the dog. Clin Sci 62: 465-469
7. Riegger, A.J.G., Liebau, G., Holzschuh, M., Witkowski, D., Steilner, H., Kochsiek, K. (1984) Role of the renin-angiotensin system in the development of congestive heart failure in the dog as assessed by chronic converting-enzyme blockade.Am J Cardiol 53: 614-618
8. Breuhaus, B.A., Saneii, H.H., Brandt, M.A., Chimoskey, J.E. (1985) Atriopeptin II lowers cardiac output in conscious sheep. Am J Physiol 249: R776-R780
9. Goetz, K.L., Wang, B.C., Geer, P.G., Sundet, W.D., Needleman, P. (1986) Effects of atriopeptin infusion versus effects of left atrial stretch in dogs which are awake. Am J Physiol 250: R221-R226
10. Lang, R.E., Thoelken, E.H., Ganten, D., Luft, F.C., Ruskoaho, H., Unger, T.H. (1985) Atrial natriuretic factor is a circulating hormone, stimulated by volume loading Nature 314: 264-266
11. Ledsome, J.R., Wilson, N., Courneya, C.A., Rankin, A.J. (1985) Release of atrial natriuretic peptide by atrial distension. Can J Physiol Pharmacol 63: 739-742
12. Bates, E.F., Shenker, Y., Grekin, R.J. (1986) The relationship between plasma levels of immunoreactive atrial natriuretic hormone and haemodynamic function in man. Circulation 73: 1155-1161
13. Kurtz, A., Bruna, R.D., Pfeilschifter, J., Taugner, R., Bauer, C. (1986) Atrial natriuretic peptide inhibits renin release from juxtaglomerular cells by a cGMP-mediated process. Proc Natl Acad Sci USA 83: 4769-4773
14. Schiffrin, E.L. (1986) Down-regulation of binding site for atrial natriuretic peptide in platelets of patients with congestive heart failure. Circulation 74 (suppl II) 1463 (abstract)
15. Hirata, Y., Tomita, M., Takada, S., Yoshimi, H. (1985) Vascular receptor binding activities and cyclic GMP responses by synthetic human and rat atrial natriuretic peptides (AMP) and receptor down-regulation by ANP. Biochem Biophys Res Commun 128: 538-546
16. Ballerman, B.J., Block, K.D., Seidman, J.G., Brenner, B.M. (1986) Atrial natriuretic peptide transcription, secretion and glomerular receptor activity during mineralocorticoid escape in the rat. Clin Invest 78: 840-843

Atrial natriuretic peptide in man

P. Weidmann, H. Saxenhofer, C. Ferrier, S. G. Shaw

Medizinische Poliklinik, University of Berne, Switzerland

The history of the atrial natriuretic peptides (ANP) begins with the observation of electron dense granules in atrial myocytes of guinea pigs (1). Similar granules were later demonstrated in human and other species (2-4), and it was noted that the granularity varies inversely with the state of hydration (5-6). This prompted the landmark study of De-Bold, showing a natriuretic and diuretic effect of extract from rat atrial tissue (7). Several groups have subsequently identified and synthesized ANPs with 21 to 28 amino acid residues (8-12), which in addition to their natriuretic effect, exhibit vasorelaxant and adrenal steroidogenesis-inhibiting (13,14) properties. Although investigations in humans began only recently, information is rapidly expanding. The following summarizes some observations on blood levels and effects of ANP in man.

Regulation of plasma ANP in normal humans

In human auricles, a prepropeptide with 151 amino acid residues is through cleavage of a 25 amino acid residue signal peptide processed to a 126 amino acid residue propeptide (10,12,15). The latter is biologically inactive and stored in the atrial myocytes (15,16). In healthy man, the biologically active α–ANP is released into the blood (17- 20); it has 28 amino acid residues and in the circulation a very short half-life of about 3 min (21,22, 23).

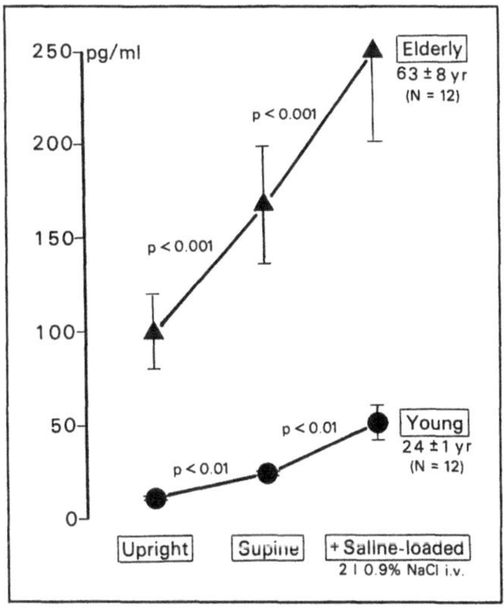

Fig. 1. Plasma immunoreactive atrial natriuretic peptide (irANP) levels in the upright and supine positions (each for 1 h), and following intravenous saline loading in the supine position in healthy young or elderly subjects (means ± SEM); from (x28)

The secretion from the heart is activated during atrial distention secondary to an increase in central blood volume and/or right or left atrial pressure (24-27). In healthy humans, plasma immunoreactive ANP (irANP) levels rise in response to intravenous loading with saline (Fig. 1) (28-30) or water (31), an acute increase in sodium intake (28,32,33), a central shift of volume produced by the change from upright to the supine position (Fig.1) (30,33,34) or by immersion into water (35,36), a vasoconstrictor hormone-induced acute increase in blood pressure (BP) and, hence afterload (Fig. 2) (37, 38), dynamic exercise (39,40), or increased heart rate induced by pacing (32,41).

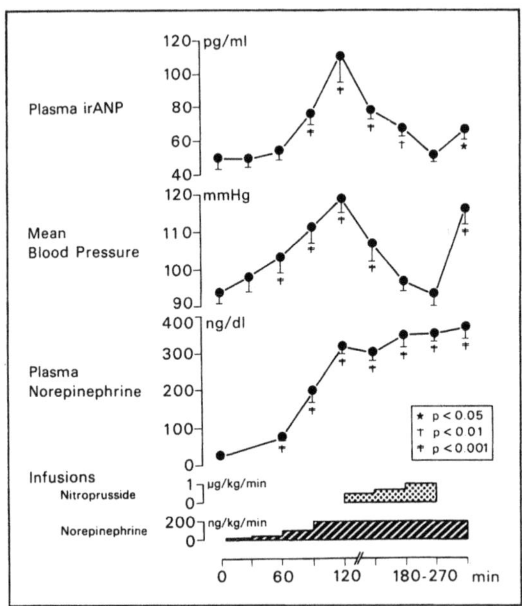

Fig. 2. Plasma irANP, mean blood presure, and plasma norepinephrine levels before and during intravenous infusions of norepinephrine alone or combined with sodium nitroprusside (means ± SEM). Note the parallel variations in irANP and blood pressure and the dissociation between irANP and norepinephrine; from (x38)

Circulating irANP concentrations also vary with aging. Levels are higher in neonates than in infants and adolescents (42). In the adult, plasma irANP concentrations increase markedly from the 3rd to the 7th decades of life (Fig. 1) (30,43). The underlying mechanism remains to be clarified.

162

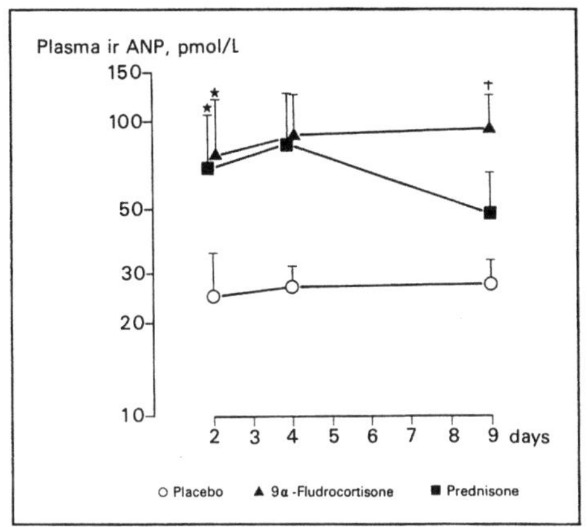

Fig. 3. Plasma irANP levels during administration of placebo, prednisone (50 mg/day), or 9α-fludrocortisone acetate (0.6 mg/day), in eight young normal men on a diet containing 130 mmol sodium and 75 mmol potassium/day (means ± SEM). Note that prednisone produced no sodium-fluid retention or increased blood pressure; escape from mineralcorticoid-induced sodium intention occured on average on day 7; from (x44)

Mineralocorticoids as well as glucocorticoids administered in high doses provoke in humans (44,45) and experimental animals (46,49) an increase in circulating irANP (Fig. 3). In rats, atrial contents of irANP (46,48,50) and ANP messenger RNA (47,48) were also elevated. This indicates an increase in cardiac ANP production and release; an additional influence of corticosteroids on the plasma clearance of ANP has not yet been excluded. The mineralocorticoid-induced rise in circulating irANP is probably at least in part a consequence of the concomitant sodium-fluid volume retention and BP elevation. However, glucocorticoid excess stimulated irANP levels without changing body sodium balance, whole blood volume or BP (44). The presence of glucocorticoid receptors in the heart (5) as well as a putative glucocorticoid receptor binding site in the second intervening sequence of the ANP gene (52,54) is consistent with the possibility that glucocorticoids can directly modulate the production and/or release of ANP. On the other hand, the heart contains no true mineralocorticoid receptors.

Certain studies *in vitro* suggested that epinephrine and arginine-vasopressin (AVP) can also activate ANP secretion from atrial myocytes (55,58). Others found no evidence of direct stimulation by catecholamines (50,59).

Effect of αANP in normal humans

Acute intravenous infusion of synthetic human αANP can produce a wide spectrum of cardiovascular, endocrine and renal interactions (Table 1). Under the acute conditions evaluated so far, BP was largely unaltered by low to modest doses of αANP (e.g. 0.007 to

Table 1. Infusion of α-Atrial Natriuretic Peptide in man

Effects			Mechanism
Blood pressure		↓	Vasodilation, inappropiate CO ↑ Blood volume ↓
Sympathetic activation → Heart rate, lipolysis		↑	Baroreflex
Plasma			
Aldosterone		↓	
Cortisol	= or	↓	Adrenal steroidogenesis ↓
Renin	to	↓ ↑	Secretion ↓ (low αANP dose) Indirect stimulation (high dose)
Arg. Vasopressin	= or	↓	Pituitary secretion?
Insulin		↑	?
Diuresis, natriuresis		↑	Renal hemodynamics (GFR ↑ , etc) Tubular (indirect and/or direct ?)

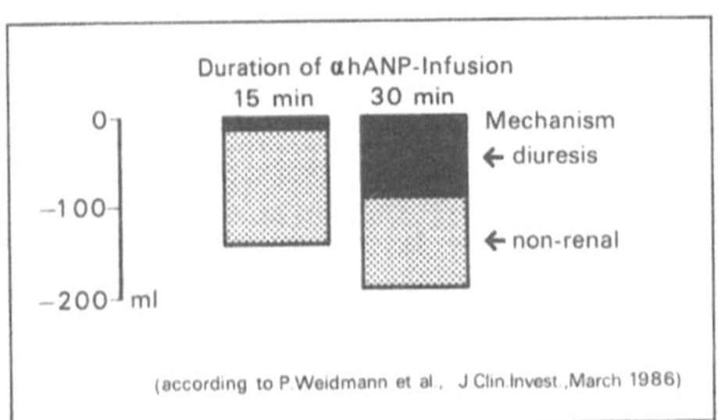

Fig. 4. Decrease in calculated plasma during infusion of human αANP (50 μg bolus followed by 0.1 μg/kg/min) in ten normal subjects. Average changes are shown; adapted (x20)

164

0.05 m/min/kg) (32,60,61), but decrease when high doses (e.g.: 0.1 to 0.2 m/min/kg) were infused (20,61,62). αANP´s hypotensive action is probably due to a decrease in total peripheral vascular resistance without a commensurate rise in cardiac output (63). ANP may modulate vascular tone through activation of specific receptors (64,66), thereby also blunting the vasoconstrictor effects of norepinephrine and angiotensin II (26,37). Nevertheless, depending on the localization and state of precontraction of the blood vessel, dilatation of constriction may result in individual regions (26,67,69). An intravascular volume contraction, which cannot be explained by concomitant diuresis alone, but is promoted by a rapid extravascular shift of fluid (20), is probably a complementary BP-lowering mechanism (Fig. 4). A relevant decrease in blood volume should reduce venous return with lowered intracardiac end-diastolic pressures and, according to the Frank-Starling relationship, have a negative influence on cardiac output. Recent observations (63) support the conclusion that this chain of hemodynamic events may counteract an appropriate adaptation of cardiac output to αANP-induced decreases in peripheral vascular resistance.

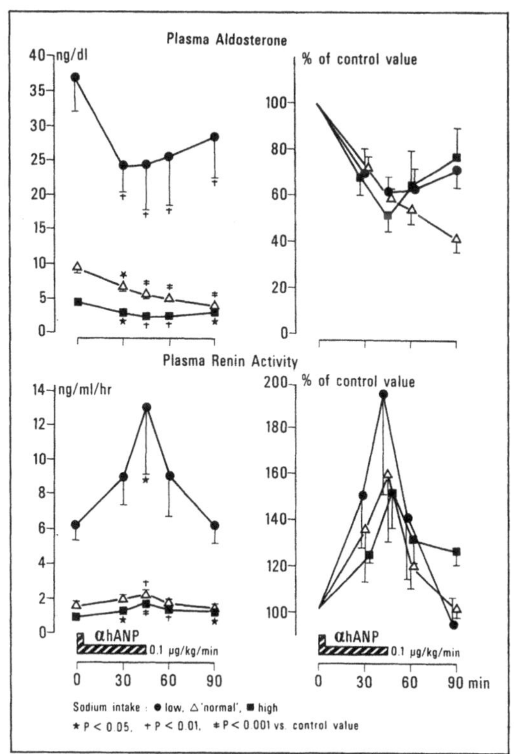

Fig.5. Effects of αANP infusion (50 µg bolus followed by 0.1 µg/kg/min) during different sodium intakes on plasma renin and aldosterone levels in 10 normal men (means ± SEM); from (x62)

The αANP induced reductions in systemic BP and volume are accompanied by a probable baroreflex-mediated activation of the sympathetic nervous system, as evidenced by rises in plasma norepinephrine, heart rate, and lipolysis (60, 62).

αANP can lower plasma aldosterone (32,62) even during a simultaneous activation of the renin-angiotensin system (Fig. 5) (62); in fact, aldosterone-responsiveness to angiotensin II is also blunted by αANP (70), while plasma cortisol levels were unchanged or decreased slightly (62). Experimental studies demonstrated that ANP may bind to specific adrenal receptors (64,71) and directly inhibit the adrenocortical steroidogenesis at a very early step prior to the metabolism of cholesterol and its side chain cleavage (13).

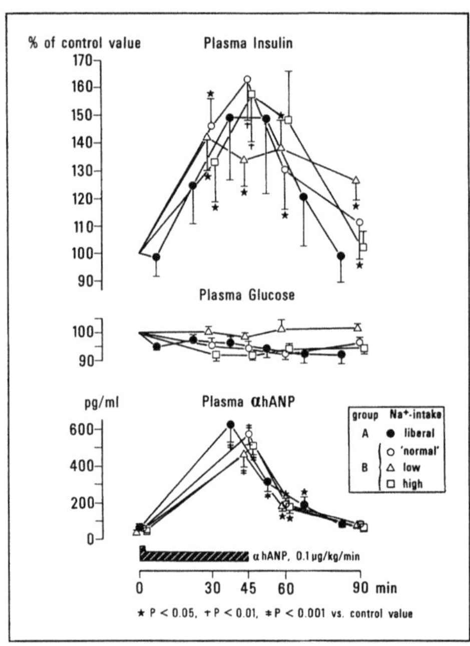

Fig. 6. Effect of αANP infusion (50 µg bolus followed by 0.1 µg/kg/min) during different sodium intakes on plasma insulin and glucose levels (means ± SEM); from (x79)

Renin release *in vitro* is inhibited by αANP (72,73). In humans, plasma renin activity (PRA) levels tended to decrease in response to low αANP doses (74), but were unchanged or even increased by high αANP infusion rates (20,32,62). In the latter setting, a direct renin-inhibitory influence of αANP is probably counteracted by stimulatory signals resulting from hemoconcentration, decreased BP and sympathetic activation (62).

Plasma AVP levels were unaltered during high dose αANP infusion in normal subjects on low, normal and high sodium intakes (20,75). Nevertheless, others reported decreased plasma AVP concentrations (76). The presence of ANP receptors in the posterior pituit-

166

ary (77) may be consistent with a functional interaction between the ANP and AVP systems (78).

In humans, intravenously infused αANP may also elevate circulating insulin by 40 to 50%, while plasma glucose levels are largely unchanged (Fig. 6) (20,79). Considering that the magnitude of this insulin-stimulation is at most minimally above the threshold to affect plasma glucose, there is no basis to propose a relavent impact of αANP on carbohydrate metabolism. However, insulin could potentially also promote renal sodium reabsorption (80). The control of biological functions usually relies on an interplay between factors with antagonistic effects. It is, therefore, of interest that αANP, which directly and through aldosterone-inhibition can enhance the natriuresis, might simultaneously activate a sodium-retaining factor such as insulin.

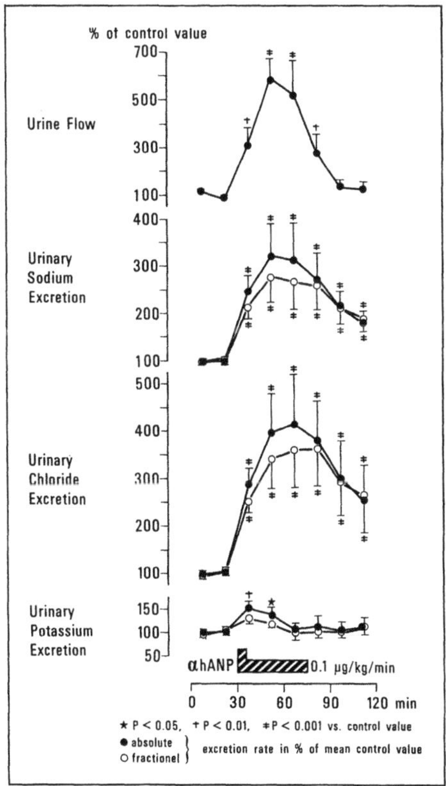

Fig. 7. Effect of αANP infusion (50 µg bolus followed by 0.1 µg/kg/min) on urine flow and monovalent electrolyte excretion rates in 10 normal subjects (means ± SEM). The mean of the two control values is taken as 100%; from (x20)

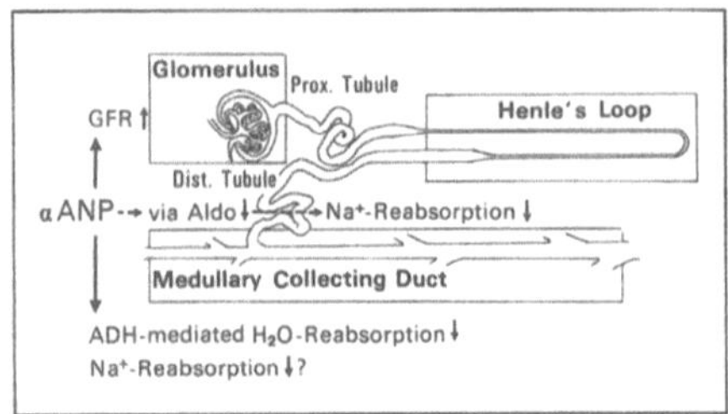

Fig. 8. Diagram of probable mechanisms of αANP-induced diuresis and natriuresis

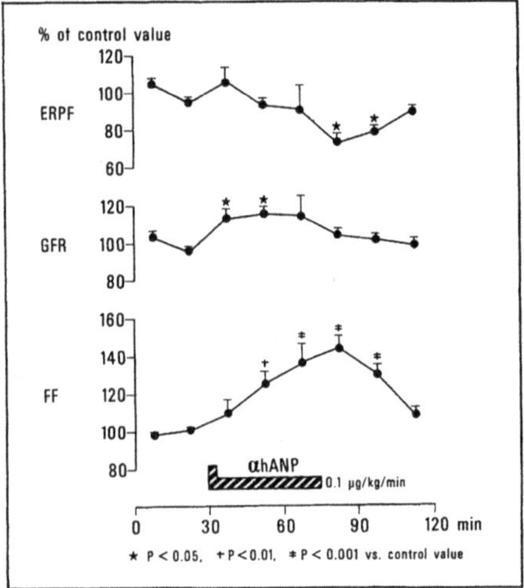

Fig. 9. Effect of αANP infusion (50 µg bolus followed by 0.1 µg/kg/min) on renal function in 10 normal subjects (means ± SEM). The mean of the two control values is taken as 100%; from (x20)

αANP infused intravenously produces in normal humans dose-dependent rises in diuresis, urinary excretion of sodium (Fig. 7), and various other electrolytes, and the clearance of free water (20,32,61,81). Compared with effects on BP, lower ANP doses are required to modify renal function (60,63). Several mechanisms seem to contribute to the excretory effects of αANP (Fig. 8). Hemodynamic changes play a role. αANP may acutely increase glomerular filtration rate (GFR), while renal blood flow is unchanged or even slightly decreased, and filtration fraction is raised markedly (Fig. 9) (20). The "isolated" rise in GFR is probably due to an ANP-induced constriction of efferent accompanied by dilatation of the afferent glomerular arterioles (82); this results in an increased glomerular filtration pressure (83,84). When GFR is augmented, it contributes importantly through increased filtered loads to ANP-mediated water and electrolyte excretion. Nevertheless, the

168

latter may occur to some degree even with unchanged GFR (20,85,86). αANP-induced changes in diuresis and natriuresis parallel to those in filtration fraction. Additional renal hemodynamic effects of ANP, such as an augmented peritubular pressure along the medullary collecting ducts (87), or a medullary redistribution of blood flow (88), could also promote natriuresis. A medullary shift of plasma flow may lead to washout of urea from the papilla, with disruption of countercurrent gradients and reduced sodium reabsorption in the thick ascending limb of the deep nephrons.

Considering possible humoral mediators, aldosterone-inhibition develops its effects on tubular sodium transport too slowly to be a relevant determinant of the initial natriuresis occurring within a few minutes after the start of infusion of synthetic αANP. However, reduced aldosterone levels, should promote sodium excretion in the established phase of increased plasma ANP concentrations. Contribution of a dopaminergic mechanism has been suspected (89), although plasma and urinary depamine levels were unchanged during αANP infusions, as were urinary levels of prostaglandins E_2 and F_{2a} (20,62,74,90). An influence of sympathetic activity has to be considered, since in the rat, sympathectomy or α-receptor blockade diminished excretory responses to ANP (91).

Direct effects of ANP on tubular transport mechanisms have also been postulated. Sodium reabsorption in the loop of Henle seems to be unaltered (92), but ANP may possibly inhibit sodium reabsorption in medullary collecting ducts (93,95) and it antagonizes the ADH-mediated tubular water permeability (96,97). The latter may be the major mechanism underlying the αANP-induced rise in free water clearance which occurred in the presence of unchanged plasma ADH levels (20).

Marked αANP-induced increases in distal tubular flow rate and sodium delivery would be exposed to stimulate potassium secretion. However, potassium excretion was increased only slightly during the first 30 min of αANP infusion and was already restored to control levels despite persistently enhanced diuresis and natriuresis between 30 and 45 min of αANP infusion (Fig. 7) (20). This suggests that potassium secretory capacity in the distal nephron is blunted during administration of αANP.

Whether ANP may directly modify proximal tubular transport is controversial. Some micropuncture studies suggested an unaltered handling of water and solutes in the proximal tubule (92,93,98,99), but observations of a raised clearance of lithium (100) and various amino acids (90) during αANP infusion in man require interpretation. Finally, the magnitude of the renal excretory effects of ANP varies in parallel with the state of hydration (62) and/or renal perfusion pressure (101). Whatever the precise interactions, the presence of ANP receptors on vascular, glomerular epithelial and endothelial and tubular cells (65,102,105), and the ANP-induced rise in particulate guanylate cyclase and cyclic guanosine monophosphate (cGMP) in glomeruli and collecting ducts (196,107), in animal kidneys studied *in vitro* are consistent with a direct influence of ANP on renal regulation at various levels. Cyclic GMP acts as the second messenger of ANP's receptor mediated actions in several (108,109), although not necessarily all (13) tissues. The cGMP stimulation is related to a rise in particulate guanylate cyclase (110). In contrast to "natriuretic hormones" that were postulated previously ("third factor", digitalis-like, or oubain-like factor), ANP does not act by modifying the Na-K-ATP′ase (111,112).

ANP-system and its possible physiological role

αANP is a promising candidate for a new and important hormonal system. It has already fulfilled two important criteria. Thus, αANP is released from a glandular site, namely the heart atria. Moreover, the moiety present in the circulation has the potential to exert in humans a remarkable spectrum of cardiovascular, endocrine, metabolic and renal action. To become established as a hormone, αANP needs to fulfill a third prerequisite, namely the demonstration that physiological maneuvers not only provoke change in ANP release or blood levels, but that these changes are in fact also modifying cardiovascular, renal, or other functions.

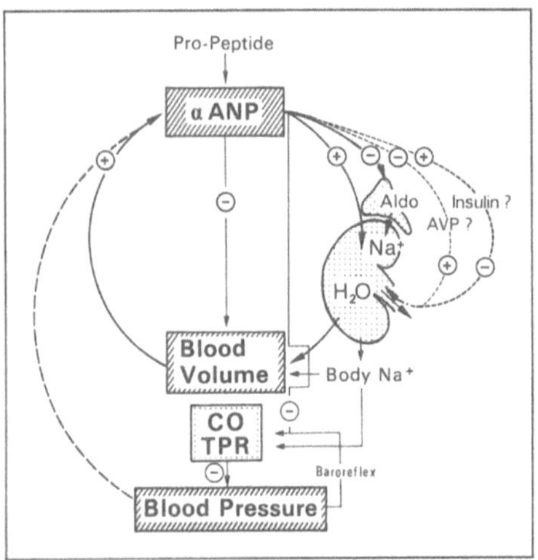

Fig. 10. Atrial natriuretic peptide system: some possible regulatory ineractions. Aldo = aldosterone, CO = cardiac output, TPR = total peripheral vascular resistance

Therefore, the schematic of the ANP-system (Fig. 10) with its possible regulatory interactions is still partly hypothetical. It is supposed that an elevated central blood volume or pressure provokes through atrial distention, the release of αANP. The latter promotes through enhanced natriuresis and diuresis and perhaps also through an extravascular shift of plasma water, a decrease in blood volume. The αANP-induced natriuresis may be facilitated by concomitant aldosterone inhibition and perhaps partly counteracted by stimulation of insulin. Furthermore, αANP may, through blood volume contraction and a decrease in total peripheral vascular resistance and/or cardiac output, lower BP. With the restoration of central blood volume and pressure to normal, the stimulus for increased ANP-release is withdrawn and plasma ANP levels return to basal levels.

In 1956, Henry and Gauer observed that distention of the left atrium causes a diuresis (113); in 1959, Anderson et al. reported that right atrial distention lowers aldosterone secretion (114). More than 20 years later, ANP has emerged to provide the yet missing link enabling the heart to modulate adrenal and renal function.

The homeostasis of physiological functions usually relies on an interplay of antagonistic factors. The possibility that glucocorticoids exert their stimulating effect on ANP secretion directly through a receptor mechanism (46,48), may be consistent with the existence of a cardio-adrenal feedback control mechanism (44). In fact, it would be functionally reasonable if the secretion of ANP which has been shown to inhibit adrenal steroidogenesis (13), might in turn be stimulated during enhanced corticosteroid activity.

Mineralocorticoid excess characteristically causes sodium retention. Natriuresis falls initially, but returns after some days to rates that are again in balance with sodium intake, thereby preventing progressively severe overhydration when mineralocorticoid excess persists. The mechanism(s) underlying this so called escape-phenomen are still incompletely understood (115). The ANP system may be a contributory factor (44,47), although in mineralocorticoid treated humans, plasma irANP already reached its highest levels before the final escape occurred (44).

Considering the potential renal and endocrine effects of αANP, it appears possible that the ANP-system also helps to mediate the compensatory increases in natriuresis and diuresis in response to sodium-volume loading, the change from standing to lying, or an acute rise in systemic pressure, and that it may perhaps even support the maintenance of excretory kidney function in aging humans. Proof for such a physiological role of αANP will require the availability of and studies with specific ANP-antagonists. Thus, it is of interest that in rats, the diuresis and natriuresis accompanying acute blood volume expansion could be blocked by monoclonal ANP-antibodies (116).

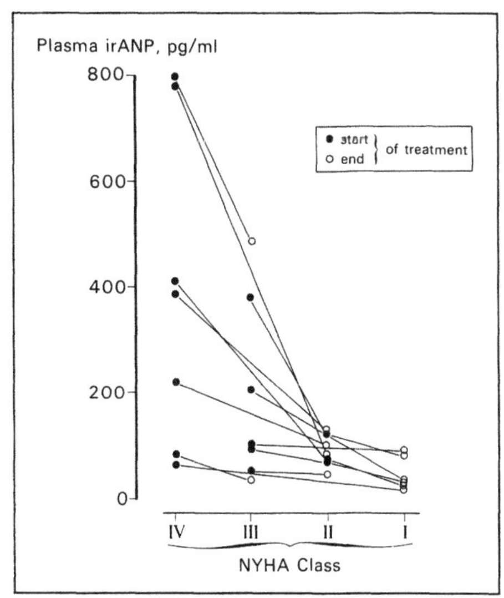

Fig. 11. Decrease in plasma irANP levels with improvement of congestive heart failure (New York Heart Association Criteria); from (x119)

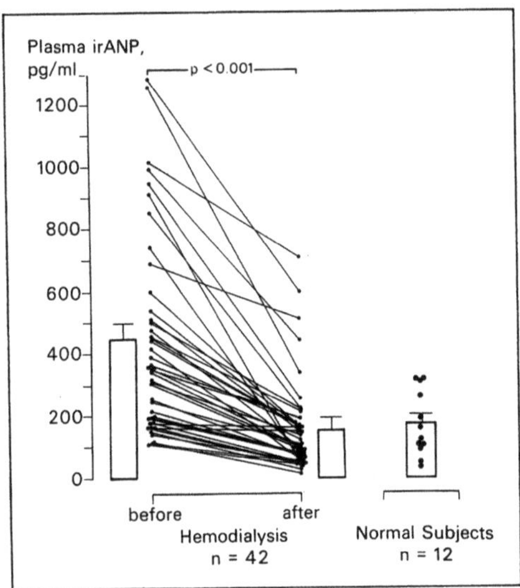

Fig. 12. Plasma irANP levels in patients with terminal renal failure before and after hemodialysis and in age-matched normal subjects. Hemodialysis-induced weight reduction averaged 2.3 kg; from (x124)

Plasma ANP levels in some pathological conditions

Plasma irANP levels are usually increased in hypervolemic states such as congestive heart failure (24,32,41,68,117-120) (Fig. 11), terminal renal failure (121-124) (Fig. 12), or the syndrome of inappropriate ADH release (SIADH) (125,126). Improvement of congestive heart failure (Fig. 11) or SIADH, and removal of excess body fluid by dialysis in patients with terminal renal failure (Fig.12), are accompanied by the return of plasma ANP towards normal values (41,119,121,123,124). High plasma irANP values were also noted in some patients with pulmonary arterial hypertension accompanying chronic obstructive lung disease (127). Circulating irANP concentrations in cardiac patients correlated positively with right and/or left atrial and pulmonary arterial pressures, and inversely with the cardiac output (24,117,120,128,129). The pathophysiological significance of the ANP system in these conditions is presently unclear. Effects of persistent high circulating ANP levels could possibly be limited by ANP receptor down-regulation. Nevertheless, should the ANP activation in congestive heart failure be biologically effective, it might antagonize the norepinephrine- and angiotensin II - mediated arteriolar vasoconstriction, antagonize the stimulated aldosterone secretion and promote natriuresis, thereby counteracting at least partly, mechanisms of increased pre- and afterloads in such patients. Regardless of its exact pathophysiological role, measurements of plasma ANP may become a valuable index for judging cardiac function and, particularly in renal failure, the state of hydration.

172

In contrast to hypervolemic states, edematous conditions with extra-cellular overhydration but often normal or even decreased blood volumes, such as the nephrotic syndrome or decompensated liver cirrhosis, seem to be usually accompanied by normal plasma irANP levels (63,130,131); in liver cirrhosis with ascites, elevated values were occasionally also reported (132). These constellations underline the role of the ANP system as a possible mechanism controlling intracardiac and/or intravascular rather than overall body-fluid balance.

Preliminary observations suggest that high altitude pulmonary edema, whose pathogenesis has been unknown, is associated with elevated plasma irANP concentrations (133, P. Bertsch et al., unpublished data). Since infused αANP can produce an extravascular shift of plasma fluid (20), a contributory role of endogenous ANP activation to the formation of high altitude and perhaps even other types of pulmonary edema deserves consideration.

Acute tachyarrhythmias usually elevate plasma irANP levels (32,41,134). The latter might possibly cause the enhanced diuresis and natriuresis, which often accompany acute increases in heart rate to >130 bpm and last at least 15 to 30 min (135).

Considering hypertensive states, high plasma irANP values have been reported in some but not all patients with untreated essential hypertension (117,136,137). Others noted largely normal values relative to the age of such patients (138,139), who also usually have normal or even slightly low blood volumes (140). A beta-blocker monotherapy elevated circulating irANP (139); therefore, a role of ANP in the antihypertensive mechanism of beta-blockers deserve consideration.

In contrast to essential hypertension, primary hyperaldosteronism is associated with a measurable excess in body sodium and interstitial as well as intravascular fluid volumes (141), and plasma irANP concentrations also seem to be high (32,138,142); removal of the aldosterone-producing adenoma normalizes these changes (143). Whether alterations in circulating irANP in chronic hypertensive states represent merely "innocent" biochemical consequences of sodium-fluid volume retention and/or high BP, or whether the ANP system may also have pathophysiological relevance, remains to be clarified.

Findings in patients with certain endocrine disturbances support the possibility that humoral factors may also modulate ANP levels. In Bartter's syndrome, plasma irANP was reported to be increased in the untreated state and restored towards normal following treatment with a cyclooxygenase inhibitor (143,144). In hypothyroidism plasma irANP may be low, perhaps due to impaired atrial release (145).

In pathological conditions, circulating irANP may not always correspond to αANP (molecular weight 3080 dalton). Higher molecular weight forms of ANP (15 000 dalton) have been noted in some patients with congestive heart failure or essential hypertension (146).

Effects of αANP-infusion in some pathological conditions

Acute renal failure is particulary interesting with regard to potential therapeutic effects of synthetic human αANP. In the rat, both the functional and histological changes of acute

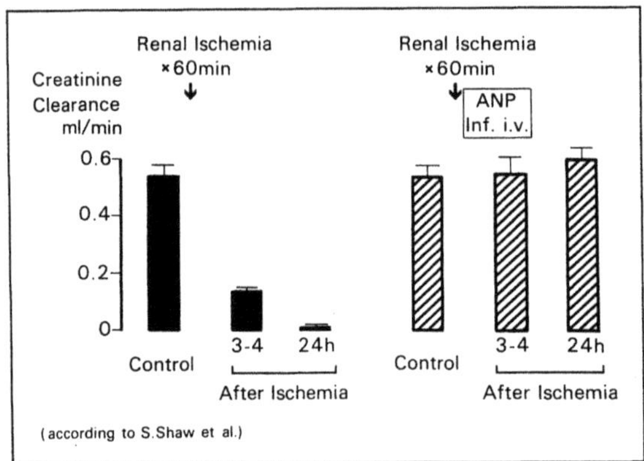

Fig. 13. Post-ischemic acute renal failure induced by 60 min of renal artery clamping in unilaterally nephrectomized rats: protective effect of intrarenal αANP infusion on creatinine clearance (n = 6, means ± SEM); from (x147)

ischemic renal failure induced by 45 to 60 min of renal arterial occlusion, could be largely prevented by a 4 h-intrarenal αANP infusion given immediately following the ischemic event (147) (Fig. 13). Others reported acute beneficial effects of a) pre-treatment with αANP on experimental norepinephrine-induced renal failure (148) and b) αANP infusions following induction of glycerol-induced acute renal failure in the rat (149). The preventive and therapeutic potential of αANP in human acute renal failure, particulary in patients undergoing major surgery or renal transplantation or in the early phase of developing impairment, awaits evaluation.

αANP has been infused in several human pathological states. In some patients with mild to moderate chronic renal impairment or with the nephrotic syndrome, αANP produced a dose-dependent increase in diuresis and natriuresis, mediated only in part by a concomitant rise in GFR (63,150). Higher αANP doses tended to lower BP; hemodynamic measurements in nephrotic patients revealed a concomitant peripheral but not renal vasodilatation, hemoconcentration resulting partly from an extravascular shift of plasma fluid, and an inappropriate adaption (rise) of cardiac output (63). The latter may be a consequence of the intravascular volume contraction, which should reduce venous return with lowered intracardiac end-diastolic pressures and, according to the Frank-Starling relationship, a negative influence on cardiac output.

Some patients with congestive heart failure responded to αANP infusions with an acute decrease in peripheral vascular resistance and preload, increased cardiac output, mild reduction in BP, lowered plasma aldosterone, and enhanced diuresis and natriuresis (68, 129).

In patients with decompensated liver cirrhosis, αANP seems to be less effective in promoting diuresis and natriuresis (151); such patients may also be particulary prone to αANP-induced hypotension (152).

174

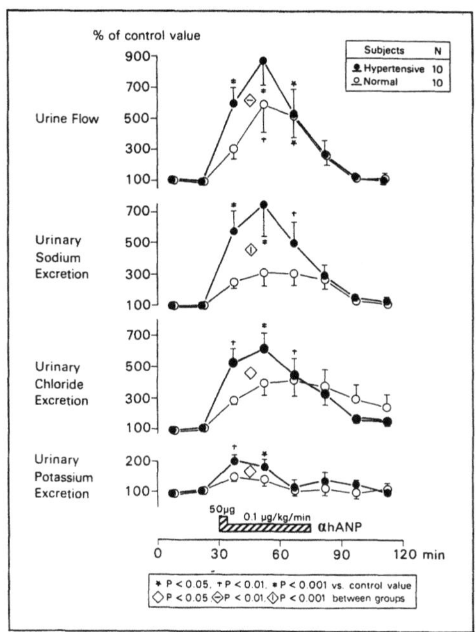

Fig. 14. Effect of αANP infusion (50 µg bolus followed by 0.1 µg/kg/min) on urine flow and monovalent electrolyte excretion in essential hypertension compared with normal subjects (means ± SEM); from (x90)

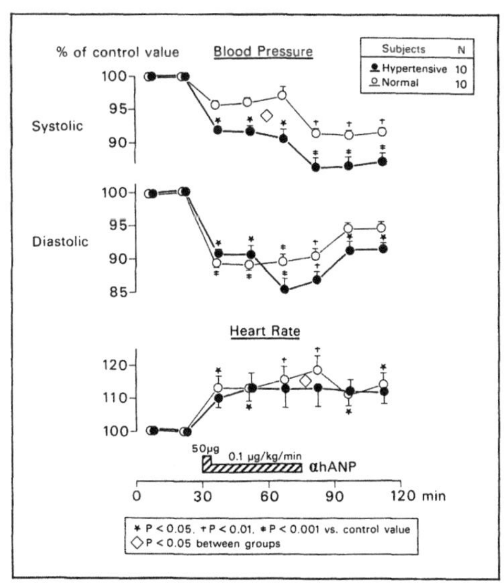

Fig. 15. Effect of αANP infusion (50 µg bolus followed by 0.1 µg/kg/min) on blood pressure (BP) and heart rate in patients with essential hypertention (basal BP 181/127 ± 9/3 mm Hg) compared with normal subjects (basal BP 127/87 ± 5/2 mm Hg) (means ± SEM); from (x90)

In patients with essential hypertension, infusions of αANP produced largely similar cardiovascular, endocrine, metabolic and renal effects as in normal humans (see above), with the exception of more markedly enhanced diuresis and electrolyte excretion in the hypertensive subjects (90) (Fig. 14). The latter could possibly be due to a higher renal perfusion pressure (101) in the hypertensive state. On the other hand, infused αANP lowered BP only slightly more in severely hypertensive than in normal subjects (90) (Fig. 15); whether this may reflect increased prior ANP-receptor occupancy due to high plasma ANP-levels in some hypertensive patients needs to be further evaluated.

ANP infusions in the reported human studies did not exceed a duration of a few hours. Therefore, the observed actions provide information on αANP´s acute pharmacological profile. Chronic effects could possibly be quite different. Prolonged administration of ANP and, if possible, the development of non-parenteral galenic preparations, will be necessary to establish the therapeutic potential of αANP in human diseases.

Acknowledgement: This work was supported in part by the Swiss National Foundation and by a research grant from Ciba-Geigy Corporation, Basel, Switzerland

References

1. Kisch, B. (1956) Electron microscopy of the atrium of the heart.I. Guinea pig. Exp Med Surg 14: 99-112
2. Jamieson, J.D., Palade, G.E. (1964) Specific granules in atrial muscle cells. J Cell Biol 23: 151-172
3. Bencosme, S.A., Berger, J.M. (1971) Specific granules in mammalian and non-mammalian vertebrate cardiocytes. In: Bajusz, E., Jasmin, G. (eds) Methods and Achievements in Experimental Pathology, 5 Karger, Basel, pp 173-213
4. Forssmann, W.G., Hock, D., Mutt, V. (1986) Cardiac hormones: morphology and biochemistry. Klin Wochenschr 64 (suppl VI): 4-12
5. Marie, J.P., Guillemot, H., Hatt, P.Y. (1976) Le degré de granulation des cardiocytes auriculaires. Etude planimétrique au cours de differents apports d´eau et de sodium chez le rat. Pathol Biol 24: 549
6. DeBold, A.J. (1979) Heart atria granularity: effects of changes in water-electrolyte balance. Proc Soc Exp Biol Med 161: 508-511
7. DeBold, A.J., Borenstein, H.B., Veress, A.T., Sonnenberg, H. (1981) A rapid and potent natriuretic response to intravenous injection of atrial myocardial extracts in rats. Life Sci 28: 89-94
8. Flynn, T.D., DeBold, M.L., DeBold, J.A. (1983) The amino acid sequence of atrial peptide with potent diuretic and natriuretic properties. Biochem Biophys Res Commun 117: 859-865
9. Currie, M-G., Geller, D.M., Cole, B.R., Siegel, N.R., Fok, K.F., Adams, S.P., Eubanks, S.R., Galluppi, G.R., Needleman, P. (1984) Purification and sequence analysis of bioactive atrial peptides (atriopeptins). Science 223: 67-69

10. Kangawa, K., Matsuo, H. (1984) Purification and complete amino acid sequence of a human atrial natriuretic polypeptide (αhANP). Biochem Biophys Res Commun 118: 131-139

11. Misono, K. S., Fukumi, H., Grammer, R. T., Inagami, T. (1984) Rat atrial natriuretic factor: complete amino acid sequence and disulfide linkage essential for biological activity. Biochem Biophys Res Commun 19: 524-529

12. Thibault, G., Carrier, F., Garcia, R., Gutkowska, J., Seidah, N. G., Chretien, M., Cantin, M., Genest, J. (1984) The human atrial natriuretic factor (hANF): purification and primary structure. Clin invest Med 7 (suppl 2): pp 59-61

13. Elliott, M. E., Goodfried, T. L. (1986) Inhibition of aldosterone synthesis by atrial natriuretic factor. Fed Proc 45: 2376 - 2381

14. Kudo, T., Baird, A. (1984) Inhibition of aldosterone production in the adrenal glomerulosa by atrial natriuretic factor. Nature 312: 756 - 757

15. Kangawa, K., Fukuda, A., Kubota, I., Hayashi, Y., Minamitake, Y., Matsuo, H. (1984) Human atrial natriuretic polypeptides (hANP): purification, structure synthesis and biological activity. J Hypertension 2 (suppl 3): pp 321 - 323

16. Kangawa, K., Fukuda, A., Matsuo, H. (1985) Structural identification of b- and g-human atrial natriuretic peptides. Nature 313: 397 - 400

17. Arendt, R. M., Stangl, E., Zähringer, J., Liebisch, D.C., Herz, A. (1985) Demonstration and characterization of a-human atrial natriuretic factor in human plasma. FEBS Lett 189: 57 - 61

18. Gutkowska, J., Bourassa, M., Roy, D., Thibault, G., Garcia, R., Cantin, M., Genest, J. (1985) Immunoreactive atrial natriuretic factor (IR-ANF) in human plasma. Biochem Biophys Res Commun 128: 1350-1357

19. Sugawara, A., Nakao, K., Moril, N., Sakamoto, M., Suda, M., Shimokura, M., Kiso, Y., Yamori, Y., Nishimura, K., Soneda, J., Ban, T., Imura, H. (1985) α–human atrial natriuretic polypeptide is released from the heart and circulates in the body. Biochem Biophys Res Commun 129: 439-446

20. Weidmann, P., Hasler, L., Gnädinger, M.P., Lang, R.E., Uehlinger, D.F., Shaw, S., Rascher, W., Reubi, F.C. (1986) Blood levels and renal effects of atrial natriuretic peptide in normal man. J Clin Invest 77: 734-742

21. Gnädinger, M.P., Lang, R.E., Hasler, L., Uehlinger, D.E., Shaw, S., Weidmann, P. (1986) Plasma kinetics of synthetic alpha-human atrial natriuretic peptide in man. Mineral Electrolyte Metab 12: 371-374

22. Nakao, K., Sugawara, A., Moril, N., Sakamoto, M., Yamada, T., Itoh, H., Shiono, S., Saito, *Y., Nishimura, K., Ban, T., Kangawa, K., Matsuo, H., Imura, H. (1986) The pharmacokinetics of a-human atrial natriuretic polypeptide in healthy subjects. Eur J Clin Pharmacol 31: 101-103

23. Yandle, T.G., Richards, A.M., Nicholls, M.G., Cuneo, R., Espiner, E.A., Livesey, J.H. (1986) Metabolic clearance rate and plasma half life of alpha-human atrial natriuretic peptide in man. Life Sciences 38: 1827-1833

24. Bates, E.R., Shenker, J., Grekin, R.J. (1986) The relationship between plasma levels of immunoreactive atrial natriuretic hormone and haemodynamic function in man. Circulation 73: 1155-1161

25. DeBold, A.J., DeBold, M.L., Sarda, I.R. (1986) Functional-morphological studies on *in vitro* cardionatrin release. J Hypertension 4 (suppl 2): S3-S7

26. Needleman, P., Greenwald, J.E. (1986) Atriopeptin: a cardiac hormone intimately involved in fluid, electrolyte and blood-pressure homeostasis. N Engl J Med 314: 828-834
27. Sato, F., Kamoi, K., Wakiya, Y., Ozawa, T., Arai, O., Ishibashi, M., Yamaji, T. (1986) Relationship between plasma atrial natriuretic peptide levels and atrial pressure in man. J Clin Endocrinol Metab 63: 823-827
28. Sagnella, G.A., Markandu, N.D., Shore, A.C., MacGregor, G.A. (1985) Effects of changes in dietary sodium intake and saline infusion on immunoreactive atrial natriuretic peptide in human plasma. The Lancet II, 1208-1211
29. Yamaji, T., Ishibashi, M., Takaku, F. (1985) Atrial natriuretic factor in human blood. J Clin Invest 76: 1705-1709
30. Haller, B.C.D., Züst, H., Shaw, S., Gnädinger, M.P., Uehlinger, D.E., Weidmann, P. (1987) Effects of posture and aging in circulating atrial peptide levels in man. J Hypertension
31. Kimura, T., Abe, K., Ota, K., Omata, K., Shoji, M., Kudo, K., Matsui, K., Inoue, M., Yasujim, M., Yoshinaga, K. (1986) Effects of acute water load, hypertonic saline infusion, and furosemide administration on atrial natriuretic peptide and vasopressin release in humans. J Clin Endocrinol Metab 62: 1003-1010
32. Espiner, E.A., Nicholls, M.G., Yandle, T.G., Crozier, I.G., Cuneo, R.C., McCormick, D., Ikram, H. (1986) Studies on the secretion, metabolism and action of atrial natriuretic peptide in man. J Hypertension 4 (suppl 2): S85-S91
33. Hollister, A.S., Tanaka, I., Imade, T., Onrot, J., Biaggioni, I., Robertson, D., Inagami, T. (1986) Sodium loading and posture modulate human atrial natriuretic factor plasma levels. Hypertension 8 (suppl II): II-106-II-111
34. Atherton, J.C., Bobinski, H., Solomon, L. (1986) Effect of posture on plasma atrial natriuretic peptide (ANP) and renal function in man. J Physiol 373: 73P
35. Anderson, J.V., Millar, N.D., O'Hara, J.P., Mackenzie, J.C., Corrall, R.J.M., Bloom, S.R. (1986) Atrial natriuretic peptide: physiological release associated with natriuresis during water immersion in man. Clin Sci 71: 319-322
36. Epstein, M., Loutzenhiser, R.D., Friedland, E., Aceto, R.M., Camargo, M.J.F., Atlas, S.A (1986) Increases in circulating atrial natriuretic factor during immersion-induced central hypervolaemia in normal humans. J Hypertension 4 (suppl.2): S 93-S 99
37. Uehlinger, D.E., Weidmann, P., Gnädinger, M.P., Shaw, S., Lang, R.E. (1986) Depressor effects and release of atrial natriuretic peptide during norepinephrine or angiotensin II infusion in man. J Clin Endocrin Metab 63: 669-674
38. Uehlinger, D.E., Zaman, T., Weidmann, P., Shaw, S., Gnädinger, M.P. (1987) Pressure-dependence of atrial natriuretic peptide during norepinephrine infusion. Hypertension (in press)
39. Tanaka, H., Shindo, M., Gutkowska, J., Kinoshita, A., Urata, H., Ikeda, M., Arakawa, K. (1986) Effect of acute exercise on plasma immunoreactive - atrial natriuretic factor. Life Sci 39: 1685-1693
40. Somers, V.K., Anderson, J.V., Conway, J., Sleight, P., Bloom, S.R. (1986) Atrial natriuretic peptide is released by dynamic exercise in man. Horm Metabol Res 18: 871-872
41. Anderson, J.V., Gibbs, J.S.R., Woodruff, P.W.R., Greco, C., Rowland, E., Bloom, S.R. (1986) The plasma atrial natriuretic peptide response to treatment of acute cardiac failure, spontaneous supraventricular tachycardia and induced re-entrant tachycardia in man. J Hypertension 4 (suppl 2): S 137-S 141

42. Weil, J., Bidlingmaier, F., Döhlemann, C., Kuhnle, U., Strom, T., Lang, R.E. (1986) Comparison of plasma atrial natriuretic peptide levels in healthy children from birth to adolescence and in children with cardiac diseases. Pediatr Res 20: 1328-1331

43. Ohashi, M., Fujio, N., Nawata, H., Kato, K.I., Ibayashi, H., Kangawa, K., Matsuo, H. (1987) High plasma concentration of human atrial natriuretic polypeptide in aged men. J Clin Endocrinol Metab 64: 81-85

44. Weidmann, P., Matter, D.R., Matter, E.E., Gnädinger, M.P., Uehlinger, D.E., Shaw, S., Hess, C. (1987) Stimulation of atrial natriuretic peptide induced by a glucocorticoid or mineralocorticoid in man. Kidney Int (in press)

45. Takeda, R., Miyamori, I., Ikeda, M. (1986) Role of atrial natriuretic polypeptides (ANP) for the pathogenesis of escape from mineralocorticoid excess. In: Proceedings of the First World Congress on Biologically Active Atrial Peptides p. 88A

46. Garcia, R., Debinski, W., Gutkowska, J., Kuchel, O., Thibault, G., Genest, J., Cantin, M. (1985) Gluco- and mineralocorticoids may regulate the natriuretic effect and the synthesis and release of atrial natriuretic factor by the rat atria in vivo. Biochem Biophys Res Commun 131: 806-814

47. Ballermann, B.J., Bloch, K.D., Seidmann, J.G., Brenner, B.M. (1986) Atrial natriuretic peptide transcription, secretion and glomerular receptor activity during mineralocorticoid escape in the rat. J Clin Invest 78: 840-845

48. Gardner, D.G., Hane, S., Trachewsky, D., Schenk, D., Baxter, J.D. (1986) Atrial natriuretic peptide mRNA is regulated by glucocorticoids in vivo. Biochem Biophys Res Commun 139: 1047-1054

49. Grekin, R.J., Ling, W.D., Shenker, Y., Bohr, D.F. (1986) Immunoreactive atrial natriuretic hormone levels increase in deoxycorticosterone acetate-treated pigs. Hypertension 8 (suppl II): II-16-II-20

50. Lachance, D., Garcia, R., Gutkowska, J., Cantin, M., Thibault, G. (1986) Mechanisms of release of atrial natriuretic factor. I Effect of several agonists and steroids on its release by atrial minces. Biochem Biophys Res Commun 135: 1090-1098

51. Funder, J.W., Duval, D., Meyer, P. (1973) The binding of tritiated dexamethasone in rat and dog heart. Endocrinology 93: 1300-1308

52. Greenberg, B.D., Bencen, G.H., Seilhammer, J.J., Lewicki, J.A., Fiddes, J.C. (1984) Nucleotide sequence of the gene encoding human atrial natriuretic factor precursor Nature 312: 656-658

53. Seidmann, C.E., Bloch, K.D., Klein, K.A., Smith, J.G. (1984) Nucleotide sequences of the human and mouse atrial natriuretic factor genes. Science 226: 1206-1209

54. Argentin, S., Nemer, M., Drouin, J., Scott, G.K., Kennedy, B.P., Davis, P.L. (1985) The gene of rat atrial natriuretic factor. J Biol Chem 260: 4568-71

55. Trippodo, N.C., MacPhee, A.A., Cole, F.E. (1984) Natriuretic response to atrial natriuretic factor enhanced by angiotensin II and vasopressin. J Hypertension 2 (suppl 3): 289-291

56. Sonnenberg, H., Veress, A.T. (1984) Cellular mechanism of release of atrial natriuretic factor. Biochem Biophys Res Commun 124: 443-449

57. Manning, P.T., Schwartz, D., Katsube, N.C., Holmberg, S.W., Needlemann, P. (1985) Vasopressin-stimulated release of atriopeptin: endocrine antagonist in fluid homeostasis. Science 229-395-397

58. Currie, M.G., Newman, W.H. (1986) Evidence for alpha-1 adrenergic receptor regulation of atriopeptin release from the isolated rat heart. Biochem Biophys Res Commun 137: 94-100

59. Garcia, R., Lechance, D., Thibault, G., Cantin, M., Gutkowska, J. (1986) Mechanism of release of atrial natriuretic factor. Biochem Biophys Res Commun 136: 510-20

60. Anderson, J., Struthers, A., Christofides, N., Bloom, S. (1986) Atrial natriuretic peptide: an endogenous factor enhancing sodium excretion in man. Clin Sci 70: 327-331

61. Biollaz, J., Nussberger, J., Porchet, M., Brunner-Ferber, F., Otterbein, E.S., Gomez, H., Waeber, B., Brunner, H.R. (1986) Four-hour infusions of synthetic atrial natriuretic peptide in normal volunteers. Hypertension 8 (suppl II): II-96-II-105

62. Weidmann, P., Hellmüller, B., Uehlinger, D.E., Lang, R.E., Gnädinger, M.P., Hasler, L., Shaw, S., Bachmann, C. (1986) Plasma levels and cardiovascular, endocrine and excretory effects of atrial natriuretic peptide during different sodium intakes in man. J Clin Endocrinol Metab 62: 1027-1036

63. Schohn, D., Gnädinger, M.P., Weidmann, P., Jahn, H., Hess, C., Shaw, S.C., Paternostro, A. (1987) Haemodynamic, endocrine and renal effects of atrial natriuretic peptide in the nephrotic syndrome. In: Proceedings of the Xth International Congress of Nephrology, London 1987. Trans Medica Europe, London, p 211

64. Hirata, Y., Tomita, M., Yoshimi, H., keda, M. (1984) Specific receptors for atrial natriuretic factor (ANF) in cultured vascular smooth muscle cells of rat aorta. Biochem Biophys Res Commun 125: 562-568

65. Napier, M.A., Vandlen, R.L., Albers-Schönberg, G., Nutt, R.F., Brady, S., Lyle, T., Winquist, R., Faison, E.P., Heinel, L.A., Blaine, E.H. (1984) Specific membrane receptors for atrial natriuretic factor in renal and vascular tissues. Proc Natl Acad Sci USA 81: 5946-5950

66. Schiffrin, E.L., St-Louis, J., Garcia, R., Thibault, G., Cantin, M., Genest, J. (1986) Vascular and adrenal binding sites for atrial natriuretic factor. Effects of sodium and hypertension. Hypertension 8 (suppl I): I-141-I-145

67. Biollaz, J., Waeber, B., Nussberger, J., Porchet, M., Brunner-Ferber, F., Otterbein, E.S., Gomez, H.J., Brunner, H.R. (1986) Atrial natriuretic peptides: reproducibility of renal effects and response of liver blood flow. Eur J Clin Pharmacol 31: 1-8

68. Cody, R.J., Atlas, S.A., Laragh, J.H., Kubo, S.H., Covit, A.B.,Ryman, K.S., Shaknovich, A., Pondolfino, K., Clark, M., Camargo, M.J.F., Scarborough, R.M. Lewicki, J.A. (1986) Atrial natriuretic factor in normal subjects and heart failure patients. Plasma levels and renal, hormonal and hemodynamic responses to peptide infusions. J Clin Invest 78: 1362-1374

69. Müller, F.B., Erne, P., Raine, A.E.G., Bolli, P., Linder, L., Resink, T.J., Cottier, C., Buehler, F.R. (1986) Atrial antipressor natriuretic peptide: release mechanisms and vascular action in man. J Hypertension 4 (suppl 2): S 1109-S 1114

70. Vierhapper, H., Nowotny, P., Waldhöusl, W. (1986)Prolonged administration of human atrial natriuretic peptide in healthy men. Reduced aldosteronotropic effect of angiotensin II. Hypertension 8: 1040-1043

71. DeLean, A., Racz, K., Gutkowska, J., Nguyen, T-T., Cantin, M., Genest, J. (1984) Specific receptor-mediated inhibitions by synthetic atrial natriuretic factor of hormone-stimulated steroidogenesis in cultured bovine adrenal cells. Endocrinology 115: 1636-1638

72. Obana, K., Naruse, M., Naruse, K., Sakurai, H., Demura, H., Inagami, T., Shizume, K. (1985) Synthetic rat atrial natriuretic factor inhibits *in vitro* and in vovo renin secretion in rats. Endocrinology 117: 1282-1284

73. Kurtz, A., Della Bruna, R., Pfeilschifter, J., Taugner, R., Bauer, C. (1986) Atrial natriuretic peptide inhibits renin release from juxtaglomerular cells by a cGMP-mediated process. Proc Natl Acad Sci USA 83: 4769-4773

74. Struthers, A.D., Anderson, J.V., Payne, N., Causon, R.C., Slater, J.D.H., Bloom, S.R. (1986) The effect of atrial natriuretic peptide on plasma renin activity, plasma aldosterone and urinary dopamine in man. Eur J Clin Pharmacol 31: 223-226

75. Gnädinger, M.P., Weidmann, P., Rascher, W., Lang, R.E., Hellmüller, B., Uehlinger, D.E. (1986) Plasma arginine-vasopressin levels during infusion of synthetic atrial natriuretic peptide on different sodium intakes in man. J Hypertension 4: 623-629

76. Fujio, N., Ohashi, M., Nawata, H., Kato, K., Ibayashi, H., Kangawa, K., Matsuo, H. (1986) α-human atrial natriuretic polypeptide reduces the plasma arginine vasopressin concentration in human subjects. Clin Endocrinol 25: 181-187

77. Januszewicz, P., Gutkowska, J., DeLean, A., Thibault, G., Garcia, R., Genest, J., Cantin, M. (1985) Synthetic atrial natriuretic factor induces release (possibly receptor-mediated) of vasopressin from rat posterior pitiutary. Soc Experimental Biol Med 178: 321-325

78. Samson, W.K. (1985) Atrial natriuretic factor inhibits dehydration and hemorrhage-induced vasopressin release. Neuroendocrinology 40: 277-279

79. Uehlinger, D.E., Weidmann, P., Gnädinger, M.P., Hasler, L., Bachmann, C., Shaw, S., Hellmüller, B., Lang, R.E. (1986) Increase in circulating insulin induced by atrial natriuretic peptide in normal humans. J Cardiovasc Phaarmacol 8: 1122-1129

80. De Fronzo, R.A. (1981) The effect of insulin on renal sodium reabsorption. Diabetologica 21: 165-171

81. Sugawara, A., Nakao, K., Moril, N., Sakamoto, M., Horil, K., Shimokura, M., Kiso, Y., Nishimura, K., Ban, T., Kihara, M., Yamori, Y., Kangawa, K., Matsuo, H., Imura, H. (1986) Significance of a-human atrial natriuretic polypeptide as a hormone in humans. Hypertension 8 (suppl I): I-151-I-155

82. Marin-Grez, M., Fleming, J.T., Steinhausen, M. (1986) Atrial natriuretic peptide causes pre-glomerular vasodilatation and post-glomerular vasoconstriction in rat kidney. Nature 324: 473-476

83. Dunn, B.R., Ichikawa, I., Pfeffer, J.M., Troy, J.L., Brenner, B.M. (1986) Renal and systemic haemodynamic effects of synthetic atrial natriuretic peptide in the anaesthetized rat. Circ Res 59: 237-246

84. Fried, T.A., McCoy, R.N., Osgood, R.W., Stein, J.H. (1986) The effects of atriopeptin II on determinants of glomerular filtration rate in the *in vitro* perfused dog glomerulus. Am J Physiol 250: F1119-F112

85. Salazar, F.J., Fiksen-Olsen, M.J., Opgenorth, T.J., Granger, J.-P., Burnett, J.C. jr., Romero, J.C. (1986) Renal effects of ANP without changes in glomerular filtration rate and blood pressure. Am J Physiol 251: F532-F536

86. Burnett, J.C. jr., Opgenorth, T.J., Granger, J.P. (1986) The renal action of atrial natriuretic peptide during control of glomerular filtration. Kidney Int 30: 16-19

87. Dunn, B.R., Troy, J.L., Ichikawa, I., Brenner, B.R. (1986) Effect of atrial natriuretic peptide (ANP) on hydraulic pressures in the rat renal papilla: implications for ANP-induced natriuresis. Kidney Int 29: 382 (abstract)

88. Borenstein, H.B., Cupples, W.A., Sonnenberg, H., Veress, A.T. (1983) The effect of a natriuretic atrial extract on renal haemodynamics and urinary excretion in anaesthetized rats. J Physiol 334: 133-140

89. Marin-Grez, M., Briggs, J.P., Schubert, G., Schnermann, J. (1985) Dopamine receptor antagonists inhibit the natriuretic factor (ANF). Life Sci 36: 2171-2176

90. Weidmann, P., Gnädinger, M.P., Ziswiler, H.R., Shaw, S., Bachmann, C., Rascher, W., Uehlinger, D.E., Hasler, L., Reubi, F.C. (1986) Cardiovascular, endocrine and renal effects of atrial natriuretic peptide in essential hypertension. J Hypertension 4 (suppl 2): S 71-S 83

91. Hathaway, S., Solomon, S. (1984) The effect of 6-hydroxydopamine on the renal response to atrial natriuretic extract and blood volume expansion. Proc West Pharmacol Soc 27: 89-93

92. Huang, C.L., Lewicki, J., Johnson, L.K., Cogan, M.G. (1985) Renal mechanism of action of rat atrial natriuretic factor. J Clin Invest 75: 769-773

93. Briggs, J.P., Steipe, B., Schubert, G., Schnermann, J. (1982) Micropuncture studies of the renal effects of atrial natriuretic substance. Eur J Physiol 395: 271-276

94. Sonnenberg, J., Honrath, U., Chong, C.K., Wilson, D.R. (1982) Atrial natriuretic factor inhibits sodium transport in medullary collecting duct. Am J Physiol 250: F 963-F 966 (1986)

95. Zeidel, M.L., Seifter, J.L., Brenner, B.M., Silva, P. (1986) Atrial peptides inhibit Na+ entry-dependent oxygen consumption in rabbit inner medullary collection duct (IMCD) cells. Clin Res 34: 731A (abstract)

96. Anderson, R.J., Dillingham, M.A. (1986) Atrial natriuretic factor inhibition of arginine vasopressin action in rabbit cortical collecting tubule. Kidney Int 29: 328

97. Samson, W.K., Vanetta, J.C. (1986) Atrial natriuretic factor inhibits vasotocin-induced water reabsorption in the toad urinary bladder. Proc Soc Exp Biol Med 181: 169-172

98. Sonnenberg, H., Cupples, W.A., DeBold, A.J., Veress, A.T. (1982) Intrarenal localization of the natriuretic effect of cardiac atrial extract. Can J Physiol Pharmacol 60: 1149-1152

99. Baum, M., Toto, R.D. (1986) Lack of a direct effect of atrial natriuretic factor in the rabbit proximal tubule. Am J Physiol 250: F66-F69

100. Burnett, J.C. jr., Granger, J.P., Opgenorth, T.J. (1984) Effects of synthetic atrial natriuretic factor on renal function and renin release. Am J Physiol 247: F 863-F 866

101. Blaine, E.H., Heinel, L.A., Schorn, T.W., Marsch, E.A., Whinnery, M.A. (1986) The character of the atrial natriuretic response: pressure and volume effects. J Hypertension 4 (suppl 2): S 17-S 24

102. Ballermann, B.J., Hoover, R.L., Karnovsky, M.J., Brenner, B.M. (1985) Physiologic regulation of atrial natriuretic peptide receptors in rat renal glomeruli. J Clin Invest 76: 2049-2056

103. Bianchi, C., Gutkowska, J., Thibault, G., Garcia, R., Genest, J., Cantin, M. (1986) Distinct localization of atrial natriuretic factor and angiotensin II binding sites in the glomerulus. Am J Physiol 251: F 594-F 602

104. Koseki, C., Hayashi, Y., Torikai, S., Furuya, M., Ohnuma, N., Imai, M. (1986) Localization of binding sites for a-rat atrial natriuretic polypeptide in rat kidney. Am J Physiol 250: F 210-F 216

105. Schinjo, M., Hirata, Y., Hagiwara, H., Akiyama, F., Murakami, K.,Kojima, S., Shimonaka, M., Inada, Y., Hirose, S. (1986) Characterization of atrial natriuretic factor receptors in adrenal cortex, vascular smooth muscle and endothelial cells by affinity labelling. Biomedical Res 7: 35-38

106. Ishikawa, S., Saito, T., Okada, K., Kuzuya, T., Kangawa, K., Matsuo, H. (1985) Atrial natriuretic factor increases cyclic GMP and inhibits cyclic AMP in rat renal papillary collecting tubule cells in culture. Biochem Biophys Res Commun 130: 1147-1153

107. Tremblay, J., Gerzer, R., Vinay, P., Pang, S.C., Béliveau, R., Hamet, P. (1985) The increase of cGMP by atrial natriuretic factor correlates with the distribution of particulate guanylate cyclase. FEBS Letters 181: 17-22

108. Matsuoka, H., Ishii, M., Sugimoto, T., Hirata, Y., Sugimoto, T., Kangawa, K., Matsuo, H. (1985) Inhibition of aldosterone production by a-human atrial natriuretic polypeptide is associated with an increase in cGMP production. Biochem Biophys Res Commun 127: 1052-1056

109. Hamet, P., Tremblay, J., Pang, S.C., Skuherska, R., Schiffrin, E.L., Garcia, R., Cantin, M., Genest, J., Palmour, R., Ervin, F.R., Martin, S., Goldwater, R. (1986) Cyclic GMP as mediator and biological marker of atrial natriuretic factor. J Hypertension 4 (suppl 2): S 49-S 56

110. Gerzer, R., Weil, J., Shom, T., Müller, T. (1986) Mechanism of atrial natriuretic factor: clinical consequences. Klin Wochenschr 64 (Suppl.VI) 21-26

111. Pollack, D.M., Mullins, M.M., Banks, R.O. (1983)Failure of atrial myocardial extract to inhibit renal Na$^+$-ATPase. Renal Physiol 6: 295-299

112. Sagnella, G.A., Nolan, D.A., Shore, A.C., MacGregor, G.A. (1985) Effects of synthetic atrial natriuretic peptides on sodium-potassium transport in human erythrocytes. Clin Sci 69: 223-226

113. Henry, J.P., Gauer, O.H., Reeves, J.L. (1956) Evidence of the atrial location of receptors influencing urine flow. Circulation Res 4: 85-90

114. Anderson, C.H., McCally, M., Farrell, G.L. (1959) The effect of atrial stretch on aldosterone secretion. Endocrinology 64: 2021

115. Higgins, J.T., Mulrow, P.J. Fluid and electrolyte disorders of endocrine diseases. In: Maxwell, M.H., Kleeman, C.R. (eds) Clinical Disorders of Fluid and Electrolyte Metabolism. McGraw-Hill, New York, pp 1291-1338

116. Hirth, C., Stasch, J.-P., John, A., Kazda, S., Morich, F., Neuser, D., Wohlfeil, S. (1986) The renal response to acute hypervolemia is caused by atrial natriuretic peptides. J Cardiovasc Pharmaco 8: 268-275

117. Arendt, R.M., Gerbes, A.L., Ritter, D., Stangl, E., Bach, P., Zähringer, J. (1986) Atrial natriuretic factor in plasma of patients with atrial hypertension, heart failure or cirrhosis of the liver. J Hypertension 4 (suppl 2): S 131-S 135

118. Lang, R.E., Dietz, R., Merkel, A., Unger, T., Ruskoaho, H., Ganten, D. (1986) Plasma atrial natriuretic peptide values in cardiac disease. J Hypertension 4 (suppl 2): S 119-S 123

119. Katoh, Y., Kurosawa, T., Takeda, S., Kurokawa, S., Sakamotom, H., Marumo, F., Kikawada, R. (1986) Atrial natriuretic peptide levels in treated congestive heart failure. Lancet I: 851

120. Raine, A.E.G., Erne, P., Buergisser, E., Mueller, F.B., Bolli, P., Burkhart, F., Buehler, F.R. (1986) Atrial natriuretic peptide and atrial pressure in patients with congestive heart failure. N Engl J Med 315: 533-537

121. Rascher, W., Tulassy, T., Lang, R.E. (1985) Atrial natriuretic peptide in plasma of volume-overloaded children with chronic renal failure. Lancet II: 303-305

122. Eisenhauer, T., Talartschik, J., Scheler, F. (1986) Detection of fluid overload by plasma concentration of human atrial natriuretic peptide (h-ANP) in patients with renal failure. Klin Wochenschr 64 (suppl VI): 73-77

123. Hasegawa, K., Matsushita, Y., Inoue, T., Morii, H., Ishibashi, M., Yamaji, T. (1986) Plasma levels of atrial natriuretic peptide in patients with chronic renal failure. J Clin Endocrinol Metab 63: 819-822

124. Gnädinger, M.P., Saxenhofer, H., Weidmann, P., Shaw, S., Uehlinger, D.E., Descoeudres, C. (1987) Plasma levels and dialysance of atrial natriuretic peptide in terminal renal failure. Kidney Int (in press)
125. Cogan, E., Debieve, M.F., Philipart, I., Pepersack, T., Abramow, M. (1986) High plasma levels of atrial natriuretic factor in SIADH. N Engl J Med 34: 1258
126. Sakamoto, H., Inoue, K., Marumo, F. (1986) High plasma atrial natriuretic peptide independent of sodium balance in SIADH. JAMA 256: 1293-1294
127. Burghuber, O.C., Hartter, E., Punzengruber, Ch., Weissel, M., Woloszczuk, W. (1986) Sekretion des humanen natriuretischen Peptids (ANP) bei Patienten mit chronischen obstruktiven Lungenkrankheiten. (COL) Acta Med Autriaca 13: 73
128. Kokubu, T., Nagae, A., Hiwada, K., Kazatani, Y., Joh, T., Matsu, O.H. (1986) Pulmonary arterial pressure and plasma concentration of atrial natriuretic factor (ANF) in patients with heart disease. Jpn Heart J 27: 791-796
129. Riegger, A.J.G., Kromer, E.P., Kochsiek, K. (1986) Atrial natriuretic peptide in patients with severe heart failure. Klin Wochenschr 64 (suppl VI): 89-92
130. Burghardt, W., Wernze, H., Diehl, K.-L. (1986) Atrial natriuretic peptide in hepatic cirrhosis: Relation to stage of disease, sympathoadrenal system and renin-aldosterone axis. Klin Wochenschr 64 (suppl VI) 103-107
131. Tulassay, T., Rascher, W., Ganten, D., Schörer, K., Lang, R.E. (1986) Atrial natriuretic peptide and volume changes in children. Clin and Exper.-Theory and Practice A8 (4/5): 695-701
132. Fernandez-Cruz, A., Marco, J., Cuadrado, L.M., Gutkowska, J., Rodriguez-Puyal, D., Caramelo, C., Lopez-Novoa, J.M. (1985) Plasma levels of atrial natriuretic peptide in cirrhotic patients. Lancet II: 1439-1440
133 Cosby, R.L., Durr, J.A., Sophocles, A., Perrinjaquet, C., Yee, B., Schrier, R. (1987) Potential role of atrial natriuretic factor (ANF) and vasopressin (AVP) in high altitude pulmonary edema. (HAPE) Kidney Int 31: 195 (abstract)
134. Kojima, S., Akabana, S., Ohe, T., Tsuchihashi, K., Yamamoto, K., Kuramochi, M., Shimomura, K., Ito, K., Omae, T. (1986) Plasma atrial natriuretic polypeptide and polyuria during paroxysmal tachycardia in Wolff-Parkinson-White syndrome patients. Nephron 44-249-252
135. Wood, P. (1963) Polyuria in paroxysmal tachycardia and paroxysmal atrial flutter and fibrillation. Br Heart J 25: 273-282
136. Richards, A.M., Nicholls, M.G., Espiner, E.A., Ikram, H., Yanddle, T.C., Joyce, S.L., Cullens, M.M. (1985) Effects of α−human atrial natriuretic peptide in essential hypertension. Hypertension 7: 812-817
137. Sagnella, G.A., Markandu, N.D., Shore, A.C., MacGregor, G.A. (1986) Raised circulating levels of atrial natriuretic peptides in essential hypertension. Lancet II: 179-181
138. Yamaji, T., Ishibashi, M., Sekihara, H., Takaku, F., Nakaoka, H., Fujii, J. (1986) Plasma levels of atrial natriuretic peptide in primary aldosteronism and essential hypertension. J Clin Endocrinol Metab 63: 815-817
139. Nakaoka, H.,Kitahara, Y., Amano, M., Imataka, K., Fujii, J. Effect of beta-adenoreceptor blockade on atrial natriuretic peptide in essential hypertension. Hypertension (in press)
140. Beretta-Piccoli, C., Weidmann, P. (1984) Circulatory volume in essential hypertension. Mineral Electrolyte Metab 10: 292-300

141. Beretta-Piccoli, C., Davies, D.L., Brown, J.J., Ferriss, B., Fraser, R., Lasaridis, A., Lever, A.F., Morton, J.J., Robertson, J.I.S., Semple, P.L. (1983) Relation of blood pressure with body sodium and plasma electrolyte in Conn's syndrome. J Hypertension 1: 197-205

142. Tunny, T.J., Higgins, B.A., Gordon, R.D. (1986) Plasma levels of atrial natriuretic peptide in man in primary aldosteronism, in Gordon's syndrome and in Bartter's syndrome. Clin Exp Pharm Phys 13: 341-345

143. Tunny, T.J., Gordon, R.D. (1986) Plasma atrial natriuretic peptide in primary aldosteronism (before and after treatment) and in Bartter's and Gordon's syndromes. Lancet I: 272-273

144. Sasaki, H., Okumura, M., Kawasaki, T., Kangawa, K., Matsuo, H., (1987) Indomethacin and atrial natriuretic peptide in pseudo-Bartter's syndrome. N Engl J Med 316: 167

145. Zimmermann, R.S., Gharib, H., Zimmermann, D., Heublein, D., Burnett, J.C. (1987) Atrial natriuretic peptide in hypothyroidism. J Clin Endocrinol Metab 64: 353-355

146. Arendt, R.M., Gerbes, A.L., Ritter, D., Stangl, E. (1986) Molecular weight heterogeneity of plasma-ANP in cardiovascular disease. Klin Wochenschr 64 (suppl VI): 97-102

147. Shaw, S.G., Weidmann, P., Hodler, J., Zimmermann, A., Paternostro, A. Atrial natriuretic peptide prevents the development of ischemic renal failure. J Clin Invest (in press)

148. Schafferhans, K., Heidbreder, E., Grimm, D., Heidland, A. (1986) Human atrial natriuretic factor prevents against norepinephrine-induced acute renal failure in the rat. Klin Wochenschr 64 (suppl VI): 73-77

149. Heidbreder, E., Schafferhans, K., Schramm, D., Götz, R., Heidland, A. (1986) Toxic renal failure in the rat: Beneficial effects of atrial natriuretic factor. Klin Wochenschr 64 (suppl VI): 78-82

150. Suda, S., Cottier, C., Saxenhofer, H., Shaw, S., Weidmann, P., Gnädinger, M.P. Atrial natriuretic peptide in mild to moderate renal failure. (unpublished data)

151. Brabant, G., Jüppner, H., Kirschner, M., Böker, K., Schmidt, F.W., Hesch, R.D. (1986) Human atrial natriuretic peptide (ANP) for the treatment of patients with liver cirrhosis and ascites. Klin Wochenschr 64 (suppl VI): 1108-111

152. Ferrier, C.P., Weidmann, P., Suchecka, K., Beretta-Piccoli, C., Gnädinger, M.P. (unpublished data)

The behavior of CDD (cardiodilatin) during graded exercise on a bicycle ergometer

U. Feuerbach, K. Völker, H. Bittner, W.G. Forssmann, W. Hollmann

Department of Cardiovascular Reasearch and Sports Medicine, DSHS Köln, Department of Anatomy and Cell Biology, University of Heidelberg, Heidelberg, FRG

Summary

Twenty healthy males, aged 16 to 55 years, were investigated for a study of cardiac hormone secretion during physical exercise. Using a bicycle ergometer the work intensity was increased gradually by 50 watts every 3 min to a maximum level. Before, during, and after the exercise the level of cardiodilatin (CDD) was determined by taking venous blood samples from a canula in the forearm. The following results were obtained:

1) The CDD-level increased in the form of a parabola related to the intensity of exercise.
2) The highest level of CDD was found among those volunteers with the highest cardio-pulmonary capacity.

Introduction

In research work on blood volume done in the 1950's and 1960's a mechanism was postulated which was thought to be in the heart (1,9,11). Marie et al. (13) and DeBold (3) postulated a substance involved in sodium and body fluid regulation.

Forssmann et al. (5), Flynn et al. (4), Cantin and Genest (2) extracted CDD/ANP (atrial natriuretic peptide) and identified its amino acid sequence. Human ANP consists of 151 amino acids and is largely identical with various animal ANPs.

The peptide affects the blood pressure regulation by increasing the diuresis-natriuresis (3,7), and by causing vasodilatation (6).

A distension of the right auricle and the therein enclosed myoendocrinal cells is considered to be the appropriate stimulus for the secretion of the cardiac hormone (10,14). In this study we wanted to investigate the following questions:

1) Does dosed physical exercise influence the CDD level in the venous blood?
2) Is there a relation in CDD level to the intensity of work and the maximum cardio-pulmonary capacity?

Methods

Twenty healthy male subjects, aged 16 to 55 years, took part in the investigation. They were grouped as follows to their maximum cardio-pulmonary capacity:

Group I	0-250 W (n=10)
Group II	0-300 W (n=6)
Group III	0-350 W (n=4)

The exercise was performed on a bicycle ergometer with increasing workload. Starting at 50 W the workload was increased by 50 W every 3 min.

The following parameters were registered at rest: 100 W, 200 W, and each additional step of workload and in the third and fifth minute of the recovery; also ECG, blood lactate (analyzed with a Boehringer Testkombination according to (8), modified by (12)), and CDD.

The blood samples for measuring the plasma CDD were taken from a canula inserted in the cubital vein. The CDD samples were analyzed by a biological assay (RIA) according to the recommendations of Forssmann (6) at the laboratory of the Department of Anatomy and Cell Biology, University of Heidelberg. The statistical analysis was done with the multifactorial analysis of variance with repetition of measurment.

Results and discussion

There was a continuous increase in CDD concentration during bicycle ergometer work. The difference between rest and maximum workload was significant ($p < 0.05$) (Fig.1).

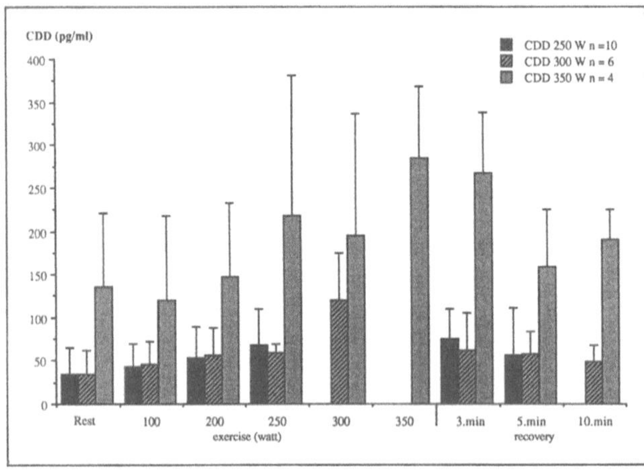

Fig. 1. CDD level in the venous blood during an increasing bicycle ergometer exercise

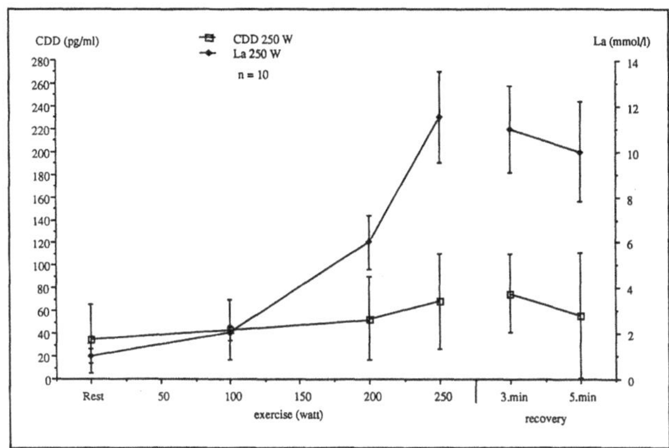

Fig. 2. CDD and blood lactate level in the blood during increasing bicycle ergometer exercise to a maximum intensity at 250 W (mean values and standard daviation)

The highest level of CDD was found in those who had the highest cardio-pulmonary capacity. This was valid at rest and at all stages of exercise. The functional curve of plasma CDD was similar to the parabolic curve of blood lactic acid in the mean values as well as in every volunteer (Fig. 2). Subjects with lower physical capacity (maximum at 250) showed a slow increase in CDD during the ergometer test and reached their maximum during recovery. The participants with a higher cardio-pulmonary capacity had a steeper increase and reached the maximum at the highest level of performance.

It is concluded that physical exercise influences the release of CDD. The higher the cardio-pulmonary capacity the steeper is the increase and the higher is the CDD level.

References

1. Bompiani, C.D., Rouiller, C., Hatt, P. (1959) Le tissue de conduction du coeur chez le rat. Etude au microscope électronique. Arch Maladie Coeur et Vaissaux 52: 1257-1274

2. Cantin, M., Genest, J. (1985) The heart and the atrial natriuretic factor. Endocrine Reviews Vol 6 Nr. 2: 107 - 127

3. DeBold, A.J., Borenstein, H.B., Veress, A.T., Sonnenberg, H. (1981) A rapid and potent natriuretic response to intravenous injection of atrial myocardial extract in rats. Life Science 28: 89 - 94

4. Flynn, T.G., DeBold, M.L., DeBold, A.J. (1983) The amino acid sequence of an atrial peptide with potent diuretic and natriuretic properties. Biochem Biophys Res Comm 117: 859 - 865

5. Forssmann, W.G., Hock, D., Lottspeich, F., Henschen, A., Kreye, V., Christmann, M., Reinecke, M., Metz, K., Carlquist, M., Mutt, V. (1983) The right auricle of the heart is an endocrine organ: Cardiodilatin as a peptide hormone candidate. Anat Embryol 168: 307- 313

6. Forssmann,W.G. (1986) Cardiac hormones: a review on the morphology, biochemistry and molecular biology of the endocrine heart. Eur J Clin Invest 16: 439 - 451

7. Genest, J. (1983) Volume hormones and blood pressure. Ann Intern Med 98: 744 - 749

8. Gutmann, J., Wahlefeld, A.D. (1974) Bestimmung mit Laktat-Dehydrogenase und NAD. In: Bergemeyer, V. (Hrsg) Methoden der enzymatischen Analyse. Verlag Chemie, Weinheim

9. Kisch, B. (1956) Electron microscopy of the atria of the heart. I. Guinea pig. Exp Surg 14: 99 - 112

10. Lang, R.E., Thölken, H., Ganten, D., Luft, F.C., Ruskoaho, H., Unger, T. (1985) Atrial natriuretic factor - a circulating hormone stimulated by volume loading. Nature 314: 264-266

11. Ledsome, J.R., Linden, R.J., O´Conner, W.J. (1961) The mechanism by which distension of the left atrium produces diuresis in anesthetized dogs. London J Physiol 159: 87

12. Marder, A., Heck, H., Föhrenbach, R., Hollmann, W. (1979) Das statische und dynamische Verhalten des Laktats und des Säure-Basen-Status im Bereich niedriger bis maximaler Azidosen bei 400 m - und 800 m Läufern bei beiden Geschlechtern nach Belastungsabbruch. Dtsch Z Sportmed 30 (7): 203 - 211 /8: 249 - 261

13. Marie, J.P., Guillemot, H., Hatt, P.Y. (1976) Le degré de granulation des cardiocytes auriculaires. Etude planimétrique au cours de différents apports d'eau et de sodium chez le rat. Path Biol 24: 549 - 554

14. Tikkanen, L., Metsärinne, K., Fyhrquist, F., Leidenius, R. (1985) Plasma atrial natriuretic peptide in cardiac disease and during infusion in healthy volunteers. Lancet II: 66 - 69

Plasma concentrations of atrial natriuretic peptide (α-h-ANP 1-28) during exercise and relationship to hemodynamic parameters in man

J. Talartschik[1], K.H. Scholz[2], T. Eisenhauer[1], K.L. Neuhaus[2]

Medical University Clinic, Depts. of Nephrology[1] and Cardiology[2], Göttingen, FRG

Summary

To elucidate the diagnostic value of Atrial Natriuretic Peptide (ANP), ANP plasma concentrations were determined by radioimmunoassay in 15 patients with coronary or valvular heart diseases who were to undergo right-heart catheterization during treadmill exercise. Continuous right atrial - (pRA), pulmonary artery - (pPA), pulmonary wedge pressures (pPC), and heart rate (HR) were correlated to ANP levels in pulmonary artery (PA), brachial artery (A), and peripheral vein (V); the best correlation was found between ANP (PA) and pPC with r=0.59 in linear regression analysis whereas no correlation was seen between HR and ANP values in PA, A and V during exercise.

With increased strain ANP concentration rose significantly in V, A and in PA (35 fmol/ml to 89 fmol/ml). ANP plasma levels were identical in A and PA, but were systematically lower in V compared to A and PA. During the first three minutes after starting treadmill exercise ANP increase in V was significantly (p<0.01) delayed, whereas ANP concentrations in A and PA rose immediately.

The data suggest that elevation of ANP during graded exertion is caused by an increase in atrial pressure and not by a rise in HR. Since there is only a quantitatively weak correlation between ANP plasma concentrations and pressures in the pulmonary circulation the predictive value of ANP levels in individual cases is limited.

Introduction

Distention of atrial cardiocytes has been demonstrated to be the main stimulus for the secretion of Atrial Natriuretic Peptide from myoendocrine cells (1,2,3). Diagnostic right heart catheterization during treadmill exercise is a well established and routinely used method to substantiate cardiac dysfunction due to coronary heart disease (CHD), cardiomyopathy (CM), hypertrophic obstructive cardiomyopathy (HOCM) and mitral or aortic valve diseases. It has been recently reported that ANP is released during physical exercise (4,5) and that ANP plasma concentrations closely correlate with atrial pressures at rest (2,6,15). The aim of this study was to find out whether pressures in the pulmonic circulation during treadmill exercise correlate with peripheral ANP plasma concentrations.

Patients and methods

Fifteen patients (four women, eleven men) undergoing diagnostic right heart catherization for clinically indicated purposes were evaluated. As depicted in Table 1 ages ranged from 40-63 years. In three patients a significant cardiac disease was excluded, five had coronary artery diseases, cardiomyopathy was present in four patients, and aortic or mitral valve regurgitation was detected in three patients. All patients were stressed in supine position by increasing degrees of strain (three minutes) until maximum exertion, determined by an increase in heart rate over 160/min., angina, pathologic ST-changes in ECG, or extreme rise or fall in blood pressure. Data were collected before (A), at each increase of exercise (B1-B5), at maximum strain (Bmax), five minutes (5M), and ten minutes (10M) after Bmax. Blood samples were drawn simultaneously from a peripheral vein (V), the brachial artery (A), and the pulmonary artery (PA). Pressures in the right atrium (pRA), in the pulmonary artery (pPA) and in the pulmonary capillary wedge position (pPC) were registered by the use of a Swan-Ganz catheter.

For the hANP-radioimmunoassay as previously described (7) blood samples were drawn into plastic tubes containing EDTA (1.5 mg/ml) and centrifuged at 3000 R/min at 4°C. Extractions were performed with reversed phase octadecylsilan cartridges (Sep-Pak C18, Waters, Massachusetts). The adsorbed peptides were eluted with 50 % Acetonitrile/H_2O containing 0.2 % Trifluoroacetic acid and were lyophilized. Recovery of synthetic ANP (5-400 fmol/ml) ranged from 78-85 %. The radioimmunoassay of h-ANP was performed with (^{125}J)-human alpha ANP and the use of a polyclonal antibody raised in rabbits (Amersham, England). Cross reactivity of synthetic h-alpha ANP (Peninsula, England) to the antibody was 100%. Dextran coated charcoal was utilized for the separation of bound and free ligands. RIA-values were reliable within a range of 3.5 - 200 fmol/assay tube and were corrected for recovery. Intra-assay variation was 5.5 % (n=12) and inter-assay variation 11.2 % (n=12).

Statistics

The paired *t*-test was used for statistical analysis. Correlation coefficients were calculated by linear regression analysis. Data are expressed as means ± standard error of the mean (SEM).

Results

With an increase in strain ANP plasma concentrations rose significantly (p<0.02) in V (mean: 31 fmol/ml [at rest] to 68 fmol/ml [Bmax]), in A (35 fmol/ml to 91 fmol/ ml), and in PA (35 fmol/ml to 89 fmol/ml) (Fig. 1). ANP values at rest were found within the normal range of healthy controls (ANP: 13-38 fmol/ml; n=52), except those patients with valvular diseases (Nos. 13-15), and with left ventricular failure due to cardiomyopathy (No. 10), (Table 1). ANP plasma levels were identical in A and PA (y=0, 99x-5.4; n=86; r=0.97), but were systematically lower in the peripheral vein

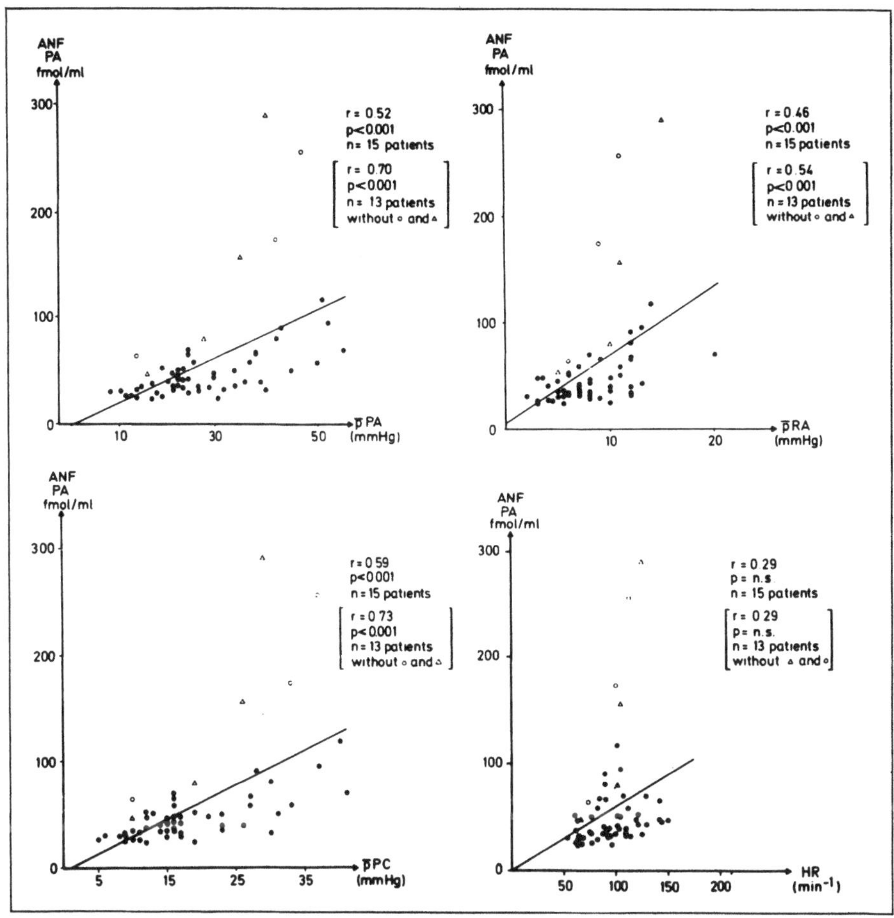

Fig. 1. ANP plasma concentrations before (A), during (B1, B2) at maximum (Bmax), and after treadmill exercise (5M, 10M) in vein, pulmonary artery and artery

(linear correlation V/A: y=0, 66x+8.5; n=86; r=0.95). Pressures in the pulmonary circulation (pRA, pPA, pPC) were significantly elevated at Bmax (p<0.001) compared to values at rest. The increase in ANP concentrations is positively correlated to a rise in right atrial (pRA), pulmonary artery (pPA), and pulmonary capillary wedge pressures (pPC), but does not depend on heart rate (HR) (Fig. 2). The best correlation was found between ANP (PA) and pPC with r=0.59 in linear regression analysis.

During the first three minutes after starting treadmill exercises (A to B1) ANP increase in V was significantly delayed, whereas ANP concentrations in the artery and PA rose immediately (Fig. 1). After exercise (5M,10M) there was no significant difference in the decrease of ANP in V, A and PA.

Table 1. Clinical data of 15 patients undergoing right heart catheterisation during treadmill exercise. CAD: coronary artery disease; CM cardiomyopathy, HOCM: hypertrophic obstructive CM

Patient No	Age (years)	sex	Diagnosis	ANF (v) at rest (fmol/ml)	ANF (v) max.exerc (fmol/ml)	max.exerc (WATT)	max. pPA (mm Hg)
1	40	f	NORMAL	26,1	44,3	150	22
2	48	m	CAD (1 vessel)	25,7	36,7	125	23
3	52	m	CAD (2 vessels)	36,2	58,3	120	25
4	58	m	CAD (1 vessel)	24,9	28,9	120	35
5	44	m	C M	25,3	30,7	125	27
6	42	m	NORMAL	27,2	44,5	75	25
7	62	f	NORMAL	25,8	34,3	75	32
8	57	m	HOCM	28,2	58,1	75	52
9	60	f	HOCM	30,9	61,6	75	38
10 △	49	f	CM, left ventricular failure	41,8	209,3	75	40
11	63	m	C M	19,5	35,4	75	50
12	50	m	CAD (2 vessels)	17,2	21,7	50	39
13 o	50	m	Mitral regurgitation CAD (2 vessels) III°	52,9	223,0	50	47
14	55	m	Aortic regurgitation III°	43,2	86,4	50	52
15	53	m	Mitral regurgitation III–IV°	44,5	46,1	25	55

Discussion

During treadmill exercise the increase in ANP plasma concentrations (A, V, PA) is closely linked with a rise in pressures in the pulmonary circulation (pRA, pPA, pPC). According to the postulated secretion of ANP by atrial stretch the predominant stimulus seems to be an increase in atrial pressure during exercise (2,4,5,6). Since it is known that both atria drain the released peptides into the coronary sinus (1,8) this may be an explanation for the quantitatively weak correlations (r=0.46) between pRA and ANP concentrations. If there is an isolated left atrial distention (e.g.: mitral regurgitation, left ventricular failure) high ANP plasma concentrations are associated with only slightly increased pressures in the right atrium (patients 10 and 13).

Our data confirm that cardiac diseases with hemodynamic changes due to vascular disorders are associated with elevated ANP concentrations compared to patients with coronary artery disease (9,10,11). The finding that an increase in heart rate alone is not responsible for ANP secretion as long as there is no increase in atrial pressure has been confirmed by other investigators (12).

ANP plasma concentrations in PA and A were almost identical, suggesting no additional secretion or inactivation during passage through the lungs and the left heart (13,15). Since ANP levels in V were systematically lower compared to A and PA an in-

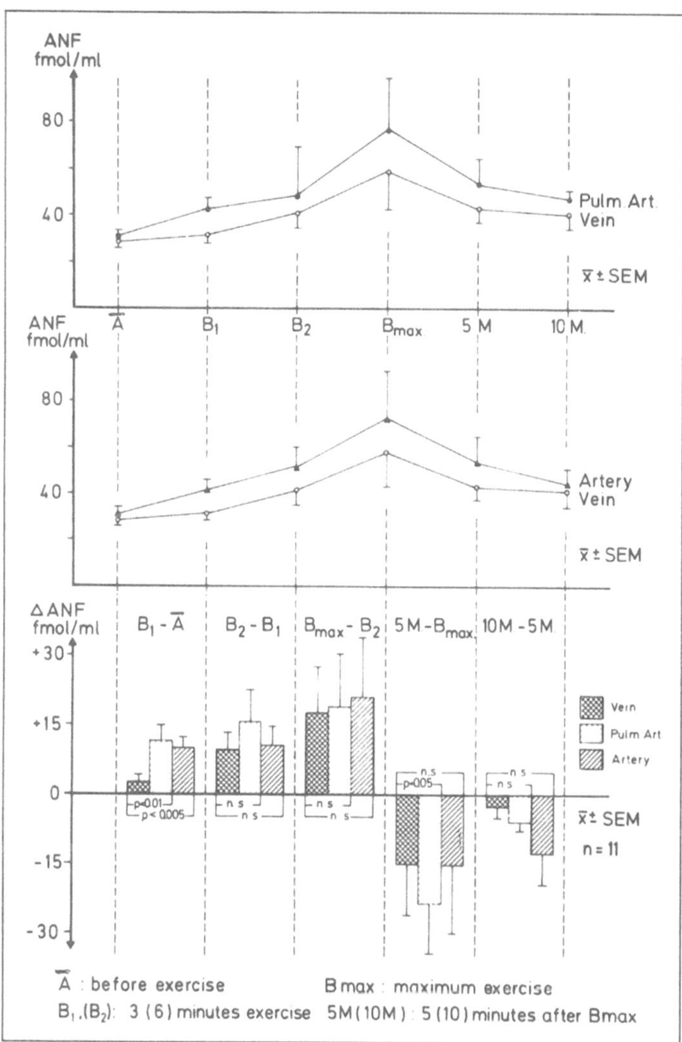

Fig. 2. ANP plasma concentrations in pulmonary artery (PA) correlated to mean pressures in PA (pPA), right atrium (pRA), capillary wedge position (pPC) and heart rate (HR) during treadmill exercise. Parameters in [ ] indicate correlations except patients No.10 and No.13

activation or absorption in the microvascular beds of the peripheral circulation may be postulated (11,14,15). A similar mechanism (e.g.: receptor binding or inactivation in kidneys, liver and peripheral tissues) may explain the findings that during the first 3 minutes after starting treadmill exercise ANP plasma levels in V remain low compared to an increase in A and PA (8,13). A venous pooling or a simple dilution effect seems unlikely if a blood turnover of at least 5 liters/min is assumed.

After peak levels at Bmax ANP decreases and is metabolized according to its half-life

(2,5-3,1 minutes) at rates similar to other vasoactive peptides, such as vasopressin and angiotensin II (1,13,14).

ANP as a non-invasive humoral marker during treadmill exercise indicates an overall increase in atrial pressure related to the type of cardiac disease, but cannot distinguish between right or left atrial distension.

References

1. Anderson, J.V., Donckier, J., McKenna, W.J., Bloom, S.R. (1986) The plasma release of atrial natriuretic peptide in man. Clin Science 71: 151-155
2. Raine, A.E.G., Erne, P., Bürgisser, E., Müller, F.B., Bolli, F., Burkart, F.,Bühler, F.R. (1986) Atrial natriuretic peptide and atrial pressure in patients with congestive heart failure. N Engl J Med 315: 533-537
3. Ledsome, J.R., Wilson, N., Courneya, C.A., Rankin, A.J. (1985) Release of atrial natriuretic peptide by atrial distention. Can J Physiol Pharmacol 63: 739-742
4. Petzl, D.H., Hartter, E., Osterode, W., Boehm, H., Woloszczuk, W. (1987) Atrial natriuretic peptide release due to physical exercise in healthy persons and in cardiac patients. Klin Wochenschr 65: (4) 194-196
5. Richards, A.M., Tonolo, G., Cleland, J.G., McIntyre, G.D., Leckie, B.J., Dargie, H.J., Ball, S.G., Robertson, J.I. (1987) Plasma atrial natriuretic peptide concentrations during exercise in sodium replete and deplete normal man. Clin Sci 72 (2): 159-164
6. Punzengruber, C., Hartter, E., Weissel, M., Burghuber, O.C., Ludvik, B., Woloszczuk, W. (1987) Periphere and rechtsatriale Plasmaspiegel des atrialen natriuretischen Peptids und ihre Beziehung zum atrialen Mitteldruck. Z Kardiol 76: 72-75
7. Eisenhauer, T., Talartschik, J., Scheler, F (1986) Detection of fluid overload by plasma concentration of human atrial natriuretic peptide (h-ANP) in patients with renal failure. Klin Wochenschr 64 (suppl VI): 68-72
8. Sato, F., Kamoi, K., Wakiya, Y., Ozawa, T., Arai, O., Ishibashi, M., Yamaji (1986) Relationship between plasma atrial natriuretic peptide levels and atrial pressure in man. J Clin Endocrinol Metab 63 (4): 823-827
9. Wambach, G., Hannekum, A., Schmidt, S., Kaufmann, W., Dalichau, H. (1987) Erhöhte Konzentrationen des atrialen natriuretischen Peptids im Plasma und im Herzvorhof bei Patienten mit Aorten- und Mitral-vitien im Vergleich zu Patienten mit koronarer Herzkrankheit. Z Kardiol 76 (2): 76-80
10. Talartschik, J., Eisenhauer, T., De Vivie, R., Talartschik, B., Scheler, F. (1987) Release of atrial natriuretic peptide during cardiac arrest and after resuscitation in 19 patients undergoing cardiac surgery with cardiopulmonary bypass. In: Brenner, B.M., Laragh, J.H. (eds) Advances in Atrial Peptide Research, Vol.2. Raven Press, New York, pp 488-492
11. Lang, R.E., Dietz, R., Merkel, A., Unger, T., Ruskoaho, H., Ganten, D. (1986) Plasma atrial natriuretic peptide values in cardiac disease. J Hypertension 4 (2): 119-123
12. Akabane, S., Kojima, S., Igarashi, Y., Kawamura, M., Matsushima, Y., Ito, K. (1987) Release of atrial natriuretic polypeptide by graded right atrial distention in anaesthetized dogs. Life Sci 40 (2): 119-125

13. Crozier, J.G., Nicholls, M.G., Ikram, H., Espiner, E.A., Yandle, T.G., Jans, S. (1986) Atrial natriuretic peptide in humans. Production and clearance by various tissues. Hypertension 8 (suppl II): II-11-II-15
14. Yandle, T.G., Richards, A.M., Nicholls, M.G., Cuneo, R., Espiner, E.A., Livesey, J.H. (1986) Metabolic clearance rate and plasma half-life of alpha-human atrial natriuretic peptide in man. Life Sci 38: 1827-1833
15. Bates, E.R., Shenker, Y., Grekin, R.J. (1986) The relationship between plasma levels of immunoreactive atrial natriuretic hormone and hemodynamic function in man. Circulation 73 (6): 1155-1161

Plasma atrial natriuretic factor responses to opiates

A. M. Vollmar*, R. M. Arendt[+], R. Schulz*
*Department of Pharmacology, Toxicology and Pharmacy of the Veterinary School, University of Munich, FRG, and [+]Medical Clinic I, Klinikum Großhadern, University of Munich, FRG

Introduction

Mechanical stimuli such as volume expansion (7) and atrial distention (3) as well as a number of drugs which directly or indirectly influence the water and electrolyte balance (6,9) stimulate the release of ANF into the circulation. Opioids are also known to influence the water balance, apparently dependent on the receptor-specificity of the opioid used. That is, kappa-opioid receptor agonists cause pronounced diuresis (8), while antidiuresis is brought about by mu-receptor agonists (11). Considering the common effect of ANF and opioids on fluid and electrolyte balance, we searched for a possible effect of mu- and kappa-opioids on the release of immunoreactive ANF (IR-ANF) into the circulation of rats.

The effects of multiple opioid receptor agonists on plasma immunoreactive atrial natriuretic factor (IR-ANF) was studied in conscious nonhydrated rats. The mu-agonist fentanyl (0.05 mg/kg, s.c.) increased plasma levels of IR-ANF up to tenfold, maximal levels were reached within 5-10 min, while U 50,488-H, a selective kappa-receptor ligand, was ineffective in this respect. Pretreatment with the narcotic antagonist naltrexone completely blocked the effect of fentanyl on plasma IR-ANF, proving the opioid receptor specificity. The quaternary antagonist N-methylnaltrexone (1 mg/kg) failed to abolish the fentanyl-induced increase of ANF, suggesting a centrally mediated action of opiates on plasma ANF.

Methods

Male Sprague-Dawley rats (Institut für wissenschaftliche Versuchstierzucht GmbH., Geretsried, FRG), weighing between 200 and 250 g were used. A standard chow diet and tap water were available *ad libitum.* Experiments were routinely started at 0900 h. Drugs were dissolved in saline and injected subcutaneously in a volume of 0.1 ml per 100 g body weight. The narcotic antagonists naltrexone and quaternary N-methylnaltrexone were given 20 min prior to the further treatment.

Animals were decapitated at different times after treatment (see results) and 3 ml blood was collected in polypropylene tubes kept on ice, each containing 3 mg sodium EDTA, 5 µg phenylmethylsulfonylfluoride (PMSF), and 1500 K.I.U. aprotinin. The blood was centrifuged (1000 xg, 10 min, 4°C) and 100 µl plasma was added to 900 µl 0.1% trifluoroacetic acid (TFA). This mixture was applied to a Sep-Pak C_{18} cartridge (Waters/ Millipore, Milford, Massachusetts, USA). After washing the loaded cartridge with 10 ml 0.1 % TFA, ANF was eluted with 1.5 ml 60 % acetonitrile in 0.1 % TFA (recovery:

75%-88%). Eluates were lyophilized and redissolved in 350 µl buffer to determine ANF by radioimmunoassay (RIA). RIA was performed as described (1).

IR-ANF rat plasma was compared to synthetic rat-ANF by HPLC. The plasma probes were extracted as described above, redissolved in 25 to 100 µl 0.01 % TFA, loaded on a C_{18} ODS ultraphere ^{TM} column (5µm, 2mm x 150mm, Beckman Instruments, San Ramon, California, USA) and eluted with a linear gradient of acetonitrile from 10% to 80% (flow rate: 0.2 ml/min). Fractions containing rat plasma ANF and synthetic ANF were determined by RIA in order to calculate the retention times.

Results

Dose-dependent effects of opioids on plasma IR-ANF: The mu-agonist fentanyl is shown in Fig. 1 to dose-dependently raise plasma IR-ANF by up to elevenfold (0.05 mg/kg, 20 min) compared to control levels. Comparable data were obtained with sufentanil, another highly selective mu-receptor agonist (data not shown). In contrast, kappa-agonist

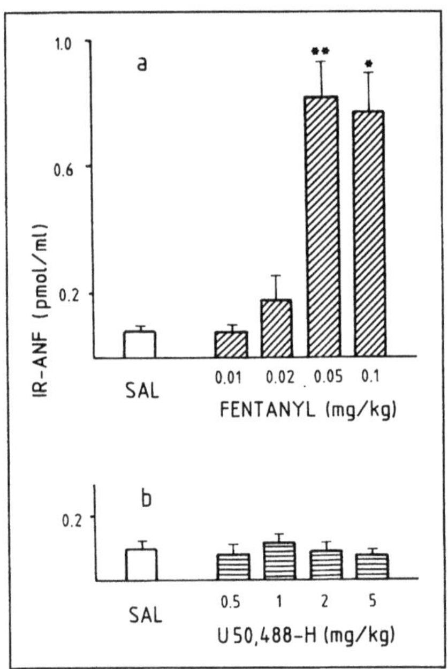

Fig. 1. Effect of the mu-opioid receptor agonist fentanyl (panel a) and the kappa-agonist U 50,488-H (panel b) on plasma IR-ANF. The responses were measured 20 min after administration of the indicated doses of agonists or saline (SAL). Values shown are means ± SEM, (n=3-6). Levels of significance refer to the difference from control. *P ≤ 0.01, ** P ≤ 0.001

U 50, 488-H failed to affect IR-ANF plasma levels, even at a dose of 5 mg/kg which induces a marked diuresis in rats (8). Ethylketocyclazocine (EKC), another kappa-agonist, also failed to increase the release of IR-ANF in the same dose range (data not shown). HPLC-analysis of IR-ANF in plasma extracts of control and fentanyl treated rats indicated that the measured IR-ANF behaved mostly like alpha-rat-ANF.

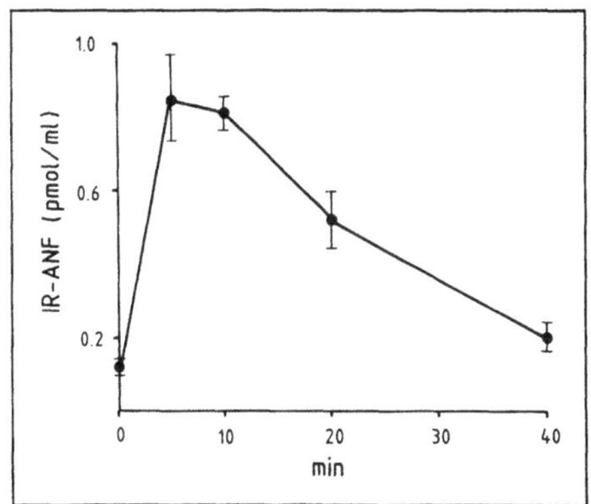

Fig. 2. Time course of change in levels of plasma IR-ANF after injection of 0.05 mg/kg fentanyl. Values shown are means ± SEM, (n=3-4). Levels of significance refer to the difference from control. * P ≤ 0.05

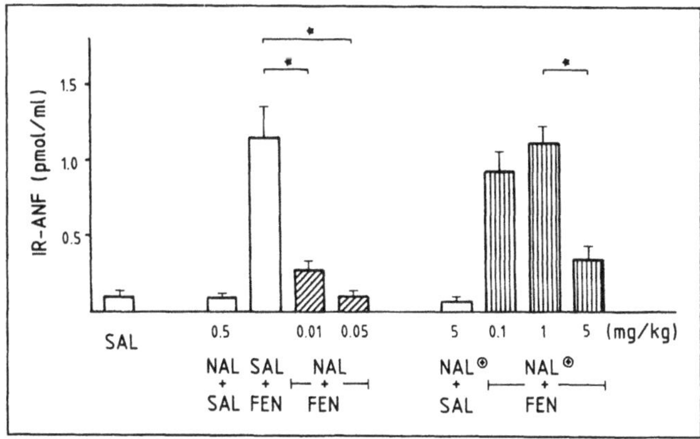

Fig. 3. Dose-dependent effect of naltrexone (NAL) and quaternary naltrexone (NAL(+)) on fentanyl (FEN)-stimulated IR-ANF release. Various doses of the antagonist or saline (SAL) were injected 20 min prior to administration of saline and 0.05 mg/kg fentanyl; Samples obtained 10 min later. Values shown are means ± SEM, (n=5-6); P≤ 0.01

Time dependency: Peak concentrations of IR-ANF were reached 5 to 10 minutes after a single injection of the maximal effective dose of fentanyl (0.05 mg/kg) (Fig. 2). After 40 min IR-ANF levels returned to near control values.

Effect of opioid antagonist on fentanyl-induced plasma IR-ANF: The tertiary narcotic antagonist naltrexone (0.05 mg/kg) given 20 min before the fentanyl injection completely blocked the stimulatory effects of fentanyl (0.05 mg/kg, 10 min) on plasma IR-ANF (Fig. 3). Naltrexone (0.5 mg/kg) did not effect IR-ANF plasma levels. For the same set of experiments the quaternary antagonist N-methylnaltrexone was used in a tenfold higher dose since it has only a tenth of the potency compared to the tertiary naltrexone (10). One mg/kg of N-methylnaltrexone was not able to prevent the increase of IR-ANF caused by fentanyl (0.05 mg/kg). Five mg/kg of quaternary naltrexone partially blocked the fentanyl effect.

Discussion

The present investigation suggests that the effect of opiates upon the release of ANF into the circulation is related to a distinct opioid receptor type. Thus, only mu-opiates are potent releasers of ANF. Kappa-receptor ligands (EKC, U 50,488-H) were without effect.

The specificity of the observed effect, the very low concentrations of the mu-opiates employed in the present study, as well as the complete blockade of their effects by the narcotic antagonist naltrexone, supports the idea that mu-opioid receptors modulate plasma ANF levels. Using quaternary antagonist N-methylnaltrexone, which does not readily cross the blood-brain barrier (2) at doses known to block mu-receptors, the IR-ANF stimulating effect of fentanyl was not affected. However, at a dose of 5 mg/kg of the quaternary antagonist, the ANF-releasing effect of fentanyl was partially abolished. This might be due to the generation of tertiary naltrexone during the course of the experiment (2). It is suggested that fentanyl acts on centrally located opioid receptors to cause an increase of plasma IR-ANF. In addition one has to take into account that opiates cause cardiovascular responses (4). We cannot rule out that, in addition to the suggested central nervous mechanism, the ANF changes observed might be due to hemodynamic changes caused by opiates (4,5).

In conclusion, our experiments suggest that ANF release into the circulation is also controlled by opioid receptors. The opioid receptors involved in this mechanism are of the mu-type and located in the central nervous system.

Acknowledgements: The excellent technical assistance of Ms. G. Hach and Ms. U. Rüberg and the secretarial assistance of Ms. H. Röder are gratefully acknowledged. This work was supported by DFG.

References

1. Arendt, R.M., Stangl, E., Zähringer, J., Liebisch, D.C., Herz, A. (1985) Demonstration and characterization of alpha-human atrial natriuretic factor in human plasma. FEBS Lett 189: 57-61
2. Brown, D.R., Goldberg, L.I. (1985) The use of quaternary narcotic antagonists in opiate research. Neuropharmacology 24: 181-191
3. Dietz, J.R. (1984) Release of natriuretic factor from rat heart-lung preparation by atrial distension. Am J Physiol 247: 1093-1096
4. Feuerstein, G. (1985) The opioid system and central cardiovascular control: Analysis of controversies. Peptides 6: 51-56
5. Johnson, M.W., Mitch, W.E., Wilcox, C.S. (1985) The cardiovascular actions of morphine and the endogenous opioid peptides. Progress in Cardiovascular Diseases, Vol. XXVII 6: 435-450
6. Lachance, D., Garcia, R., Gutkowska, J., Cantin, M., Thibault, G. (1986) Mechanisms of release of atrial natriuretic factor. 1. Effect of several agonist and steroids on its release by atrial minces. Biochem Biophys Res Comm 135: 1090-1098
7. Lang, E., Thölken, H., Ganten, D., Luft, F.C., Ruskoaho, H., Unger, T. (1985) Atrial natriuretic factor - a circulating hormone stimulated by volume loading. Nature 314: 264-265
8. Leander, J.D. (1983) Further study of kappa opioids on increased urination. J Pharmacol Exp Ther 227: 35-41
9. Manning, P.T., Schwartz, D., Katsube, N.C., Holmberg, S.W., Needleman, P. (1985) Vasopressin-stimulated release of atriopeptin: Endocrine antagonist in fluid homeostasis. Science 229: 395-397
10. Schulz, R., Wüster, M., Herz, A. (1979) Centrally and peripherally mediated inhibition of intestinal motility by opioids. Naunyn-Schmiedeberg´s Arch Pharmacol 308: 255-260
11. Walker, L.A., Murphy, J.C. (1984) Antinatriuretic effect of acute morphine administration in conscious rats. J Pharmacol Exp Ther 229: 404-408

Atrial natriuretic peptide in term newborn infants

A. Timnik, J. Weil, M. Brügmann, T. Strom, *M. Haufe, *J.M. Heim, *R. Gerzer

Departments of Pediatrics and *Medicine, University of Munich, FRG

Summary

The aim of this study was to measure plasma and urinary levels of atrial natriuretic peptide (ANP) and cyclic guanosine 3´, 5´monophosphate (cGMP) in term newborn infants (n=25) aged one to six days.

Plasma ANP and cGMP levels were significantly higher in umbilical venous blood and in peripheral venous blood of infants aged two to four days than in older control infants aged 30 days to 12 months (n=24, p<0.01). There was a linear correlation between the two parameters (r=0.71, p<0.001).

Furthermore, in contrast to ANP, a linear correlation was found between plasma and urinary cGMP levels (r=0.69, p<0.02). Therefore, urinary cGMP - but not ANP - excretion seems to reflect plasma cGMP concentrations to a certain extent. Additionally, an age dependence of urinary cGMP excretion was found that showed significantly higher values in infants aged one to three days than in infants aged four to six days (p<0.01). There was no influence of elevated ANP or cGMP levels on urinary sodium excretion or urine volume.

Introduction

Only few studies have been performed in the early human postnatal period that reported elevated plasma levels of atrial natriuretic peptide (ANP) (10,12). However, it is not clear how target organs, such as kidney or blood vessels, react to these elevated levels during this period of life. Since the effect of ANP on target cells is mediated, at least in part, by cyclic guanosine 3´, 5´ monophosphate (cGMP), this nucleotide seems to be a marker for the cellular effect of ANP (1,2,3). Furthermore, in newborn infants no studies have been published measuring urinary excretion of ANP and cGMP. Urinary measurements may be of value in infancy, since blood sampling is often difficult or not possible due to ethical reasons.

Additionally, the influence of ANP on the regulation of water and electrolyte homeostasis is still unclear in newborn infants (8). In the present study plasma ANP and cGMP concentrations were measured simultaneously. Furthermore, urinary excretions of ANP and cGMP were repeatedly measured throughout the first six days of life and correlated to plasma levels, to urinary sodium excretion, and urine volume.

This study was supported by grants from the Deutsche Forschungsgemeinschaft (We 1130/1-1 and Ge 399/3-1).

Subjects and methods

We investigated term newborn infants (n=25) after uncomplicated spontaneous vaginal delivery; birth weights ranged between 3120g - 3780g. Five minute Apgar scores were 9 or 10 in all infants. Blood samples were taken from umbilical veins after early clamping of the cord, and from peripheral venous blood between the second and fourth day of life. Peripheral venous blood was taken from control reconvalescent infants (n=24) aged 30 days to 12 months, admitted to the hospital for uncomplicated pediatric diseases, such as social and behavioral problems, or infections of the upper airways. In all infants studied, blood samples were taken at the occasion of routine clinically indicated blood sampling.

We also took venous blood from women after uneventful pregnancy at term (n=20) 5 to 10 minutes after delivery and from healthy non-pregnant women aged 21 to 36 years (n=16).

From 17 out of 25 newborns, urine was collected for 8 h (2200 h - 0600 h) throughout the first 6 days of life.

Levels of ANP and cGMP in plasma and in urine were measured by radioimmunoassay as recently described (3,4,9).

Urinary sodium excretions were measured by routine laboratory methods. For statistical evaluations the Wilcoxen U-Test for non-dependent parameters was applied.

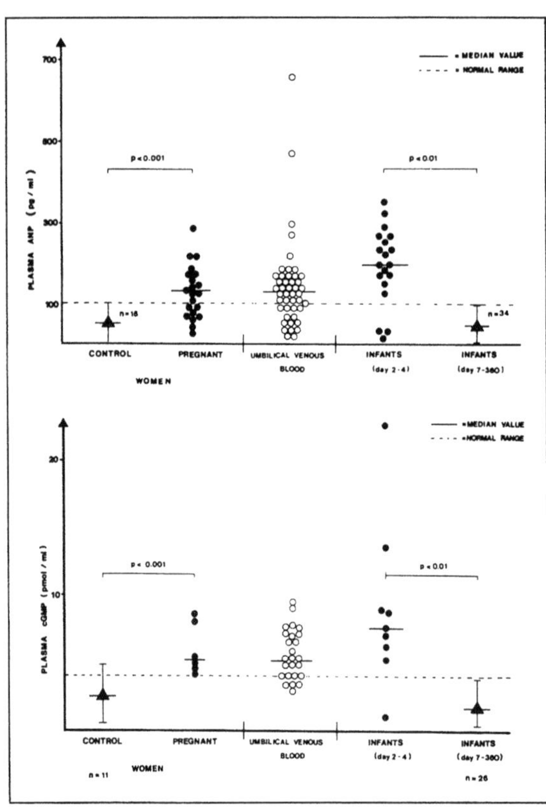

Fig. 1. Plasma ANP (top) and cGMP (bottom) levels in non-pregnant, healthy women, and in pregnant women immediately after delivery, as well as in umbilical venous blood and in peripheral blood of infants ages two to four days and of older control infants

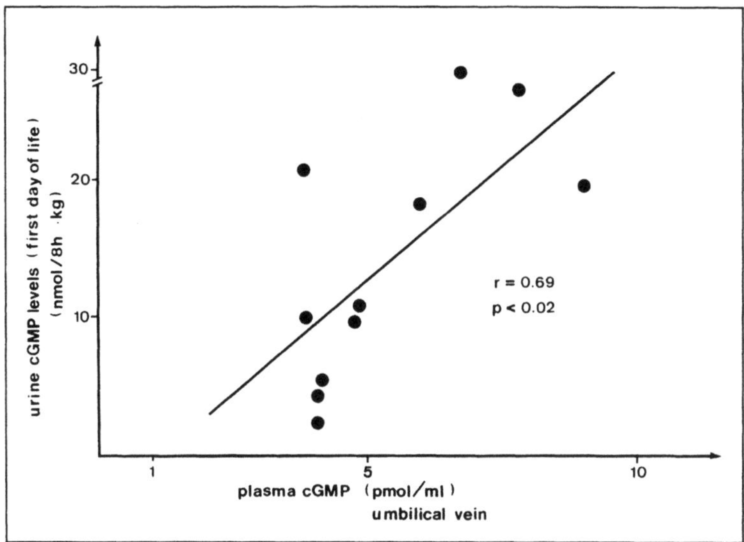

Fig. 2. Correlation between plasma and urinary cGMP levels

Results

Figure 1 shows plasma ANP and cGMP levels in umbilical venous blood and in peripheral venous blood of newborn infants aged two to four days and of control infants aged 30 days up to 1 year. ANP and cGMP plasma levels of women immediately after delivery and of nonpregnant healthy women are also depicted. In the pregnant women significantly higher ANP (range: 17.4 - 298.3 pg/ml, median value: 138.6 pg/ml) and cGMP levels (4.1 - 8.9 pmol/ml; median value: 6.1 pmol/ml) were found than in healthy nonpregnant women (median values: ANP 51.3 pg/ml; cGMP 3.4 pmol/ml; p < 0.001 for ANP and cGMP). Significantly higher levels of both parameters were measured in umbilical venous blood (median values: ANP 136.2 pg/ml; cGMP 6.1 pmol/ml), and in newborn infants aged two to four days (median values: ANP 184.2 pg/ml; cGMP 7.9 pmol/ml) than in older control infants (median values: ANP 48.3 pg/ml; cGMP 2.8 mol/ml; p < 0.01 for both parameters).

There was a linear correlation between ANP and cGMP plasma levels in all infants studied (data not shown, r=0.71, p < 0.001).

In Fig. 2 the correlation between plasma cGMP levels in umbilical venous blood and urinary cGMP levels in newborn infants aged one day of life is depicted. There was a linear correlation between the two parameters (r=0.69; p<0.02). No correlation was found between plasma and urinary ANP levels.

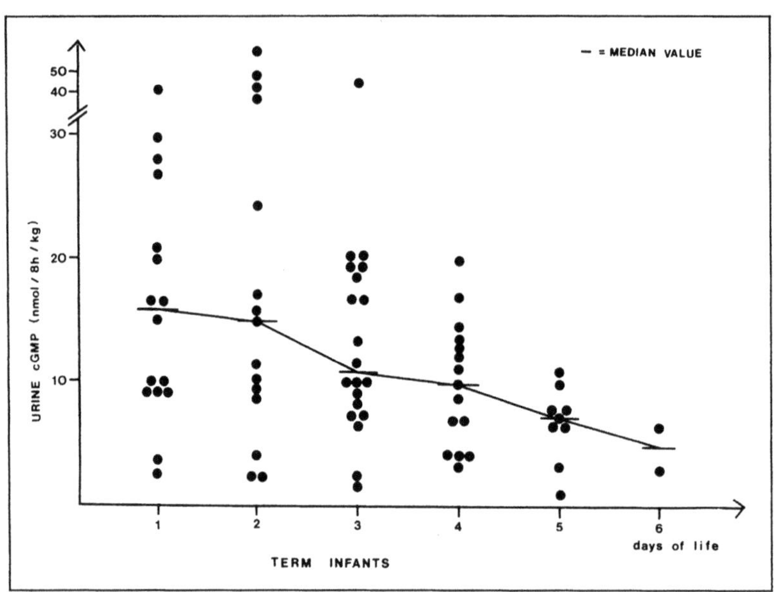

Fig. 3. Urinary excretion of cGMP in newborn infants throughout the first six days of life

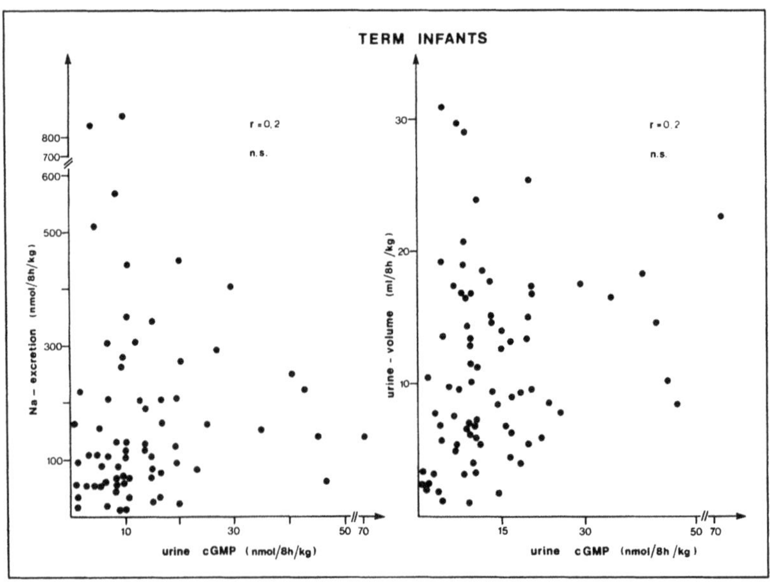

Fig. 4. Correlation between urinary cGMP excretion and sodium excretion or urine volume in newborn infants ages one to six days of life

208

In Fig. 3 the daily urinary cGMP excretion is shown from newborn infants throughout the first six days of life. The cGMP excretion was significantly higher during days 1 to 3 than during days 4 to 6 of life ($p < 0.01$). No age dependence of urinary ANP excretion was detectable in newborn infants in the early neonatal period.

As shown in Fig. 4 no correlation was found between urinary cGMP excretion and sodium excretion or urine volume.

Discussion

To our knowledge this is the first study reporting plasma and urinary ANP and cGMP levels in the early neonatal period.

We found elevated plasma ANP and cGMP concentrations in umbilical venous blood and peripheral venous blood of term newborn infants aged one to four days of life compared to older infants. Similar to our previous study in children suffering from heart failure (11), there was a linear correlation between the two parameters in plasma, which gives further evidence that cGMP is a marker for the cellular response to ANP (3). In contrast to ANP only the urinary cGMP excretions reflected plasma cGMP concentrations to a certain extent. There was a significant correlation between plasma and urinary cGMP levels, although a considerable variation of the individual values occurred. Therefore, urinary cGMP- but not ANP- excretion may be of value in studying the changes of the ANP/cGMP system during the early neonatal period.

This assumption is supported by the second part of this study showing an age dependency of urinary cGMP excretion with higher values in newborn infants aged one to three days than in older infants aged four to six days. These results of urinary cGMP excretion are in good agreement with the changes of plasma cGMP levels we found in the newborn infants studied. We could not detect an enhanced natriuresis or diuresis in infants with high ANP and cGMP levels. Our results are in accordance with the study of Tulassay et al. (8) who could not detect a correlation between plasma ANP levels and sodium excretion in preterm infants. This is a surprising finding, since ANP is thought to possess natriuretic and diuretic properties. The lack of effect of ANP in newborn infants may be explained either by a down-regulation of specific ANP receptors or by counteraction of other volume regulating hormones, such as vasopressin or the renin-angiotensin system.

Measuring the ANP binding sites on platelets (6) from umbilical venous blood, we could not detect a reduced number of binding sites in term newborn infants compared to subjects with normal ANP levels. Thus, these experiments did not support the hypothesis of ANP receptor "down-regulation" in the early neonatal period (7).

On the other hand, the latter hypothesis of simultaneous activation of the natriuretic as well as of the sodium-retaining hormonal system is supported by previous results, showing elevated plasma vasopressin and aldosterone levels in term newborn infants during the first days of life (5).

Further studies with simultaneous measurements of plasma ANP and cGMP and of sodium retaining hormones (e.g., vasopressin) may help to clarify the physiological role of ANP in water and electrolyte homeostasis during the human postnatal period.

Acknowledgement: The authors would like to thank Misses Gabi Heilig and Misses Barbara Schütte for skillful technical assistance.

The α-hANP antibody was kindly provided by R.E. Lang, Dept. of Pharmacology, University of Heidlelberg, FRG.

References

1. Gerzer, R., Weil, J., Strom, T., Müller, T. (1986) Mechanisms of action of actrial natriuretic factor: clinical consequences. Klin Wochenschr 64 (suppl VI): 21-26
2. Gerzer, R., Heim, J.M., Schütte, B., Weil, J. (1987) Cellular mechanism of action of actial natriuretic factor Klin Wochenschr 65 (suppl VIII): 109-114
3. Heim, J.M., Gottmann, K., Weil, J., Haufe, M.C., Gerzer, R. (1987) Is cyclic GMP a useful marker for ANF action ? Z Kardiol (in press)
4. Lang, R.E., Thölken, H., Ganten, D., Luft, F.C., Ruskoaho, H., Unger, T. (1985) Atrial natriuretic factor - a circulating hormone stimulated by volume loading. Nature 314: 264-266
5. Leake, R.D., Weitzman, R.E. (1980) The fetal-maternal neurohypophysiological system. In: Tulchinsky, D., Ryan, K.J. (eds) Maternal-fetal Endocrinology, 1st edn WB Sanders, Philadelphia, pp 233-241
6. Strom, T., Weil, J., Bidlingmaier, F. (1987) Platelet receptors for atrial natriuretic peptide in man. Life Sci 40: 769-773
7. Strom, T., Weil, J., Timnik, A., Knorr, D., Bidlingmaier, F. (1987, Abstract) Binding sites for atrial natriuretic peptide (ANP) on platelets in patients with high plasma ANP levels. Pediatr Res (in press)
8. Tulassay, T., Rascher, W., Seyberth, H.J., Lang, R.E., Toth, M., Sulyok, E. (1986) Role of atrial natriuretic peptide in sodium homeostasis in premature infants. J Pediatr 109: 1023-1027
9. Weil, J., Lang, R.E., Suttmann, H., Rampf, U., Bidlingmaier, F., Gerzer, R. (1985) Concomittant increase of plasma atrial natriuretic peptide and cyclic GMP during volume loading. Klin Wochenschr 63: 1265-1268
10. Weil, J., Bidlingmaier, F., Döhlemann, C., Kuhnle, U., Strom, T., Lang, R.E. (1986) Comparison of plasma atrial natriuretic peptide levels in healthy children from birth to adolescence and in children with cardiac diseases. Pediatr Res 20: 1328-1331
11. Weil, J., Gerzer, R., Strom, T., Lang, R.E., Döhlemann, C., Knorr, D., Bidlingmaier, F. (1987) Increased plasma cyclic guanosine monophosphate concentrations in children with high levels of circulating atrial natriuretic peptide Pediatrics 80: 545-548
12. Yamaji, T., Hirai, N., Ishibashi, M., Takaku, F., Yanaihara, T., Nakayama, T. (1986) Atrial natriuretic peptide in umbilical cord blood: evidence for circulating hormone in human foetus. J Clin Endocrinol Metab 63: 1414-1417

Effect of an ANP bolus injection on endocrine and renal regulation in patients with liver cirrhosis and ascites

G. Brabant, H. Stegner[1], A. Lueg, D. Stolze, R. Brunkhorst[2], F.W. Schmidt[3], R.D. Hesch

Depts. Clinical Endocrinology, Nephrology[2] and Gastroenterology[3], School of Medicine, Hanover, FRG, and University Clinic Eppendorf[1], Hamburg , FRG

Summary

Basal levels of atrial natriuretic peptide (ANP), aldosterone, renin, and vasopressin were investigated in patients with liver cirrhosis in 10 patients with ascites following a bolus injection of 33µg ANP. This dose of ANP significantly increased the fractional excretion of sodium but not of potassium. In comparison to normal controls ANP levels showed significantly lower maximum levels after bolus injection, but the plasma disappearance rate appeared to be unaltered. Aldosterone plasma levels were significantly decreased whereas the decrease of renin levels showed a high individual variation. Vasopressin was significantly elevated but did not change following ANP injection. ANP was measured in ascites of five patients simultaneously to the plasma determinations. The results suggest a free distribution of the peptide into the ascites. This phenomenon might explain differences in the regulation and biological action of ANP in patients with cirrhosis and ascites as compared to normal controls or patients without ascites.

Introduction

The pathophysiological changes leading to ascites formation in patients with liver cirrhosis are still incompletely understood. As it is well-known that the capacity to excrete sodium and water is impaired in these patients, alterations in a number of endocrine systems might be involved. Apart from the well established changes in the activity of the renin-aldosterone system in cirrhosis (8), prostaglandins (9), the renal kallikreinkinin system (28), the sympathoadrenal system (1,19) and vasopressin (2) have been discussed. After the characterization of a human atrial natriuretic peptide (ANP) as a selectively natriuretic and diuretic agent, a number of authors investigated the pathophysiological role of this peptide in patients with liver cirrhosis. Plasma levels were never found to be decreased as expected, but most authors reported elevated plasma concentrations (12,13, 14,17,23). There is no explanation yet for this phenomenon. Gerbes et al. (this symposium) demonstrated identical concentrations in cirrhotics and controls and were unable to find significant amounts of different molecular forms of circulating ANP using a highly specific HPLC method. Detailed investigations on the plasma disappearance rate of the peptide are still lacking.

The interaction of ANP with the renin-aldosterone system have been shown by a number of authors (5,25,26). A modulatory role in the release of vasopressin has been de-

"

monstrated for the peptide. Other reports point towards a direct influence of vasopressin on ANP secretion and discuss a potential interaction of ANP and vasopressin in the regulation of vascular smooth muscle (7,18,20).

We could previously demonstrate in a number of patients that pharmacological doses of ANP are effective in inducing natriuresis and diuresis in patients with liver cirrhosis (3). Under these conditions it was tempting to investigate the impact of a bolus injection of ANP on the renin-aldosterone system and on vasopressin levels in patients with cirrhosis to gain further insight into the potential interaction of these regulatory systems of sodium and water excretion under pathological conditions of the disease.

Patients and methods

Basal ANP levels were determined in 55 patients with liver cirrhosis from various causes (post hepatitis, ethyl-toxic, primary biliary cirrhosis, and undefined). In all these patients aldosterone and renin were determined.

Ten patients with liver cirrhosis and ascites (five female, five male patients, six postinfectious and four with ethylic cirrhosis) were subjected to an intravenous bolus injection of 33 µg ANP (Bissendorf Peptide, Wedemark, FRG). All these patients were under continuous therapy with spironolactone (200 mg) and butizide (20 mg), as well as salt (2g/d) and water (1.5 l/d) restriction. All patients gave written consent according to the principles of §II of the Declaration of Helsinki. Apart from clinical parameters such as blood pressure and heart rate, which under these doses have been shown to be unaltered, (4), the changes in plasma concentrations of ANP, renin, aldosterone and vasopressin were measured in comparison to the control period. Blood samples were drawn at -10, -5, 0 min and in 10 min intervals thereafter for two hours. In four patients ANP plasma levels were determined in 20 second intervals for the first two min to better define the plasma disappearance rate of the peptide. In five patients parallel to the plasma measurements, ANP was determined in the ascites at -10, 0, 5, 10, 30 and 60 min. All tests were performed at 1400 h. The changes in ANP plasma levels were compared to a group of five normal controls.
Fractional excretion of sodium and potassium was measured in 30 min increments starting at -30 min until the end of the experiments.

ANP was measured by a specific direct radioimmunoassay as published elsewhere (17). Renin and aldosterone levels were determined by commercially available radioimmunoassay kits (Gammacoat I^{125} plasma renin activity RIA, Travenol, München; Aldosterone RIA Coat-A-count I^{125}, Biermann, Bad Nauheim). Vasopressin was measured by a modified method described by Skowsky et al. (27). Statistical evaluation was performed by the paired t-test.

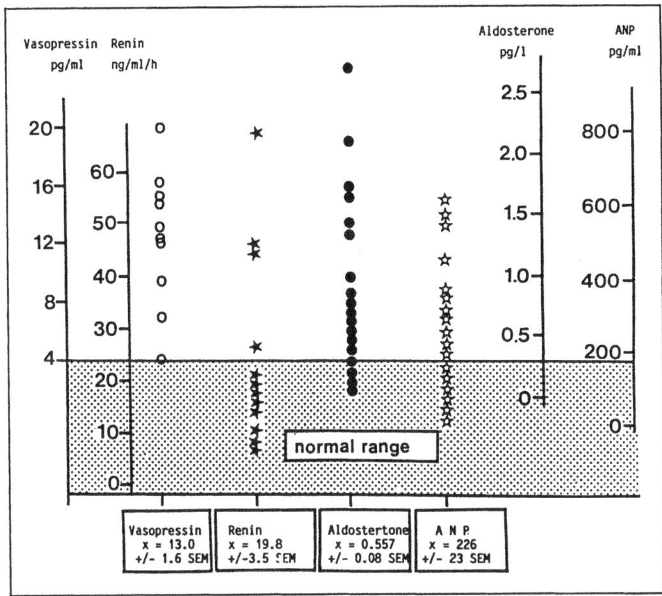

Fig. 1. Basal plasma levels of ANP, aldosterone, renin and vasopressin in patients with liver cirrhosis of various causes and degrees. Mean levels and SEM are given

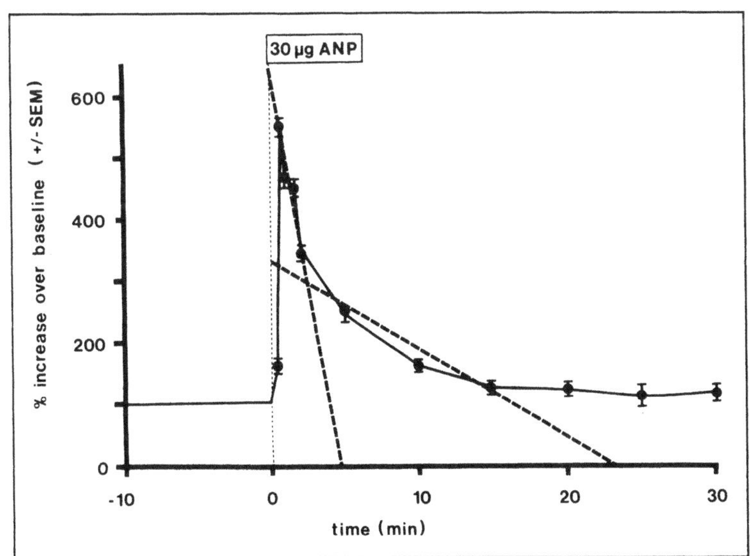

Fig. 2. Plasma disappearance of ANP after bolus injection. Mean of four patients with liver cirrhosis and ascites (in % of basal values) when ANP levels were sampled every 20 sec during the first 2 min

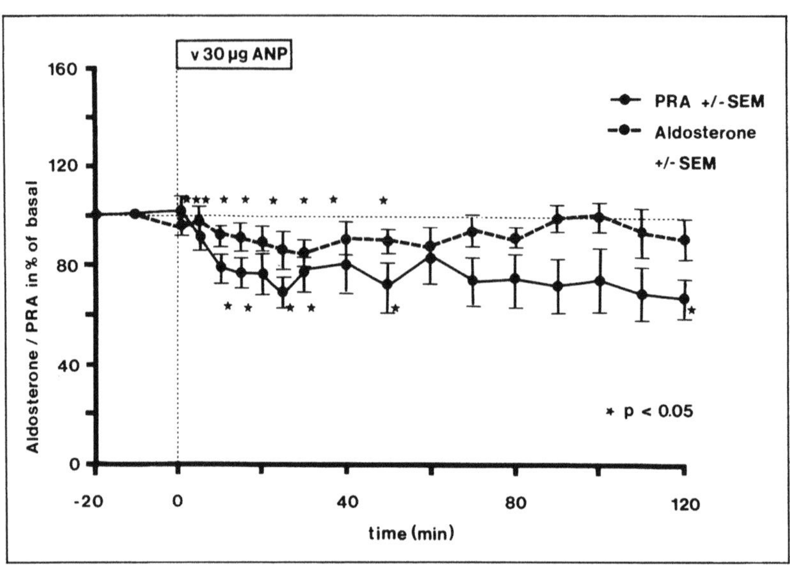

Fig. 3. Mean plasma aldosterone and renin levels following a bolus injection of 33 µg ANP (in % of basal values) in 10 patients with liver cirrhosis and ascites. Statistical evaluation was performed on the absolute values

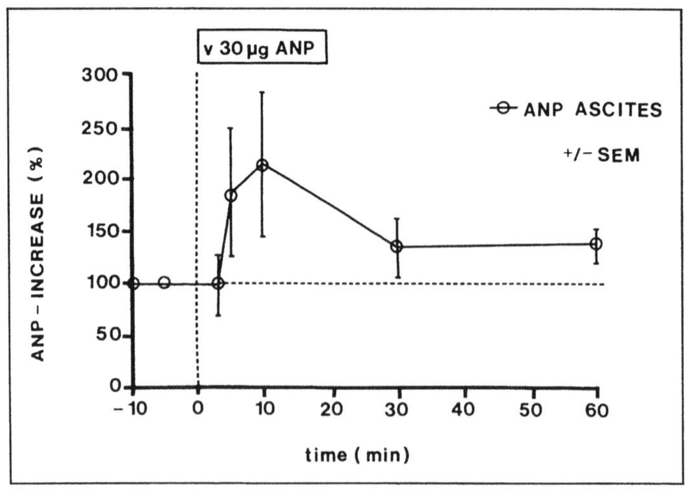

Fig. 4. Alterations of ANP in ascites following a bolus injection of 33 µg ANP. Mean of five patients with liver cirrhosis (in % of basal values)

214

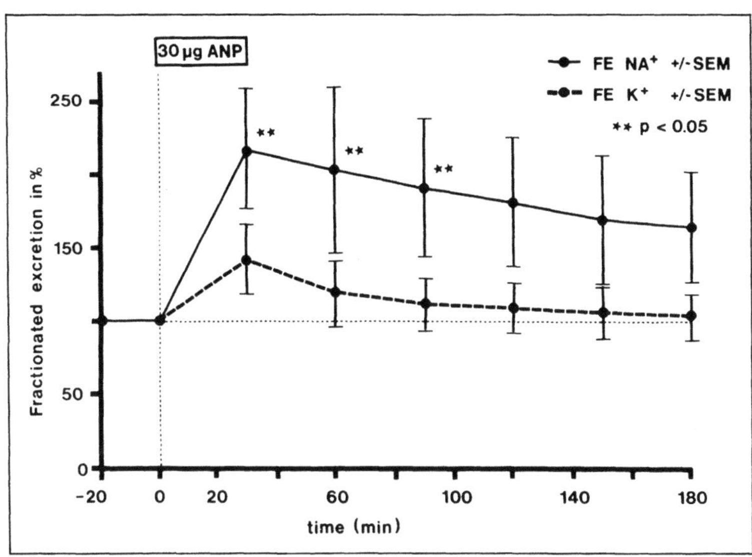

Fig. 5. Mean fractional excretion of sodium and potassium in 10 patients with liver cirrhosis and ascites (in % of basal values). Statistical evaluation was performed on the absolute values

Results

ANP basal levels varied over a wide range, but averages were significantly higher (226 +/- 23 pg/ml, +/- SEM) in the total group (55 patients) than with the assay for normal controls (104+/- 8 pg/ml) (Fig.1); aldosterone and renin levels were equally elevated but there was no correlation to ANP levels individually; the same results could be demonstrated in 10 patients for vasopressin (13 +/-1.6 pg/ml) where again no correlation of individual ANP and vasopressin levels were found.

Following bolus injection of ANP, basal plasma levels of the peptide rose from 258 +/- 25 (SEM) pg/ml to 1539 +/- 322 pg/ml after 1 min. As shown in Fig. 2 the fast component of the plasma disappearance rate was very short (2.1 +/- 1.1 min) and the slow component was 13 +/- 2.5 min. The maximum increase of ANP levels was much higher in controls (4844 +/- 236 pg/ml) as compared to the group of patients, but no difference in either components of the plasma disappearance rates could be demonstrated between patients and normal controls. Following an ANP bolus, renin slightly decreased showing a high individual variation. In contrast, aldosterone decreased significantly during the first 90 min. After that period aldosterone plasma levels returned to baseline values (Fig. 3). Increased basal plasma vasopressin levels were unaltered by ANP bolus injection.

Measurement of ANP in the ascites revealed detectable concentrations in all patients (139 +/- 34 pg/ml). As shown in Fig. 4, ANP ascites levels increased in all patients following ANP bolus injection.

In confirmation of previous results ANP significantly increased the fractional excretion of sodium but not potassium during the first 90 min. after the ANP bolus with a maximum of 129% after 30 min. (Fig. 5).

Discussion

For a long time a potential role of a selective natriuretic factor in the hormonal dysregulation leading to ascites formation in patients with liver cirrhosis (6) has been speculated. In contrast to expectations, measurement of ANP in plasma showed not lowered but normal or elevated levels of the hormone. Methodological problems appear to be unlikely as elevated plasma levels are reported by measurement with the direct RIA as well as with extraction methods (10,14,15). It is conceivable that changes in the metabolism of ANP may account for these elevations. As shown in animal models (21) and in man (22) the plasma half-life of ANP is very short with a fast component of about 2 min and a slow component of 13 min. The results of our study exactly agree with these findings and there was no difference noticeable between healthy controls and patients with liver cirrhosis and ascites. However, the maximum levels of ANP after bolus injections were much higher in normal controls than in cirrhotic patients.

For the first time to our knowledge we could demonstrate significant amounts of ANP in the ascites of patients with liver cirrhosis. Following injection of a pharmacological dose of ANP these concentrations increased, suggesting a free distribution of the hormone into the ascites. This observation fits well to the impaired increase of ANP plasma levels in ascites patients and to the findings of Gerbes et al. These authors reported on an impaired increase of plasma ANP concentrations after stimulation of endogenous secretions by head out water immersion only in patients with ascites but not in normal controls or cirrhotics without ascites. This concept might help to explain that the same endogenous stimulus is less effective in patients with cirrhosis and ascites as compared to normal controls or cirrhotics without ascites. Free distribution of ANP between the plasma and ascites may further explain why patients with high circulating ANP concentrations respond to a given pharmacological stimulus (3). ANP in ascites may be degraded with a much slower rate, thus the large pool of ANP in ascites may redistribute and the continuous exposure to the peptide may lead to ANP receptor "down-regulation" (4).

The results of the ANP-induced changes in the renin-aldosterone axis confirmed other *in vivo* and *in vitro* studies on the regulatory influence of ANP. In humans, however, the results have been controversial. This might be explained by the pulsatile nature of secretion of both hormones (11). As aldosterone secretion is shown to be inhibited on several levels it is conceivable that ANP significantly inhibited only the stimulated levels of aldosterone.

An interaction of ANP and vasopressin has been discussed by several authors (7,16,18, 20,24). In patients with liver cirrhosis it is well known that vasopressin plasma levels are elevated by non-osmotic stimuli (1). Our study confirms these findings in a group of 10 patients. A bolus injection of ANP, sufficient to significantly increase the fractional sodium excretion, however, did not measurably influence the elevated vasopressin levels. These data and the lack of correlation between basal plasma ANP and vasopressin levels are not in favor of a significant interaction of the two peptides.

In summary, the multiple changes in the endocrine regulation in patients with liver cirrhosis and ascites could be confirmed in our patients. Even though a number of alterations of ANP has been shown, the importance of this peptide for the dysregulation of salt and water excretion in cirrhosis has to be further defined. It can be speculated, however, that the mechanism of distribution and redistribution between plasma and ascites may contribute to the impaired responses to ANP demonstrated in patients with severe forms of decompensated liver cirrhosis.

Acknowledgement: The excellent technical assistance of Mrs. H. Renftel-Heine and Miss S. Kaesler are greatfully acknowledged. ANP was a gift of Bissendorf Peptide, Wedemark, FRG. This work was supported by DFG grant Bra 915/1-1.

References

1. Bichet, D.G., Szatalowicz, V., Chaimovitz, C., Schrier, R.W. (1982) Role of vasopressin in abnormal water excretion in cirrhotic patients. Ann Intern Med 96: 413-417
2. Bichet, D.G., Van Putten, V.J., Schrier, R.W. (1982) Potential role of increased sympathetic activity in impaired sodium and water excretion in cirrhosis. N Engl J Med 307: 1552-57
3. Brabant, G., Jüppner, H., Böker, K., Schmidt, F.W., Hesch, R.D. (1986) Human atrial natriuretic peptide (hANP) in the treatment of liver cirrhosis with ascites. Fed Proc 45: 1024
4. Brabant, G., Jüppner, H., Kirschner, M., Böker, K., Schmidt, F.W., Hesch, R.D. (1986) Human atrial natriuretic peptide (ANP) for the treatment of patients with liver cirrhosis and ascites. Klin Wochenschr 64 (Suppl VI): 108-111
5. Burghardt, W., Wernze, H., Diehl, K.L. (1986) Atrial natriuretic peptide in hepatic cirrhosis: relation to stage of disease, sympathoadrenal system and renin aldosterone axis. Klin Wochenschr 64 (Suppl IV): 103-107
6. Buckalew, V.M., Gruber, K.A. (1983) Natriuretic hormone. In: Epstein, M. (ed) The kidney in liver disease. Elsevier, New York, pp: 479-500
7. Dillingham, M.A., Anderson, R.G. (1986) Inhibition of vasopressin action by atrial natriuretic factor. Science 231: 1572-1573
8. Epstein, M. (1983) The renin-angiotensin-aldosterone system in liver disease. In: Epstein, M. (ed) The kidney in liver disease, pp 353-376 2nd edn. Elsevier, New York
9. Epstein, M., Lifschitz, M., Preston, S. (1979) Augmentation of renal prostaglandin E in decompensated cirrhosis. Kidney Int 16: 920
10. Fernandez-Cruz, A., Marco, J., Cuadrado, L.M., Gutkowska, J., Rodriguez-Puyol, D., Caramelo, C., López-Novoa, J.M. (1985) Plasma levels of atrial natriuretic peptide in cirrhotic patients. Lancet II: 1439-1440
11. Follenius, M., Brandenburger, G. (1987) Episodic renin secretion. In: Wagner T.O.F., Filicori, M. (eds) Episodic hormone secretion: from basic science to clinacal application, p 97
12. Fyhrquist, F., Tötterman, K.J., Tikkanen (1985) Infusion of atrial natriuretic peptide in liver cirrhosis with ascites. Lancet II: 1439

13. Henricksen, J.H., Schuetten. H.J., Bendtsen, F., Warberg, J. (1986) Circulating atrial natriuretic peptide (ANP) and central blood volume (CBV) in cirrhosis. Liver 6: 361-368

14. Gerbes, A.L., Arendt, R.M., Ritter, D., Jüngst, D., Zähringer, J., Paumgartner, G. (1985) Plasma atrial natriuretic factor in patients with cirrhosis. N Engl J Med 313: 1609-1610

15. Gerzer, R., Idzikowski, M., Weil, J., Heim, J.M., Birkner, B., Loeschke, K. (1987) Effekte des atrialen natriuretischen Faktor bei Patienten mit Leberzirrhose und Aszites Gastroenterology 25: 36-37

16. Itoh, H., Nakao, K., Yamada, T., Morii, N., Shiono, S., Sugawara, A., Saito, Y., Mukoyama, M., Arai, H., Katsuura, G., Eigyo, M., Matsushita, A., Imura, H., (1987) Modulatory role of vasopressin in the secretion of atrial natriuretic peptide in conscious rats. Endocrinol 120: 2186-2188

17. Jüppner, H., Brabant, G., Kapteina, U., Kirschner, M., Klein, H., Hesch, R.D. (1986). Direct radioimmunoassay for human atrial natriuretic peptide (hANP) and its clinical evaluation. Biochem Biophys Res Commun 139: 1215-1223

18. Kimura, T., Abe, K., Ota, K., Omata, K., Shoji, M., Kudo, K., Matsui, K., Inoue, M., Yasujima, M., Yoshinaga, K. (1986). Effects of acute water load, hypertonic saline infusion and furosemide administration on atrial natriuretic peptide and vasopressin release in humans. J Clin Endocrinol Metab 62: 1003-1010

19. Krakoff, L.R. (1983) The sympathoadrenal system in liver disease. In: Epstein, M. (ed.) The kidney in liver disease, p. 501 - 514, 2nd edn. Elsevier, New York

20. Manning, P.T., Schwartz, D., Katsube, N.C., Holmberg, S.W., Needleman, P. (1985) Vasopressin-stimulated release of atriopeptin: endocrine antagonists in fluid homeostasis. Science 229: 395-398

21. Murthy, K.K., Thibault, G., Garcia, R., Gutkowska, J., Genest, J., Cantin, M. (1986) Degradation of atrial natriuretic factor in the rat. Biochem J 240: 461-469

22. Nakao, K., Sugawara, A., Morii, N., Sakamoto, M., Yamada, T., Itoh, H., Shiono, S., Saito, Y., Nishimura, K., Ban, T., Kangawa, K., Matsuo, H., Imura, H. (1986) The pharmacokinetics of alpha human atrial natriuretic polypeptide in healthy subjects. Eur J Pharmacol 31: 101-103

23. Nozuki, M., Mouri, T., Itoh, K., Takahashi, K., Totsune, K., Saito, T., Yoshinaga, K. (1986) Plasma concentrations of atrial natriuretic peptide in various diseases. Tohoku J Exp Med. 148: 439-447

24. Ogawa, K., Arnolda, L.F., Woodcock, E.A., Hiwatari, M., Johnston, C.I. (1987) Lack of atrial natriuretic peptide on vasopressin release. Clin Sci 72: 525-530

25. Ring-Larsen, H., Hendriksen, J.H., Wilken, C., Clausen, J., Pals, H., Christensen, N.J. (1986) Diuretic treatment in decompensated cirrhosis and congestive heart failure: effect of posture. Br Med J 292: 1351-1353

26. Rosoff, L., Zia, P., Reynolds, T., Horton, R. (1975) Studies of renin and aldosterone in cirrhotic patients with ascites. Gastroenterology 69: 698-705

27. Skowsky, W.R., Rosenbloom, A.A., Fisher, D.A. (1974) Radioimmunoassay of arginine vasopressin in serum: development and application. J Clin Endocrinol Metab 38: 278-287

28. Zipser, R.D., Kerlin, P., Hoefs, J.C., Zia, P., Barg, A. (1980) Renal kallikrein excretion in alcoholic cirrhosis: relation to other vasoactive systems. Am J Gastroenterol 75: 183-187

Atrial natriuretic peptide (Alpha- h -ANP) in 157 children with congenital heart disease

T. Eisenhauer[1], J. Talartschik[1], G. Eigster [2]

University Clinic, Depts of Nephrology[1] and Pediatric Cardiology[2], Göttingen, FRG

Summary

Atrial natriuretic peptide plasma concentrations were determined in a group of 157 children with congenital heart diseases undergoing cardiac catheterisation. Valvular heart diseases associated with typical atrial distension such as tricuspid atresia, transposition of great arteries and arterial septal defect (n = 32) were found to have elevated ANP plasma concentrations in 91.4%; fifty-six children of all evaluated patients had clinical signs and symptoms of congestive heart failure. Fifty-one of these patients (91%) showed elevated ANP levels. Seventy patients with indication for cardiac surgery were found to have elevated ANP plasma concentrations in over 81% (n = 54). Heart diseases with hemo-dynamically significant left to right shunt volume (n = 32) (Qp:Qs more than 1.5) were associated with elevated ANP levels (>30 fmol/1) in 87.5%. No correlation was found between ANP values and the degree of the pressure gradient across the valve in children with stenotic valvular diseases such as a pulmonary or aortic stenosis. Our data suggest that elevated ANP levels in children with congenital heart diseases indicate atrial distention due to either a cardiac disorder with significant volume load or congestive heart failure.

Introduction

The recently discovered hormone ANP plays a keyrole in volume and electrolyte homeostasis. Elevation of atrial pressure and atrial distension seem to be the principal stimuli of ANP secretion (1,2,3). Different clinical entities like chronic renal insufficiency, volume overload and congestive heart failure have been reported to be associated with elevated ANP plasma concentrations (4-7). High ANP levels in children with congenital heart diseases have first been described by Lang et al., (8).

To elucidate possible correlations of the ANP plasma concentration with atrial pressure, the type of heart disease, and the degree of cardiac insufficiency, 157 consecutive pediatric patients with congenital heart diseases undergoing cardiac catheterization were investigated.

Table 1. Elevated (>30 fmol/ml) and normal (<30 fmol/ml) ANP concentrations in 157 children with different congenital heart diseases, congestive heart failure and indication for surgery.

	total	ANP < 30fmol/ml			ANP > 30fmol/ml		
		n	CHF n	OP n	n	CHF n	OP n
1. VSD	23	9	1	o	14	10	11
2. Fallot-Tetra	8	1	o	1	7	o	5
3. PDA	4	1	o	1	3	2	3
4. AV-Kanal	14	4	2	1	10	10	10
5. TGA	21	2	o	2	19	9	11
6. Coarctatio	11	6	o	2	5	2	2
7. ASD	6	1	o	o	5	o	5
8. v. Pst.	16	8	o	2	8	1	1
9. v. Aost.	15	8	o	1	7	2	2
10. subv. Ao. St.	8	2	o	1	1	o	o
11. suprav. Ao. St.	3	2	o	1	1	o	o
12. SV	8	4	2	2	4	1	2
13. TA	8	o	o	o	8	5	2
14. Varia (mit Vitium)	7	o	o	o	7	6	1
15. Varia (ohne Vitium)	5	1	o	o	4	1	1
total ⟶	157	51	5	13	106	51	57
		≙100%	9,8%	25,5%	≙100%	48,1%	53,8%

Patients and methods

Patients

Pediatric patients (95 boys and 62 girls) with congenital heart diseases who underwent cardiac catheterization, age one day to twenty years, were evaluated. Cardiac catheterization was carried out according to standard techniques.

Depending on the age and disease of the patient up to eight samples for determination of ANP were drawn from peripheral vein, superior, or inferior caval vein, right atrium, left atrium, right and left ventricle, and aorta (total 499). Table 1 shows the number and distribution of the different congenital heart diseases of our patients.

Methods

By previous extraction with reversed-phase octadecylsilan columns (Sep-Pak C18), elution with 50% acetonitrile/H_2O containing 0.2% triflouroacetic acid and lyophylisation, ANP was determined by radioimmunoassay as described (9).

According to our normal range of healthy controls (N = 50) and data in the literature (7, 8) ANP values below 30 fmol/ml were considered normal.

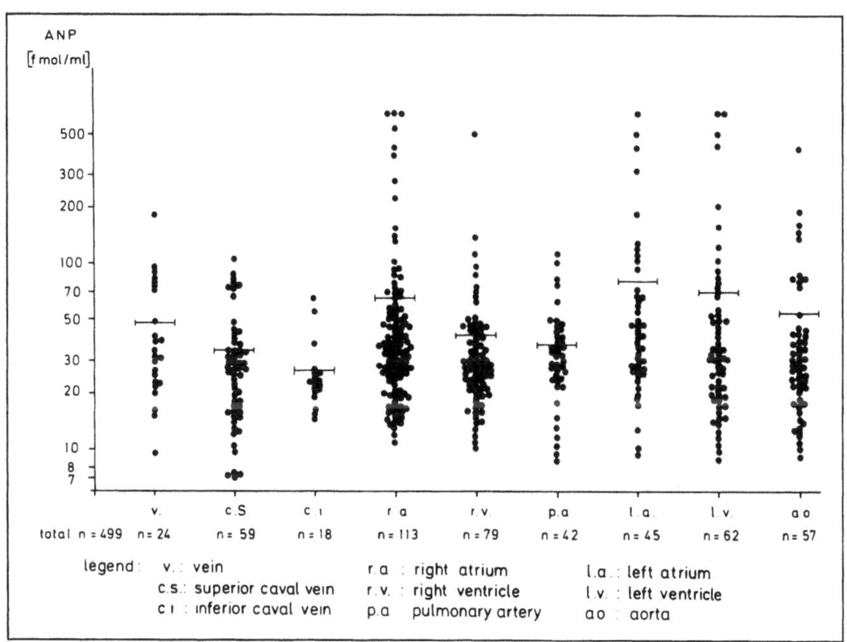

Fig.1. ANP plasma concentrations in 157 children with congenital heart diseases undergoing right/left heart catheterization.

Results

Figure 1 illustrates the distribution of ANP plasma concentrations according to the site of blood sampling in all patients. Highest ANP levels were observed in both atria and in the left ventricle. In peripheral veins and superior and inferior caval veins ANP concentrations were substantially lower as compared to the arteries.

Table 2. Heart diseases with hemodynamically significant left to right shunt volume (Qp:Qs>1.5) and ANP plasma concentrations.

total n = 47	Qp : Qs < 1,5 n = 15 100 %	Qp : Qs > 1,5 n = 32 100 %
ANP < 30 fmol/ml	n = 11 73,3%	n = 4 12,5%
ANP > 30 fmol/ml	n = 4 26,7%	n = 28 87,5%

Qp : flow rate pulmonic circulation (l/min)
Qs : flow rate systemic circulation (l/min)

*1. VSD n= 23   4. AV-Canal n=14
3. PDA n= 4   7. ASD   n= 6

No specific correlation between a single cardiac disease and ANP plasma concentrations was detected.

Cardiac defects with considerable atrial distension like transposition of great arteries, arterial septal defect, and tricuspid atresia were associated in 91.4% (N = 32) with elevated ANP plasma concentrations.

In children with aortic or pulmonary stenosis no correlation between the pressure gradient across the valve and ANP values was found. Heart diseases with hemodynamically significant left to right shunt volume (Qp:QS>1.5) as ventrical septal defect, persistent ductus arteriosus, AV channel and arterial septal defect showed elevated ANP values in 87.5% (N:32), whereas in diseases with insignificant shunt volume ANP values were elevated in only 26.7% (Table 2).

A weak but significant linear correlation was found between mean right atrial pressures and ANP plasma concentrations in the right ventricle and pulmonary artery (r = 0.44; p<0.001).

Discussion

Although we found no specific distribution pattern of ANP characteristic of a specific cardiac disease in our patients, determination of this new hormone may be of diagnostic and prognostic value in children with congenital heart failure.

As expected, heart diseases with typical atrial distention were found to have substantially elevated ANP plasma concentrations. High ANP values were also found in congenital heart diseases with significant left to right shunt volume (Qp:QS>1.5), reflecting atrial stretch and volume overload of the right heart.

As described by others, even in this inhomogeneous group of patients, we found a quantitative correlation between right atrial pressure and ANP concentrations in the right heart.

Since the observation that ANP plasma concentrations in the left atrium were higher than in the pulmonary artery (Fig. 1), and since it cannot be explained by the hemodynamics of the cardiac disease itself, an extracardial source of ANP production is suggested as described by (10), who demonstrated in an animal model ANP secretion by the lung. The finding that ANP values of patients on extracorporeal circulation during cardiac operations remain still in a subnormal level, although the cardioplegic heart is totally removed from circulation, lends further support to the hypothesis of extracardial ANP production (11).

Our patients with clinical signs and symptoms of congestive heart failure (91%) showed elevated ANP plasma concentrations (Table 1).

Cardiac catheterization and selective angiocardiography remain the cornerstones of hemodynamic evaluation in children with congenital heart disease. Periodic control of ANP plasma concentrations together with other noninvasive techniques, may provide the clinician with a simple additional diagnostic tool to monitor right atrial distension. Increase of ANP plasma concentrations could reflect either volume overload or progressive congestive heart failure.

The finding that of our seventy patients with indication for cardiac surgery fifty-four (81%) had elevated ANP plasma concentrations underlines the potential prognostic value of this new hormone to estimate the severity of congenital heart diseases in children.

References

1. Lang, R.E., Thoelken, H., Ganten, D., Luft, F.C., Ruskoaho, H., Unger, T. (1985) Atrial natriuretic factor - a circulating hormone stimulated by volume loading. Nature 314: 264-266

2. Needlemann, P., Adams, S.P., Sole, B.R., Currie, M.G., Geller, D.M., Michener, M.L., Saper, C.B., Schwartz, D., Standaert, D.G. (1985) Atriopeptins as cardiac hormones. Hypertension 7: 469-482

3. Tikkanen, I., Fyhrquist, F., Metsärinne, K., Leidenius, R. (1985) Plasma atrial natriuretic peptide in cardiac disease and during infusion in healthy volunteers. Lancet 2: 66-69

4. Raine, A.E.G., Erne, P. et al. (1986) Atrial natriuretic peptide and atrial pressure in patients with congestive heart failure. New Engl J Med 315: 533-537

5. Nakaoka, H., Imataka, K. et al. (1985) Plasma levels of atrial natriuretic factor in patients with congestive heart failure. New Engl J Med 313: 892-993

6. Riegger, A.J.G., Kromer, E.P., Kochsiek. (1985) Der natriuretische Vorhoffaktor bei schwerer Herzinsuffizienz. DMW 110: 1607-1610

7. Rascher, W., Tulassay, T., Lang, R.E. (1985) Atrial natriuretic peptide in plasma of volume-overloaded children with chronic renal failure. Lancet 10: 303-305

8. Lang, R.E., Unger, T. et al. (1985) Alpha atrial natriuretic peptide concentrations in plasma of children with congenital heart diseases. Br Medm J 11: 1241

9. Eisenhauer, T., Talartschik, J., Scheler, F. (1986) Detection of fluid overload by plasma concentration of human atrial natriuretic peptide (h-ANP) in patients with renal failure. Klin Wochenschr (Suppl. VI): 68-72

10. Gutkowska, J., Cantin, M., Genest, J., Sirois, P. (16th 21st May, 1987) Release of atrial natriuretic factor - like immunoreactivity from the isolated perfused rat lung. 2nd World Congress on Biologically Active Atrial Peptides, New York, N.Y.

11. Talartschik, J., Eisenhauer, T., de Vivie, R., Talartschik, B., Scheler, F. (1987) Release of atrial natriuretic peptide (α-H-ANP 1-28) during cardiac arrest and after resuscitation in 19 patients undergoing cardiac surgery with cardiopulmonary bypass. Proceedings of 2nd World Congress on Biologically Active Atrial Peptides. New York (in press)

Receptor binding of cardiac hormones to dog kidney with congestive heart failure

M. Gagelmann, K. Forssmann, A. Whalley, G. A. J. Riegger*, W.G. Forssmann

Department of Anatomy and Cell Biology, University of Heidelberg, FRG, and *Medical University Clinik, Würzburg, FRG

Introduction

During chronic and acute cardiac failure, plasma levels of cardiodilatin/atrial natriuretic peptide (CDD/ANP) are elevated, depending on the severity of the disease. Compared to normal subjects, however, CDD/ANP infusion in high doses in patients with chronic heart failure showed no significantly increased urine or sodium excretion rates (1,6). It was reported that in platelets from patients with congestive heart failure, binding sites for CDD/ANP are lower than in controls (9). Down-regulation of CDD/ANP receptors in cultured vascular smooth muscle cells during incubation with ANP has been reported (3), and Takayanagi and collaborators (11) observed that the number of ANP receptors was reduced in aortic smooth muscle of spontaneous hypertensive rats during hypertension. The loss of the biological activity of CDD/ANP with respect to diuresis during congestive heart failure may be due to a receptor down-regulation in the kidney. Therefore, we investigated binding of CDD/ANP in an animal model of congestive heart failure.

Material and methods

Heart failure with a chronic low output was induced by right ventricular pacing (240 beats/min; 10 to 18 days) in female mongrel dogs. Implantation of the pacemaker was done under general anesthesia. Control and cardiomyopathic dogs were nephrectomized under barbiturate anesthesia. The renal vein, renal artery, and ureter were cannulated for *in vitro* perfusion. After perfusion of the kidney with a solution containing procaine-HCl (2), the experiment began with a wash of Tyrode solution until a continuous venous and urine outflow was achieved. The Tyrode solution (47mM KCl, 128mM NaCl, 14.4 mM NaHCO$_3$, 1.2mM MgCl$_2$, 0.1mM Na$_2$Ca-EDTA, 10mM glucose, 2.5mM CaCl$_2$, pH 7.2) was kept at 37°C and rinsed with carbogen. The perfusion pressure was 80 mmHg. In the second step [125]I- hANP (NEN) was added to the perfusion solution (5500cpm/ml) for 5 to 6 min. Finally, the kidney was washed to remove nonspecifically bound radioactivity. In some experiments binding of the iodine labelled tracer was analyzed after 5 min of infusion of unlabelled CDD-28/ANP. The venous and urine outflows were collected for radioactivity measurements. For the study of [125]I- hANP distribution kidneys were cut longitudin- ally and sliced (0.4mm). Tissue samples from the outer cortex, the cortex, the outer and inner medulla, including the papilla were collected, blotted, and counted.

Wash
αhANP
Pretreatment
125J-αhANP
Perfusion
Wash
Urine Outflow (ml/min)
Venous Outflow (ml/min)
Venous (▲—▲) and Urine (△···△) Radioactivity (cpm/ml × 10⁻³)
Time (min)
Relative Distribution of 125J-αhANP in the Kidney (%)
Control Dogs
αhANP (0.75 µg/ml)
Pretreatment
Cardiomyopathic Dogs
αhANP (0.25 µg/ml)
Subcapsular Cortex
Cortex
Outer Medulla
Inner Medulla and Papilla

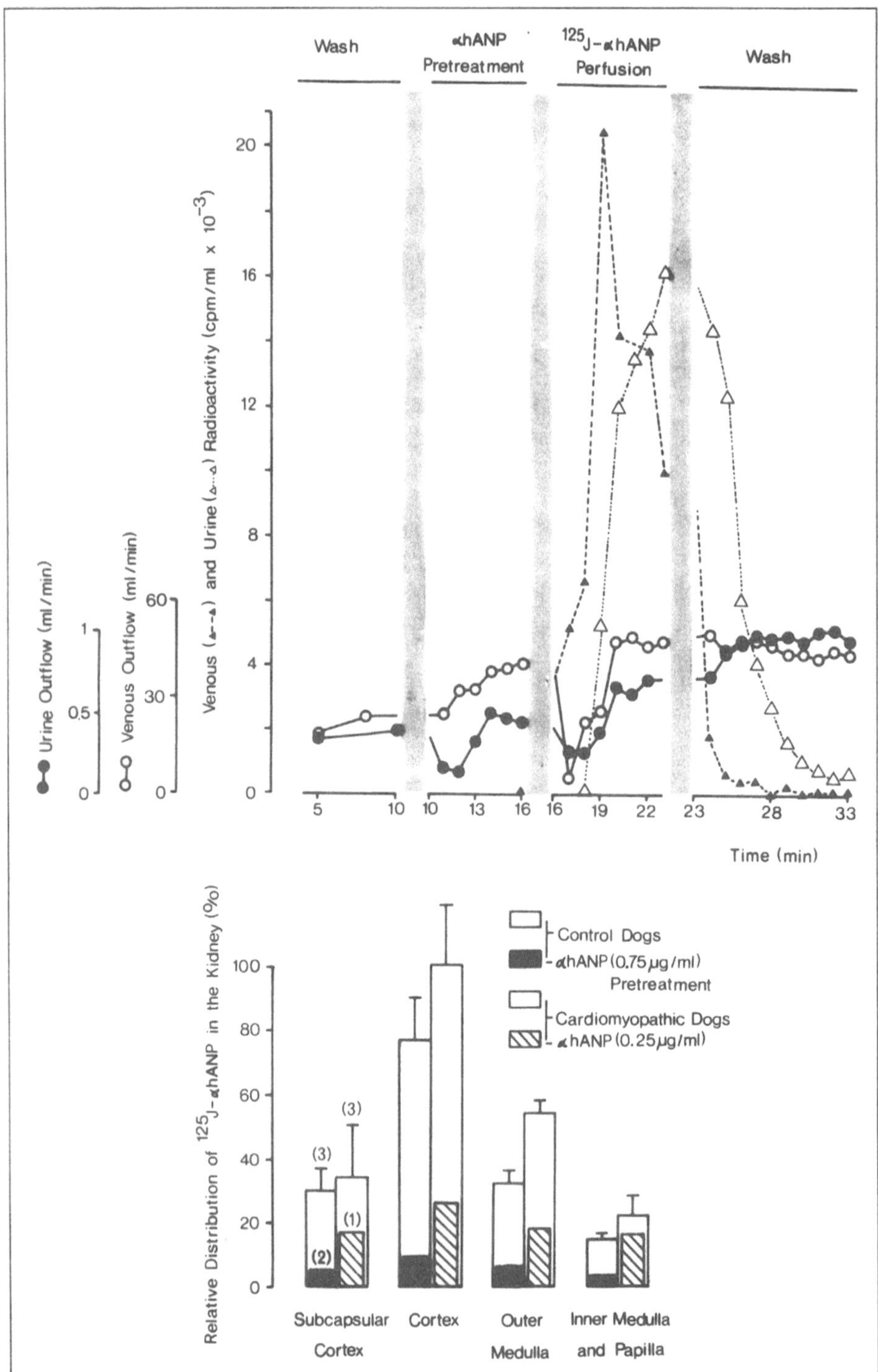

226

Fig. 1. Time course of an *in vitro* perfusion of a kidney of a control dog (upper) and relative distribution of ^{125}I- hANP binding in the kidneys of cardiomyopathic and control animals (lower); *Upper panel:* the urine and venous outflow was determined during a wash with Tyrode solution followed by pretreatment with 0.75µg/ml hANP (208ml) and perfusion with 265ml ^{125}I- hANP (5500cpm/ml). Unbound radioactivity was removed by the wash. (Dashed bars indicate change of solution). For further analysis kidneys were frozen in liquid nitrogen and stored at -70°C; *Lower panel:* frozen kidneys were sliced longitudinally (0.4mm) and the tissue samples were excised. Values (cpm/blot weight) are normalized to the maximal mean value in the cortex of cardiomyopathic dogs (100%). Values are mean values ± SEM, the number of animals is indicated in parentheses. In some experiments the kidneys were pretreated with unlabelled CDD/ANP followed by perfusion with ^{125}I- hANP

Results and discussion

A typical experiment is presented in Fig. 1 (upper panel) showing the time course of an *in vitro* perfusion of a kidney from a control dog. A constant venous and urine outflow was achieved during this wash. Pretreatment of the kidney with 0.75µg/ml CDD-28 in the next step was associated with increased venous and urine flow. This is probably due to a constriction of the efferent and dilatation of the afferent glomerular artery and an increase in glomerular capillary permeability. After pretreatment with unlabelled CDD/ANP, and subsequent treatment with a ^{125}I- hANP-containing solution, the procedure was followed by a wash to remove nonspecifically bound tracer. Compared to the venous outflow the radioactivity in the urine was retarded (2 min).

The relative distribution of ^{125}I- hANP binding to kidneys from control as well as from cardiomyopathic dogs is demonstrated in Fig. 1 (lower panel). In both, the regional distribution was similar, showing a maximal value in the cortex. The binding to kidneys of cardiomyopathic dogs is not significantly changed compared to controls. Additionally, studies were conducted to determine if binding of ^{125}I- hANP decreased upon pretreatment with unlabelled CDD/ANP. This is shown in Fig.1 (lower panel). After preloading with CDD/ANP, binding of ^{125}I- hANP was poor in the kidneys of cardiomyopathic and control dogs. The markedly reduced binding of ^{125}I- hANP in the presence of unlabelled CDD/ANP indicates the specificity.

In conclusion it may be suggested, that a receptor down-regulation in the kidneys of dogs with experimental congestive heart failure seems not to be the mechanism for the altered excretory parameters of the kidney. It seems likely that the changed glomerular filtration and filtration fraction (7) may be explained by mechanisms other than receptor binding of CDD/ANP to the kidney. Clinical analysis in the living experimental animals in comparison to isolated perfused kidney suggests defects in kidney function which may possibly be due to alterations in the intracellular mechanism. However, the small number of experiments in this study and the high variability impairs the interpretation. The problem is even more accentuated in light of reports from Scarborough et al. (8) and Takayanagi et al. (11) on two different ANP receptors in cultured vascular smooth muscle cells, and in the adrenal cortex - and the coupling of only one to the cGMP metabolism. However, existence of a single class of high-affinity sites and high density of ANP receptors have been reported for the rat kidney cortex while low-affinity sites may be related to degradation processes (5). Recently, Ishikawa et al. (4), demonstrated the presence of only a single class of binding sites for alpha-ANP in the human cortical membranes. Thus, further studies on the involvement of ANP receptors in the kidney during congestive heart failure are required.

References

1. Crozier, I.G., Nicholls, M.G., Ikram, H., Espiner, E.A., Comez, H.J., Warner, N.J. (1986) Hemodynamic effects of atrial peptide infusion in heart failure. Lancet 2: 1242
2. Forssmann, W.G., Pickel, V., Reinecke, M., Hock, D., Metz, J. (1981) Immuno-histochemistry and immunocytochemistry of nervous tissue. In: Heym, C., Forssmann, W.G. (eds) Techniques in neuroanatomical research. Springer, Berlin Heidelberg New York, p 171
3. Hirata, Y., Tomita, M., Takada, S., Yoshimi, H. (1985) Vascular receptor binding activities and cyclic GMP responses by synthetic human and rat atrial natriuretic peptides (ANP) and receptor down-regulation by ANP. Biochem Biophys Res Commun 128: 538
4. Ishikawa, Y., Umemura, S., Yasuda, G., Uchino, K., Shindou, T., Minamizawa, K., Toya, Y., Kaneko, Y. (1987) Identification of an atrial natriuretic peptide specific receptor in human kidney. Biochem Biophys Res Commun 147: 135
5. Napier, M.A., Vandlen, R.L., Albers-Schöberg, G., Nutt, R.F., Brady, S., Lyle, T., Winquist, R., Faison, E.P., Heinel, L.A., Blaine, E.H. (1984) Specific membrane receptors for atrial natriuretic factor in renal and vascular tissues. Proc Natl Acad Sci USA 81: 5946
6. Riegger, A.J. G., Kromer, E.P., Kochsiek, K. (1986) Human atrial natriuretic peptide: plasma levels, hemodynamic, hormonal and renal effects in patients with severe congestive heart failure. J Cardiovasc Pharmacol 8: 1107
7. Riegger, A.J.G., Elsner, D., Muders, F., Pascher, E., Kromer, E.P. (1988) Endocrine heart in experimental cardiomyopathy. In: Functional morphology of the endocrine heart. pp 153-159. Forssmann, W.G., Schnermann, D.D., Alt, J. (eds) Steinkopff Verlag Darmstadt
8. Scarborough, R.M., Schenk, D.B., McEnroe, G.A., Arfsten, A., Kang, L.-L., Schwartz, K., Lewicki, J.A. (1986) Truncated atrial natriuretic peptide analogs. J Biol Chem 261: 12960
9. Schiffrin, E.L. (1986) Down-regulation of binding sites for atrial natriuretic peptide in platelets of patients with congestive heart failure. Circulation 74 (suppl II): II-463 (abstract)
10. Takayanagi, R., Imada, T., Grammer, R.T., Misono, K.S., Naruse, M., Inagami, T. (1986) Atrial natriuretic factor in spontaneously hypertensive rats: concentration changes with the progression of hypertension and elevated formation of cyclic GMP. J Hypertension 4: S303
11. Takayanagi, R., Snajdar, R.M., Imada, T., Tamura, M., Pandey, K.N., Misono, K.S., Inagami, T. (1987) Purification and characterization of two types of atrial natriuretic factor receptors from bovine adrenal cortex: guanylate cyclase-linked and cyclase-free receptors. Biochem Biophys Res Commun 144: 244

Characterization of the ANF system in patients with cirrhosis of the liver

A.L. Gerbes, *R.M. Arendt, *E. Stangl, V. Gülberg, T. Sauerbruch, D. Jüngst, G. Paumgartner

Depts. of Medicine II and *I, Klinikum Großhadern, University of Munich, Munich, FRG

Summary

Defects of the ANF system in cirrhosis could not be demonstrated in terms of an absolute deficiency of plasma levels or evidently major abnormalities of processing in patients with cirrhosis. In the present study the responsiveness of the ANF system to acute volume stimulation by water immersion, the diuretic and natriuretic effects of ANF infusion were examined. Stimulation of ANF release by 1 h immersion was significantly blunted in 10 cirrhotic patients with ascites (increase of plasma ANF by 46 ± 18%), whereas 11 cirrhotics without ascites showed a 104 ± 16% increase, similar to the 117 ± 29% stimulation in 25 healthy controls. Immersion increased urinary volume by 3.6 ± 0.6, 2.0 ± 0.8, and 0.7 ± 0.4 ml/min, and urinary sodium excretion by 146 ± 38, 75 ± 43, and 43 ± 19 µmol/min in controls, cirrhotics without ascites, and cirrhotics with ascites, respectively. Infusion of ANF for 30 min prompted an increase in diuresis and natriuresis in seven cirrhotic patients, which was less marked in patients with ascites as compared to patients without. Thus, the stimulus-response coupling for ANF may be impaired in patients with cirrhosis and ascites.

Introduction

The pathophysiology of renal sodium retention and ascites formation in patients with cirrhosis of the liver has become rather perplexing. There is increasing evidence that activation of the renin-aldosterone system is just one of several factors involved in the impaired volume regulation of cirrhosis. For many years, a deficiency of a putative natriuretic hormone in cirrhosis has been postulated, but never satisfactorily demonstrated. Therefore, investigations on the role of the novel atrial natriuretic factor (ANF) (7) in cirrhosis have been anticipated.

The first communication on ANF plasma levels in cirrhosis showed that there is no absolute deficiency of this novel natriuretic and diuretic hormone in patients with cirrhosis (9). These findings have been confirmed by several groups, reporting ANF plasma levels in patients with cirrhosis and ascites equal to or higher than normal (11).

However, the immunoreactive ANF in plasma of patients with cirrhosis might include different molecular species with altered biological activity, as has been suggested in patients with congestive heart failure (3). By high performance gel permeation chromatography only negligible amounts of immunoreactivity coeluting with precursor forms of

ANF 99-126 have been detected (2). Thus, no evidence for major abnormalities of processing of ANF was found in cirrhosis.

Neither basal plasma levels nor characterization of immunoreactivity necessarily reflect the functional status or the compensatory reactivity of the ANF system. Therefore, in the present study responsiveness of the ANF system to volume stimulation was investigated. Head out of water immersion (WI) in a thermoneutral bath has been shown to induce central hypervolemia with atrial distention and to prompt natriuresis and diuresis (4). Epstein and others have demonstrated in numerous investigations that WI is a useful tool for the study of volume regulation (5): it increases central volume by shifting blood from peripheral vessels thus obviating the necessity of infusing volume expanders that might alter plasma composition. We demonstrated that WI rapidly increases ANF plasma levels in healthy human subjects (10). In the present study WI was used to investigate the response of ANF to acute volume stimulation in patients with cirrhosis.

Infusion of ANF has been shown to induce diuresis and natriuresis in healthy subjects (15). The initial observation of a diuretic effect of ANF in a patient with cirrhosis and refractory ascites (8) prompted us to investigate the renal response to ANF infusion in cirrhotic patients.

Patients and methods

Water immersion: 25 healthy controls and 21 patients with cirrhosis, 10 with ascites and 11 without ascites were investigated after informed consent had been obtained. There was no evidence for cardiovascular, renal, or pulmonary disease in control or patients. No diuretics had been given to any subject for one week preceeding the study. Subjects were on a hospital diet containing approximately 150 meq sodium/day and were prohibited alcohol, tobacco, tea, and coffee the day before and during the experiment. In the morning after complete emptying of the bladder, a catheter was placed in a forearm vein, subjects were given 400 ml of water orally and were seated next to a tank with thermoneutral ($34.5 \pm 0.2°C$) water. After 1 h the subjects were immersed up to their necks in the water bath, maintaining the same position for 1 h. This was followed by another hour of sitting outside the tank. Throughout the investigation 200 ml/hour of water were given orally to ensure adequate urine flow. Urine obtained by spontaneous emptying of the bladder was collected before (0), after 60 min (60 min) and 60 min subsequent to (120 min) the end of immersion. Blood samples were obtained at 0.60 and 120 min.

ANF infusion: ANF (Bissendorf, West Germany) was infused (50ng/kg/min) into a forearm vein of seven patients with cirrhosis, three without and four with ascites in the supine position, (the patients bladder were emptied 1h before starting the infusion) urine was collected immediately before, 1 and 2 h after the start of the infusion. Plasma samples were collected before, 30 min and 2 h after start of the infusion. ANF was determined in extracted plasma samples as described before (2). Plasma renin activity was measured by RIA (16). Data are given as mean and standard error. Data were evaluated statistically by paired or unpaired Students *t*-test.

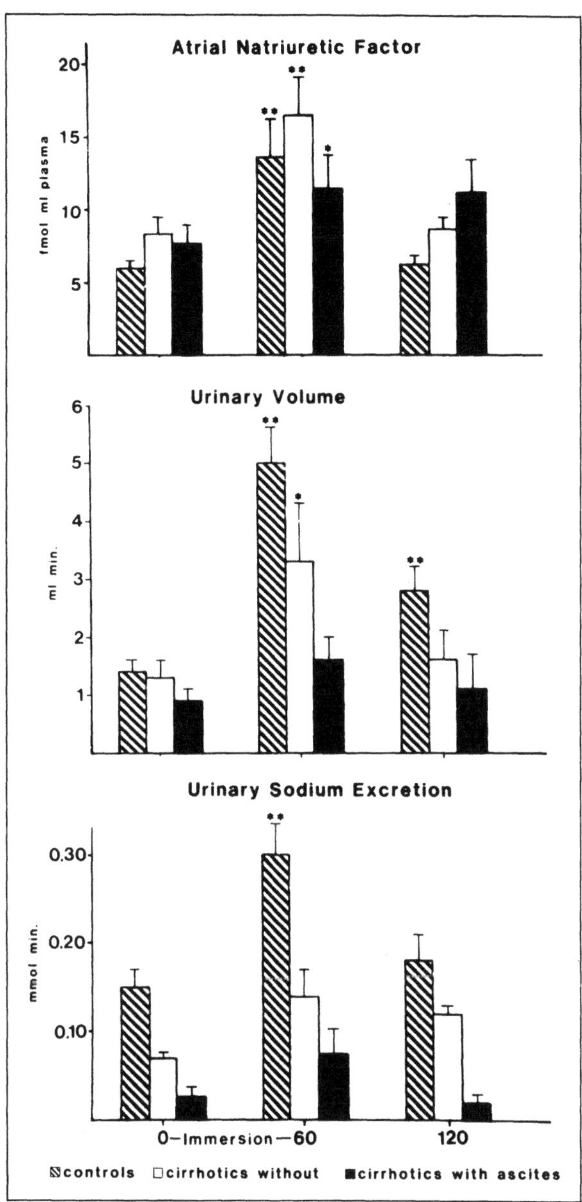

Fig. 1. Plasma levels of Atrial Natriuretic Factor, urinary volume and urinary sodium excretion before (0 min), after 1 hour (60 min) and 1 hour subsequent to (120 min) the end of water immersion in 25 healthy controls, 11 cirrhotic patients without and 10 cirrhotic patients with ascites. * p < 0.05, ** p < 0.01 as compared to baseline levels

Results

Volume stimulation by water immersion caused a rise of plasma ANF from 6.0 ± 0.6 to 13.6 ± 2.6 fmol/ml in healthy subjects and from 8.5 ± 1.3 to 16.5 ± 2.6 fmol/ml in cirrhotic patients without ascites; increases were significant at the p < 0.01 level. In cirrhotics with ascites stimulation of ANF from 7.7 ± 1.3 to 11.4 ± 2.3 fmol/ml was significantly (p < 0.05) blunted as compared to cirrhotics without ascites (Fig. 1).

The renal response to immersion is illustrated in Fig. 1. Mean increases in urinary volume were found to be 3.64 ± 0.60 ml/min in controls, 2.02 ± 0.81 ml/min in cirrhotics without ascites, and only 0.68 ± 0.35 ml/min (not significantly different from baseline values) in cirrhotics with ascites. Similar differences were observed in urinary sodium excretion: immersion induced a rise by 146 ± 38 μmol/min in controls, by 75 ± 43 μmol/min in cirrhotics without ascites, and by only 43 ± 19 μmol/min in cirrhotics with ascites. Increases in both cirrhotic groups did not reach the level of significance.

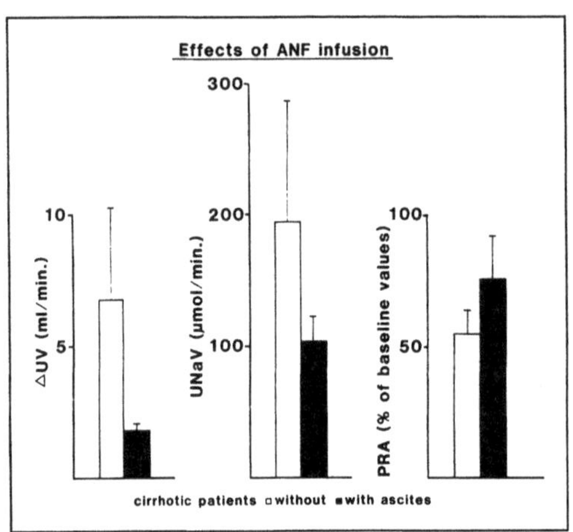

Fig. 2. Increase of diuresis and natriuresis 1 hour after the beginning of a 30 minute infusion of ANF and suppression of plasma renin activity after 30 minutes of infusion in 3 cirrhotic patients without and 4 cirrhotic patients with ascites

Infusion of ANF markedly stimulated diuresis and natriuresis in the hour subsequent to the beginning of the infusion. After another hour renal parameters returned to baseline levels. In patients without ascites the effects were stronger than in patients with ascites (Fig. 2): urinary volume increased by 6.8 ± 3.6 as compared to 1.8 ± 0.3 ml/min, sodium excretion increased by 195 ± 93 as compared to 105 ± 19 μmol/min. Plasma renin activity in patients without ascites decreased from 13 ± 11 to 7 ± 5 ng AI/ml/h; and a similar suppression from 27 ± 21 to 16 ± 10 ng AI/ml/h was seen in patients with ascites.

Discussion

The observation that water immersion significantly increases ANF plasma levels in a large number of healthy subjects is in accordance with previous results from our laboratory (10) and has been confirmed by other investigations (1,6,13,14). Similar increases were observed in patients with cirrhosis of the liver without ascites. In cirrhotics with ascites, however, stimulation of ANF release into the plasma was found to be significantly reduced after 1 h of immersion. This might be due to a decreased volume stimulus in these patients. However, plasma renin activity as an indicator of centrally effective volume decreased in cirrhotics with ascites by a ratio not different from that in cirrhotics without ascites or controls (unpublished observations). Furthermore, central hemodynamics and intracardiac pressures are influenced by water immersion independent of the degree of ascites (12). Thus, a blunted increase of atrial pressure as a stimulus for ANF release does not seem a likely explanation for the observed blunted ANF stimulation in these patients. Therefore, a defective ANF synthesis or release might be suspected in these patients.

The blunted ANF stimulation was paralleled by reduced increases of urinary sodium excretion and urinary volume following immersion in cirrhotics with ascites. While this might suggest a (patho-)physiological role of ANF in volume regulation, no correlations of either basal or stimulated ANF with renal response could be found. In view of the involvement of several other hormonal systems in volume regulation, however, this finding is not surprising.

Furthermore, at comparable baseline plasma levels of ANF, excretion of sodium was diminished in cirrhotic patients as compared to controls, and in patients with ascites as compared to patients without ascites. The renal response to stimulation by water immersion was less marked than the increase of ANF in the cirrhotic groups. These findings might be consistent with the increased activitiy of sodium retaining principles, counteracting the renal action of ANF, or blunted responsiveness of the kidney to ANF in cirrhosis.

Infusion of ANF at a rather low dose induced a marked diuresis and natriuresis in patients with cirrhosis, with a less marked response of the patients with ascites. This might have been due to the lowering of systematic blood pressure by ANF, causing a reflectory increase of renin activity, counteracting the renal effects of ANF. However, with the dose applied no marked decreases of blood pressure were seen and plasma renin activity was suppressed in both cirrhotic groups. Thus, a reduced activity of the renal receptor-effector coupling must be taken into consideration in patients with decompensated cirrhosis.

This study demonstrates a blunted response of ANF release to water immersion in patients with cirrhosis of the liver and might indicate a reduced renal response to ANF in patients with ascites, thus, suggesting a role for this novel hormone in the impairment of acute volume regulation in cirrhosis.

Acknowledgement: This work was supported by a grant from the Friedrich-Baur-Stiftung of the Faculty of Medicine, University of Munich (head: Prof. Dr. Buchborn).

References

1. Anderson, J., Struthers, A., Christofides, N., Bloom, S. (1986) Atrial natriuretic peptide: an endogenous factor enhancing sodium excretion in man. Clin Sci 70: 327-331
2. Arendt, R.M., Gerbes, A.L., Ritter, D., Stangl, E. (1986) Molecular weight heterogeneity ofplasma-ANF in cardiovascular disease. Klin Wochenschr 64 (Suppl VI): 97-102
3. Arendt, R.M., Gerbes, A.L., Ritter, D., Stangl, E. (1987) Differential processing of plasma-ANF in cardiovascular disease. In: B.M. Brenner, J.H. Laragh (eds): Biologically Active Atrial Peptides. Raven Press New York, pp 544-547
4. Behn, C., Gauer, O.H., Kirsch, K., Ekert, P. (1969) Effects of sustained intrathoracic vascular distention on body fluid distribution and renal excretion in man. Pflügers Arch ges Physiol 31: 1233-135
5. Epstein, M. (1978) Renal effects of head-out water immersion in man: implications for an understanding of volume homeostasis. Physiol Rev 58: 529-581
6. Epstein, M., Loutzenhiser, R., Friedland, E., Aceto, R.M., Camargo, M.I.F., Atlas, S.A. (1987) Relationship of increased plasma atrial natriuretic factor and renal sodium handling during immersion-induced central hypervolemia in normal humans. J Clin Invest 79: 738-745
7. Forssmann, W.G. (1986) Cardiac hormones. Review on the morphology, biochemistry and molecular biology of the endocrine heart. Eur J Clin Invest 16: 439-451
8. Fyhrquist, F., Tötterman, K.J., Tikkanen, I. (1985) Infusion of atrial natriuretic peptide in liver cirrhosis with ascites. Lancet II : 1439
9. Gerbes, A.L., Arendt, R.M., Ritter, D., Jüngst, D., Zähringer, J., Paumgartner, G. (1985) Plasma atrial natriuretic factor in patients with cirrhosis. N Engl J Med 313: 1609-1610
10. Gerbes, A.L., Arendt, R.M., Schnizer, W., Silz, S., Jüngst, D., Zähringer, J., Paumgartner, G. (1986) Regulation of atrial natriuretic factor release in man: effect of water immersion. Klin Wochenschr 64: 666-667
11. Gerbes, A.L., Arendt, R.M., Paumgartner, G. (1987) Editorial review. Atrial natriuretic factor-possible implications in liver disease. J Hepatology 5: 123-132
12. Nicholls, U.M., Shapiro, M.D., Groves, B.S., Schrier, R.W. (1986) Factors determining renal response to water immersion in non-excretor cirrhotic patients. Kidney Int 30: 417-421
13. Ogihara, T., Shima, J., Hara, H. et al. (1986) Significant increase in plasma immunoreactive atrial natriuretic peptide concentration during head-out water immersion. Life Sci 38: 2413-2418.
14. Pendergast, D.R., DeBold, A.J., Pazik, M., Hong, S.K. (1987) Effect of head-out immersion on plasma atrial natriuretic factor in man (42497). Proc Soc Exp Biol Med 184: 429-35
15. Weidmann, P., Hasler, L., Gnädinger, M.P., Lang, R.E., Uehlinger, D.E., Shaw, S., Rascher, W., Reubi, F.C. (1986) Blood level and renal effects of atrial natriuretic peptide in normal man. J Clin Invest 77: 734-742
16. Wernze, H., Spech, H.I., Müller, G. (1978) Studies on the activity of the renin-angiotensin-aldosterone system (RAAS) in patients with cirrhosis of the liver. Klin Wochenschr 36: 389-397

The effect of human atrial natriuretic peptide on renal and cardiovascular responses to insulin-induced hypoglycemia in healthy man

E. Jungmann, C. Konzok, E. Höll, W. Fassbinder*, K. Schöffling

Center of Internal Medicine, Johann Wolfgang Goethe-University, Frankfurt, FRG, and
*III. Department of Internal Medicine, City Hospital, Fulda, FRG

Summary

In healthy men insulin-induced hypoglycemia offers the opportunity to examine under nearly physiologic conditions the potential effects of human atrial natriuretic peptide (human ANF- (99-126), hANP) on catecholamine- or vasopressin-related events. Thus, evidence is presented that 100 µg hANP given as an i.v. bolus injection does not inhibit the counter-regulatory increase in epinephrine and norepinephrine concentrations in plasma in response to decreased blood glucose levels. However, hANP is found to inhibit, by direct or indirect mechanisms, the epinephrine-mediated increase in heart rate and systolic blood pressure and to antagonize the action predominantly of norepinephrine on diastolic blood pressure. Whereas the sodium-retaining effect of insulin-induced hypoglycemia is completely reversed by hANP, its anti-diuretic effect is only partially suppressed. It appears likely that under physiologic conditions the antagonizing action of hANP on the renal response to vasopressin is more prominent than its potential capacity to inhibit vasopressin secretion. Moreover, hANP is found to increase the effectiveness of exogenous insulin. This is probably predominantly due to its inhibiting effect on the hepatic insulin break-down.

In endocrinological research, insulin-induced hypoglycemia is a well-defined and thoroughly studied experimental model for investigating the effects of metabolic stress on hormonal systems. The concomitant stimulation by insulin-induced hypoglycemia of the release of vasopressin, epinephrine and norepinephrine is part of its most prominent hormonal characteristics (1,2). Therefore, insulin-induced hypoglycemia offers the unique opportunity to examine in healthy man under nearly physiological conditions whether human atrial natriuretic peptide (human ANF-(99-126), hANP) is capable of influencing the release or the effects of raised concentrations of these hormones on renal and cardiovascular indexes.

Methods

After having given fully informed consent eight male healthy volunteers, all normal-weighted medical students aged 23 - 40 years, were studied. All subjects continued their normal diets, without any restriction on sodium or potassium intake. All subjects were requested to abstain from caffeine, smoking and strenuous physical activity for at least 24 h prior to the experiments. None of the subjects had taken any medication. All subjects underwent four random experiments, which were separated by intervals of at least one week. The subjects fasted for 10 h and remained seated from 0800 h to 1200 h apart from brief periods of standing to urinate. Urine was collected from 0800 h to 1000 h (base-line samples) and from 1000 h to 1200 h. Both at 0800 h and at 1000 h, subjects drank 300 ml tap water to ensure adequate urine flow. At 0800 h an indwelling teflon cannula (Braunula[R], Braun Melsungen) was inserted into the antecubital vein. At 1000 h insulin, hANP, or a similar looking placebo injection was given intravenously over 1 minute in a double-blinded manner according to the following injection scheme:

I: 100 mg hANP (Bissendorf Peptide) + 0.125 μ/kg body weight human regular insulin (Actrapid[R], Novo)

II: hANP + 3.125 ml/kg body weight physiologic saline instead of insulin

III: insulin + 2 ml physiologic saline as placebo instead of hANP

IV: control experiment (placebo injections only)

All injections were well tolerated, side-effects or adverse reactions were not registered. Venous samples were obtained before and 30, 45, 60 and 120 min after the injections for radioenzymatic determination of epinephrine and norepinephrine (3) and for the radio-immunological measurement of insulin (Pharmacia) and glucagon (Serono). Blood glucose was measured enzymatically (Boehringer Mannheim). At 0800 h and 1200 h, in addition, venous creatinine levels were determined from samples (Beckman Instruments). Urine volume was measured immediately, and samples were taken to determine sodium, potassium, chloride, and creatinine concentrations (Beckman Instruments). Blood pressure was recorded with a mercury manometer. All data are expressed as means $\pm$ standard error of the mean. The effects of the experimental procedures on urine volume, and sodium, and chloride excretion are presented by the percent change of these parameters from respective base-line values.

Results and discussion

Thirty minutes after the injection of insulin + hANP, higher insulin levels were measured than after injection of insulin alone ($p < 0.05$). This increase in insulin levels is apparently due to an inhibition of hormone break-down, probably within the liver (4). At the same time, the minimum level of blood glucose were reduced from 36 ± 4 mg/dl to 24 ± 5 mg/dl ($p < 0.01$).

As expected from aggravated hypoglycemia, the glucagon response to insulin + hANP was greater than that to insulin alone. The counter-regulatory response of epinephrine levels to hypoglycemia in the insulin + hANP experiment was likewise greater than that in the experiment studying the responsiveness to insulin alone (792 ± 188 vs. 576 ± 118 ng/1, $p < 0.05$). This difference is also easily explained by more severe hypoglycemia following injection of insulin + hANP. Norepinephrine responsiveness remained unchanged by hANP. Thus, we cannot present any evidence that hANP is capable of influencing catecholamine stimulation by metabolic stress in human subjects. An inhibitory effect of hANP on catecholamine release in man was concluded from a preliminary study in cultured human pheochromocytoma cells (6). This conclusion was consistent with observations in experimental animals, both under *in vivo* and *in vitro* conditions (6,7). In contrast, reports on the potential effects of hANP on catecholamine responsive-

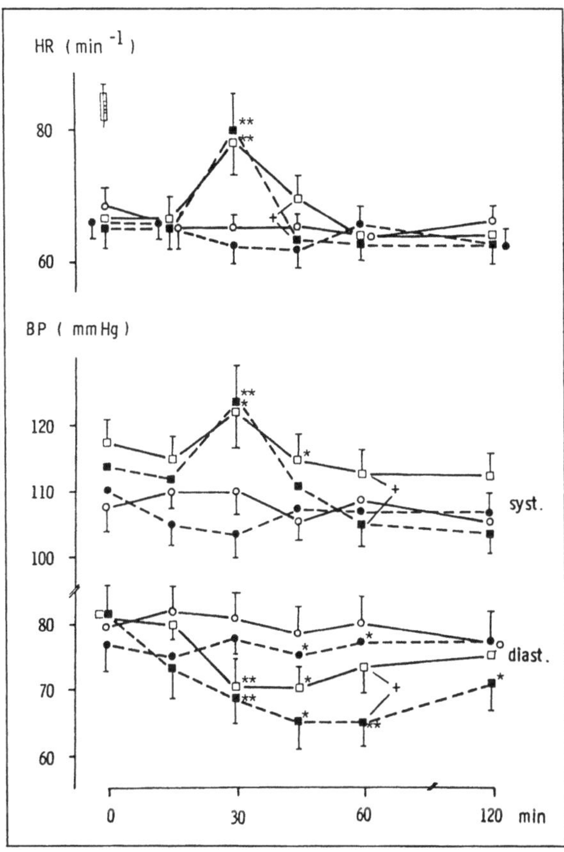

Fig. 1. The effect of 100mg hANP i.v. on heart rate (HR) and blood pressure (BP) responses to insulin-induced hypoglycemia (0.125 U/kg b.wt. i.v.) in eight healthy volunteers (open circles, placebo, filled circles, hANP, open squares, insulin, filled squares insulin + hANP; * p < 0.05 vs. placebo; ** p < 0.01 vs. placebo; + p < 0.05)

ness in healthy man were conflicting: epinephrine levels were found to be decreased as well as unchanged by hANP, norepinephrine levels were reported to be either increased or unchanged by hANP administration (8).

Thirty minutes after the injection of insulin + hANP, responses of the heart rate and systolic blood pressure were considerably smaller than expected from the severity of the concomittant hypoglycemia or epinephrine stimulation (Fig. 1). Thereafter, the responsiveness of both parameters was completely blocked by hANP. In contrast, hANP appeared to potentiate the response of diastolic blood pressure to hypoglycemia, most impressively 15 - 60 min after the injections. Therefore, we can confirm the previously suggested antagonistic properties of hANP not only in regard to the action of norepinephrine, which predominantly governs the maintenance of diastolic blood pressure

237

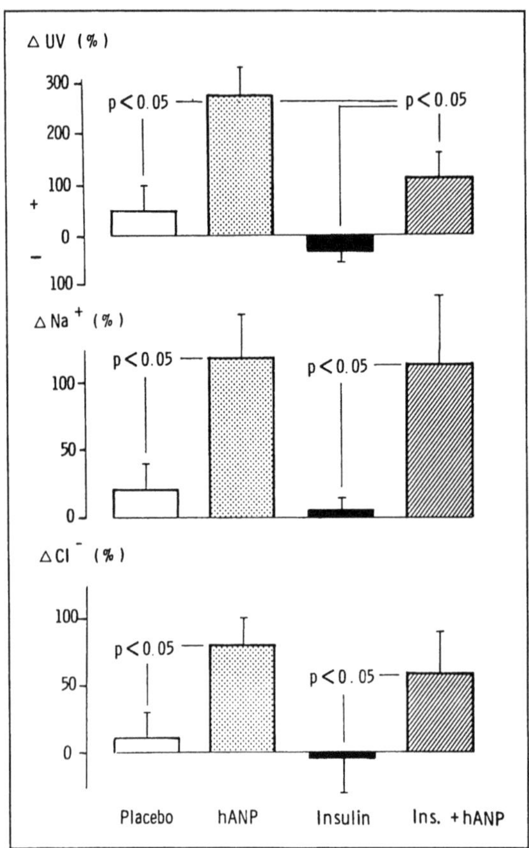

Fig. 2. The effect of 100 µg hANP i.v. on renal responses to insulin-induced hypoglycemia (o.125 U/kg b. wt.): changes from respective base-line levels of urinary volume (UV), sodium (Na^+), and chloride (Cl^-) excretion

during insulin-induced hypoglycemia, (9) but also in regard to epinephrine-mediated tachycardia and increase in cardiac output (6,8,10).

In response to insulin-induced hypoglycemia, vasopressin levels were found to be increased from base-line by about 100 % ($p < 0.05$) resulting in a reduction in urinary volume by 72 ± 3 % ($p < 0.01$) and in urinary sodium excretion by 31 ± 14 % ($p < 0.05$) (Fig. 2). There is good evidence that hANP is capable of antagonizing the vasopressin action in man (11). Whereas hANP was repeatedly shown in animal studies to inhibit vasopressin secretion (12), studies undertaken to confirm this action of hANP in man produced conflicting results (8). The effect of hANP on renal responses to insulin-induced hypoglycemia (fig. 2) closely resembles its effect on the action of exogenous vasopressin (11). In man, therefore, the effect of hANP on vasopressin action on the kidney function during insulin-induced hypoglycemia is considered to be more prominent than its potential inhibitory action on vasopressin stimulation.

References

1. Rosak, C. (1987) Regulation der Hypoglykämie. Fortschr Med 105: 230-232
2. Baylis, P-H., Zerbe, R.L., Robertson, G.L. (1981) Arginine vasopressin response to insulin-induced hypoglycemia. J Clin Endocr Metab 53: 935-940
3. Da Prada, M., Zürcher, G. (1976) Simultaneous radioenzymatic determination of plasma and tissue adrenalin, noradrenalin and dopamine within the femtomole range. Life Sci 19: 1161-1174
4. Jungmann, E., Höll, E., Konzok, C., Schirmer, U., Walter-Schräder, M.C., Althoff, P.H., Schöffling, K. (1988) Zur Wirkung des humanen atrialen natriuretischen Peptides auf Insulinspiegel und Insulinsekretion. Med Klin 83: 325-326
5. Jungmann, E., Walter-Schrader, M.C., Haak, T., Fassbinder, W., Rosak, C., Wambach, G., Althoff, P.H., Schöffling, K. (1988) Impaired renal responsiveness to human atrial natriuretic peptide (hANP) in normotensive patients with Type 1 diabetes mellitus. Klin Wochenschr 66: 527-532
6. Kuchel, O., Debinski, W., Racz, K., Buu, N.T., De Léan, A., Garcia, R., Cantin, M., Genest, J. (1987) Atrial natriuretic factor is an inhibitory modulator of peripheral sympathoadrenomedullary activity. In: Brenner, B.M., Laragh, J.H. (eds.) Biologically active atrial peptides. Raven Press, New York
7. Holtz, J., Sommer, O., Bassenge, E. (1987) Inhibition of sympathoadrenal activity by atrial natriuretic factor in dogs. Hypertension 9: 350-354.
8. Nicholls, M.G., Richards, A.M. (1987) Human studies with atrial natriuretic factor. Endocrinology and Metabolism Clinics of North America 16: 199-223
9. Liang, C.S., Doherty, J.U., Faillace, R., Maekawa, K., Arnold, S., Gavras, H., Hood, W.B. (1982) Insulin infusion in conscious dogs. J Clin Invest 69: 1321-1336
10. Winquist, R.J. (1987) Pharmacologic effects of atrial natriuretic peptide. Endocrinology and Metabolism Clinics of North America 16: 163-182
11. Vierhapper, H., Waldhäusl, W. (1987) Effects of human atrial natriuretic peptide on dDAVP-induced antidiuresis in man. Acta endocrinol (suppl 283): 15
12. Samson, W.K. (1987) Atrial natriuretic factor and the central nervous system. Endocrinology and Metabolism Clinics of North America 16: 145-162